T0141854

Intelligent Systems Reference Library

Volume 109

Series editors

Janusz Kacprzyk, Polish Academy of Sciences, Warsaw, Poland
e-mail: kacprzyk@ibspan.waw.pl

Lakhmi C. Jain, Bournemouth University, Fern Barrow, Poole, Australia, and
University of Canberra, Canberra, Australia
e-mail: jainlc2002@yahoo.co.uk

About this Series

The aim of this series is to publish a Reference Library, including novel advances and developments in all aspects of Intelligent Systems in an easily accessible and well structured form. The series includes reference works, handbooks, compendia, textbooks, well-structured monographs, dictionaries, and encyclopedias. It contains well integrated knowledge and current information in the field of Intelligent Systems. The series covers the theory, applications, and design methods of Intelligent Systems. Virtually all disciplines such as engineering, computer science, avionics, business, e-commerce, environment, healthcare, physics and life science are included.

More information about this series at http://www.springer.com/series/8578

Achim Zielesny

From Curve Fitting
to Machine Learning

An Illustrative Guide to Scientific Data
Analysis and Computational Intelligence

Second Edition

 Springer

Achim Zielesny
Institut für biologische und chemische
    Informatik
Westfälische Hochschule
Recklinghausen
Germany

ISSN 1868-4394   ISSN 1868-4408   (electronic)
Intelligent Systems Reference Library
ISBN 978-3-319-81313-4   ISBN 978-3-319-32545-3   (eBook)
DOI 10.1007/978-3-319-32545-3

Printed on acid-free paper

This Springer imprint is published by Springer Nature
The registered company is Springer International Publishing AG Switzerland

To my parents

Preface

Preface to the first edition

The analysis of experimental data is at heart of science from its beginnings. But it was the advent of digital computers in the second half of the 20th century that revolutionized scientific data analysis twofold: Tedious pencil and paper work could be successively transferred to the emerging software applications so sweat and tears turned into automated routines. In accordance with automation the manageable data volumes could be dramatically increased due to the exponential growth of computational memory and speed. Moreover highly non-linear and complex data analysis problems came within reach that were completely unfeasible before. Non-linear curve fitting, clustering and machine learning belong to these modern techniques that entered the agenda and considerably widened the range of scientific data analysis applications. Last but not least they are a further step towards computational intelligence.

The goal of this book is to provide an interactive and illustrative guide to these topics. It concentrates on the road from two-dimensional curve fitting to multidimensional clustering and machine learning with neural networks or support vector machines. Along the way topics like mathematical optimization or evolutionary algorithms are touched. All concepts and ideas are outlined in a clear cut manner with graphically depicted plausibility arguments and a little elementary mathematics. Difficult mathematical and algorithmic details are consequently banned for the sake of simplicity but are accessible by the referred literature. The major topics are extensively outlined with exploratory examples and applications. The primary goal is to be as illustrative as possible without hiding problems and pitfalls but to address them. The character of an illustrative cookbook is complemented with specific sections that address more fundamental questions like the relation between machine learning and human intelligence. These sections may be skipped without affecting the main road but they will open up possibly interesting insights beyond the mere data massage.

All topics are completely demonstrated with the aid of the computing platform Mathematica and the Computational Intelligence Packages (CIP), a high-level function library developed with Mathematica's programming language on top of Mathematica's algorithms. CIP is open-source so the detailed code of every method is freely accessible. All examples and applications shown throughout the book may be used and customized by the reader without any restrictions. This leads to an interactive environment which allows individual manipulations like the rotation of 3D graphics or the evaluation of different settings up to tailored enhancements for specific functionality.

The book tries to be as introductory as possible calling only for a basic mathematical background of the reader - a level that is typically taught in the first year of scientific education. The target readerships are students of (computer) science and engineering as well as scientific practitioners in industry and academia who deserve an illustrative introduction to these topics. Readers with programming skills may easily port and customize the provided code. The majority of the examples and applications originate from teaching efforts or solution providing. The outline of the book is as follows:

- The introductory **chapter 1** provides necessary basics that underlie the discussions of the following chapters like an initial motivation for the interplay of data and models with respect to the molecular sciences, mathematical optimization methods or data structures. The chapter may be skipped at first sight but should be consulted if things become unclear in a subsequent chapter.
- The main chapters that describe the road from curve fitting to machine learning are chapters 2 to 4. The curve fitting **chapter 2** outlines the various aspects of adjusting linear and non-linear model functions to experimental data. A section about mere data smoothing with cubic splines complements the fitting discussions.
- The clustering **chapter 3** sketches the problems of assigning data to different groups in an unsupervised manner with clustering methods. Unsupervised clustering may be viewed as a logical first step towards supervised machine learning - and may be able to construct predictive systems on its own. Machine learning methods may also need clustered data to produce successful results.
- The machine learning **chapter 4** comprises supervised learning techniques, in particular multiple linear regression, three-layer feed-forward neural networks and support vector machines. Adequate data preprocessing and their use for regression and classification tasks as well as the recurring pitfalls and problems are introduced and thoroughly discussed.
- The discussions **chapter 5** supplements the topics of the main road. It collects some open issues neglected in the previous chapters and opens up the scope with more general sections about the possible discovery of new knowledge or the emergence of computational intelligence.

The scientific fields touched in the present book are extensive and in addition constantly and progressively refined. Therefore it is inevitable to neglect an awful lot of important topics and aspects. The concrete selection always mirrors an author's preferences as well as his personal knowledge and overview. Since the missing parts unfortunately exceed the selected ones and people always have strong feelings about what is of importance the final statement has to be a request for indulgence.

Recklinghausen, April 2011 *Achim Zielesny*

Preface to the second edition

The first edition was friendly reviewed as a useful introductory cookbook for the novice reader. The second edition tries to keep this character and resists the temptation to heavily expand topics or lift the discussion to more subtle academic levels. Besides numerous minor additions and corrections throughout the whole book (together with the unavoidable introduction of some new errors) the only substantial extension of the second edition is the addition of Multiple Polynomial Regression (MPR) in order to support the discussions concerning the method crossover from linear and near-linear up to highly non-linear machine learning approaches. As a consequence several examples and applications have been reworked to improve readability and line of reasoning. Also the construction of minimal predictive models is outlined in an updated and more comprehensible manner.

The second edition is based on the extended version 2.0 of the Computational Intelligence Packages (CIP) which now allows parallelized calculations that lead to an often considerably improved performance with multiple (or multicore) processors. Specific parallelization notes are given throughout the book, the description of CIP is accordingly extended and reworked examples and applications make now use of the new functionality.

With this second edition the book hopefully strengthens its original intent to provide a clear and straight introduction to the fascinating road from curve fitting to machine learning.

Recklinghausen, February 2016 *Achim Zielesny*

Acknowledgements

Certain authors, speaking of their works, say, "My book", "My commentary", "My history", etc. They resemble middle-class people who have a house of their own, and always have "My house" on their tongue. They would do better to say, "Our book", "Our commentary", "Our history", etc., because there is in them usually more of other people's than their own.

Pascal

Acknowledgements to the first edition

I would like to thank Lhoussaine Belkoura, Manfred L. Ristig and Dietrich Woermann who kindled my interest for data analysis and machine learning in chemistry and physics a long time ago.

My mathematical colleagues Heinrich Brinck and Soeren W. Perrey contributed a lot - may it be in deep canyons, remote jungles or at our institute's coffee kitchen. To them and my IBCI collaborators Mirco Daniel and Rebecca Schultz as well as the GNWI team with Stefan Neumann, Jan-Niklas Schäfer, Holger Schulte and Thomas Kuhn I am deeply thankful.

The cooperation with Christoph Steinbeck was very fruitful and an exceptional pleasure: I owe a lot to his support and kindness.

Karina van den Broek, Mareike Dörrenberg, Saskia Faassen, Jenny Grote, Jennifer Makalowski, Stefanie Kleiber and Andreas Truszkowski corrected the manuscript with benevolence and strong commitment: Many thanks to all of them.

Last but not least I want to express deep gratitude and love to my companion Daniela Beisser who not only had to bear an overworked book writer but supported all stages of the book and its contents with great passion.

Every book is a piece of collaborative work but all mistakes and errors are of course mine.

Acknowledgements to the second edition

Kolja Berger implemented the code for parallelized CIP calculations with strong commitment and Marie Theiß carefully rechecked the updated CIP version 2.0 commands throughout the whole book: Many thanks to both of them.

For additional support and helpful discussions I am deeply grateful to Karina van den Broek, Karolin Kleemann, Christoph Steinbeck, Andreas Truszkowski, my mathematical colleagues Heinrich Brinck and Soeren W. Perrey, and my companion Daniela Beisser who again accompanied all efforts with great encouragement and passion.

As a final remark I simply have to repeat myself: Every book is a piece of collaborative work but all mistakes and errors are again mine.

Contents

Chapter 1
Introduction

This chapter discusses introductory topics which are helpful for a basic understanding of the concepts, definitions and methods outlined in the following chapters. It may be skipped for the sake of a faster passage to the more appealing issues or only browsed for a short impression. But if things appear dubious in later chapters this one should be consulted again.

Chapter 1 starts with an overview about the interplay between data and models and the challenges of scientific practice especially in the molecular sciences to motivate all further efforts (section 1.1). The mathematical machinery that plays the most important role behind the scenes is dedicated to the field of optimization, i.e. the determination of the global minimum or maximum of a mathematical function. Basic problems and solution approaches are briefly sketched and illustrated (section 1.2). Since model functions play a major role in the main topics they are categorized in an useful manner that will ease further discussions (section 1.3). Data need to be organized in a defined way to be correctly treated by corresponding algorithms: A dedicated section describes the fundamental data structures that will be used throughout the book (section 1.4). A more technical issue is the adequate scaling of data: This is performed automatically by all clustering and machine learning methods but may be an issue for curve fitting tasks (section 1.5). Experimental data experience different sources of error in contrast to simulated data which are only artificially biased by true statistical errors. Errors are the basis for a proper statistical analysis of curve fitting results as well as for the assessment of machine learning outcomes. Therefore the different sources of error and corresponding conventions are briefly described (section 1.6). Machine learning methods may be used for regression or classification tasks: Whereas regression tasks demand a precise calculation of the desired output values a classification task requires only the correct assignment of an input to a desired output class. Within this book classification tasks are tackled as adequately coded regression tasks which is sketched in a specific section (1.7). The Computational Intelligence Packages (CIP) offer a largely unified structure for different types of calculations which is summarized in a following section to make their use more intuitive and less subtle. In addition a short description of Mathematica's top-down programming and proper initialization is provided (section 1.8). This

© Springer International Publishing Switzerland 2016
A. Zielesny, *From Curve Fitting to Machine Learning*, Intelligent Systems
Reference Library 109, DOI 10.1007/978-3-319-32545-3_1

chapter ends with a note on the reproducibility of calculations reported throughout the book (section 1.9).

1.1 Motivation: Data, models and molecular sciences

Essentially, all models are wrong, but some are useful.

G.E.P. Box

Science is an endeavor to understand and describe the real world out there to (at best) alleviate and enrich human existence. But the structures and dynamics of the real world are very intricate and complex. A humble chemical reaction in the laboratory may already *involve perhaps 10^{20} molecules surrounded by 10^{24} solvent molecules, in contact with a glass surface and interacting with gases ... in the atmosphere. The whole system will be exposed to a flux of photons of different frequency (light) and a magnetic field (from the earth), and possibly also a temperature gradient from external heating. The dynamics of all the particles (nuclei and electrons) is determined by relativistic quantum mechanics, and the interaction between particles is governed by quantum electrodynamics. In principle the gravitational and strong (nuclear) forces should also be considered. For chemical reactions in biological systems, the number of different chemical components will be large, involving various ions and assemblies of molecules behaving intermediately between solution and solid state (e.g. lipids in cell walls)* [Jensen 2007]. Thus, to describe nature, there is the inevitable necessity to set up limitations and approximations in form of simplifying and idealized models - based on the known laws of nature. Adequate models neglect almost everything (i.e. they are, strictly speaking, wrong) but they may keep some of those essential real world features that are of specific interest (i.e. they may be useful).

The dialectical interplay of experiment and theory is a key driving force of modern science. Experimental data do only have meaning in the light of a particular model or at least a theoretical background. Reversely theoretical considerations may be logically consistent as well as intellectually elegant: Without experimental evidence they are a mere exercise of thought no matter how difficult they are. Data analysis is a connector between experiment and theory: Its techniques advise possibilities of model extraction as well as model testing with experimental data.

Model functions have several practical advantages in comparison to mere enumerated data: They are a comprehensive representation of the relation between the quantities of interest which may be stored in a database in a very compact manner with minimum memory consumption. A good model allows interpolating or extrapolating calculations to generate new data and thus may support (up to replace) expensive lab work. Last but not least a suitable model may be heuristically used to explore interesting optimum properties (i.e. minima or maxima of the model function) which could otherwise be missed. Within a market economy a good model is simply a competitive advantage.

The ultimate goal of all sciences is to arrive at quantitative models that describe nature with a sufficient accuracy - or to put it short: to calculate nature. These calculations have the general form

answer $= f(\text{question})$ or output $= f(\text{input})$

where input denotes a question and output the corresponding answer generated by a model function f. Unfortunately the number of interesting quantities which can be directly calculated by application of theoretical ab-initio techniques solely based on the known laws of nature is rather limited (although expanding). For the overwhelming number of questions about nature the model functions f are unknown or too difficult to be evaluated. This is the daily trouble of chemists, material's scientists, engineers or biologists who want to ask questions like the biological effect of a new molecular entity or the properties of a new material's composition. So in current science there are three situations that may be sensibly distinguished due to our knowledge of nature:

- **Situation 1:** The model function f is theoretically or empirically known. Then the output quantity of interest may be calculated directly.
- **Situation 2:** The structural form of the function f is known but not the values of its parameters. Then these parameter values may be statistically estimated on the basis of experimental data by curve fitting methods.
- **Situation 3:** Even the structural form of the function f is unknown. As an approximation the function f may be modelled by a machine learning technique on the basis of experimental data.

A simple example for situation 2 is the case that the relation between input and output is known to be linear. If there is only one input variable of interest, denoted x, and one output variable of interest, denoted y, the structural form of the function f is a straight line

$y = f(x) = a_1 + a_2 x$

where a_1 and a_2 are the unknown parameters of the function which may be statistically estimated by curve fitting of experimental data. In situation 3 it is not only the values of the parameters that are unknown but in addition the structural form of the model function f itself. This is obviously the worst possible case which is addressed by data smoothing or machine learning approaches that try to construct a model function with experimental data only.

Situations 1 to 3 are widely encountered by the contemporary molecular sciences. Since the scientific revolution of the early 20th century the molecular sciences have a thorough theoretical basis in modern physics: Quantum theory is able to (at least in principle) quantitatively explain and calculate the structure, stability and reactivity of matter. It provides a fundamental understanding of chemical bonding and molecular interactions. This foundational feat was summarized in 1929 by Paul A. M. Dirac with famous words: *The underlying physical laws necessary for the mathematical theory of a large part of physics and the whole of chemistry are thus completely known* ... it became possible to submit molecular research and development (R&D) problems to a theoretical framework to achieve correct and satisfactory solutions - but unfortunately Dirac had to continue ... *and the difficulty is only that the exact application of these laws leads to equations much too complicated to be soluble.* The humble "only" means a severe practical restriction: It is in fact only the smallest quantum-mechanical systems like the hydrogen atom with one single proton in the nucleus and one single electron in the surrounding shell that can be treated by pure analytical means to come to an exact mathematical solution, i.e. by solving the Schroedinger equation of this mechanical system with pencil and paper. Nonetheless Dirac added an optimistic prospect: *It therefore becomes desirable that approximate practical methods of applying quantum mechanics should be developed, which can lead to an explanation of the main features of complex atomic systems without too much computation* [Dirac 1929]. A few decades later this hope begun to turn into reality with the emergence of digital computers and their exponentially increasing computational speed: Iterative methods were developed that allowed an approximate quantum-mechanical treatment of molecules and molecular ensembles with growing size (see [Leach 2001], [Frenkel 2002] or [Jensen 2007]). The methods which are ab-initio approximations to the true solution of the Schroedinger equation (i.e. they only use the experimental values of natural constants) are still very limited in applicability so they are restricted to chemical ensembles with just a few hundred atoms to stay within tolerable calculation periods. If these methods are combined with experimental data in a suitable manner so that they become semi-empirical the range of applicability can be extended to molecular systems with several thousands of atoms (up to more than a hundred thousand atoms by the writing of this book [Clark 2010/2015]). The size of the molecular systems and the time frames for their simulation can be even further expanded by orders of magnitude with mechanical force fields that are constructed to mimic the quantum-mechanical molecular interactions so that an atomistic description of matter exceeds the million-atoms threshold. In 1998 and 2013 the Royal Swedish Academy of Sciences honored these scientific achievements by awarding the Nobel prize in chemistry with the prudent comment in 1998 that *Chemistry is no longer a purely experimental science* (see [Nobel Prize 1998/2013]). This atomistic theory-based treatment of molecular R&D problems corresponds to situation 1 where a theoretical technique provides a model function f to "simply calculate" the desired solution in a direct manner.

Despite these impressive improvements (and more is to come) the overwhelming majority of molecular R&D problems is (and will be) out of scope of these atomistic computational methods due to their complexity in space and time. This is especially true for the life and the nano sciences that deal with the most complex natural and artificial systems known today - with the human brain at the top. Thus the molecular sciences are mainly faced with situations 2 and 3: They are a predominant area of application of the methods to be discussed on the road from curve fitting to machine learning. Theory-loaded and model-driven research areas like physical chemistry or biophysics often prefer situation 2: A scientific quantity of interest is studied in dependence of another quantity where the structural form of a model function f that describes the desired dependency is known but not the values of its parameters. In general the parameters may be purely empirical or may have a theoretically well-defined meaning. An example of the latter is usually encountered in chemical kinetics where phenomenological rate equations are used to describe the temporal progress of the chemical reactions but the values of the rate constants - the crucial information - are unknown and may not be calculated by a more fundamental theoretical treatment [Grant 1998]. In this case experimental measurements are indispensable that lead to xy-error data triples (x_i, y_i, σ_i) with an argument value x_i, the corresponding dependent value y_i and the statistical error σ_i of the y_i value (compare below). Then optimum estimates of the unknown parameter values can be statistically deduced on the basis of these data triples by curve fitting methods. In practice a successful model function may at first be only empirically constructed like the quantitative description of the temperature dependence of a liquid's viscosity (illustrated in chapter 2) and then later be motivated by more theoretical lines of argument. Or curve fitting is used to validate the value of a specific theoretical model parameter by experiment (like the critical exponents in chapter 2). Last but not least curve fitting may play a pure support role: The energy values of the potential energy surface of hydrogen fluoride could be directly calculated by a quantum-chemical ab-initio method for every distance between the two atoms. But a restriction to a limited number of distinct calculated values that span the range of interest in combination with the construction of a suitable smoothing function for interpolation (shown in chapter 2) may save considerable time and enhance practical usability without any relevant loss of precision.

With increasing complexity of the natural system under investigation a quantitative theoretical treatment becomes more and more difficult. As already mentioned a quantitative theory-based prediction of a biological effect of a new molecular entity or the properties of a new material's composition are in general out of scope of current science. Thus situation 3 takes over where a model function f is simply unknown or too complex. To still achieve at least an approximate quantitative description of the relationships in question a model function may be tried to be solely constructed with the available data only - a task that is at heart of machine learning. Especially quantitative relationships between chemical structures and their biological activities or physico-chemical and material's properties draw a lot of attention: Thus QSAR (Quantitative Structure Activity Relationship) and QSPR (Quantitative Structure Property Relationship) studies are active fields of research in the life, material's and nano sciences (see [Zupan 1999], [Gasteiger 2003], [Leach 2007] or [Schneider 2008]). Cheminformatics and structural bioinformatics provide a bunch of possibilities to represent a chemical structure in form of a list of numbers (which mathematically form a vector or an input in terms of machine learning, see below). Each number or sequence of numbers is a specific structural descriptor that describes a specific feature of a chemical structure in question, e.g. its molecular weight, its topological connections and branches or electronic properties like its dipole moments or its correlation of surface charges. These structure-representing inputs alone may be analyzed by clustering methods (discussed in chapter 3) for their chemical diversity. The results may be used to generate a reduced but representative subset of structures with a similar chemical diversity in comparison to the original larger set (e.g. to be used in combinatorial chemistry approaches for a targeted structure library design). Alternatively different sets of structures could be compared in terms of their similarity or dissimilarity as well as their mutual white spots (these topics are discussed in chapter 3). A structural descriptor based QSAR/QSPR approach takes the form

$$\text{activity/property} = f(\text{descriptor1}, \text{descriptor2}, \text{descriptor3}, ...)$$

with the model function f as the final target to become able to make model-based predictions (the methods used for the construction of an approximate model function f are outlined in chapter 4). The extensive volume of data that is necessary for this line of research is often obtained by modern high-throughput (HT) techniques like the biological assay-based high-throughput screening (HTS) of thousands of chemical compounds in the pharmaceutical industry or HT approaches in materials science all performed with automated robotic lab systems. Among others these HT methods lead to the so called BioTech data explosion that may be thoroughly exploited for model construction. In fact HT experiments and model construction via machine learning are mutually dependent on each other: Models deserve data for their creation as well as the mere heaps of data produced by HT methods deserve models for their comprehension.

With these few statements about the needs of the molecular sciences in mind the motivation of this book is to show how situations 2 (model function f known, its parameters unknown) and 3 (model function f itself unknown) may be tackled on the road from curve fitting to machine learning: How can we proceed from experimental data to models? What conceptual and technical problems occur along this path? What new insights can we expect?

1.2 Optimization

```
Clear["Global`*"];
<<CIP`Graphics`
```

At the beginning of each section or subsection the global Clear command clears all earlier variables and definitions and thus cares for a proper initialization. Then the necessary CIP packages are loaded, e.g. the Graphics package for this section. A proper initialization prevents possible code interferences due to earlier definitions. Note that Mathematica has a top-down programming style: Once a variable is assigned it keeps its value.

Optimization means a process that tries to determine the optima, i.e. the minima and maxima of a mathematical function. A plethora of important scientific problems can be traced back to an issue of optimization so they are essentially optimization problems. Optimization tasks also lie at heart of the road from curve fitting to machine learning: The methods discussed in later chapters will predominantly use mathematical optimization techniques to do their job. It should be noticed that the following optimization strategies are also utilized for the (common) research situation where no direct path to success can be advised and a kind of educated trial and error is the only way to progress.

A mathematical function may contain ...

- ... **no optimum at all**. An example is a 2D straight line, a 3D plane (illustrated below) or a hyperplane in many dimension. But also non-linear functions like the exponential function may not contain any optimum.

```
pureFunction=Function[{x,y},1.0+2.0*x+3.0*y];
xRange={-0.1,1.1};
yRange={-0.1,1.1};
labels={"x","y","z"};
CIP`Graphics`Plot3dFunction[pureFunction,xRange,yRange,labels]
```

All CIP based calculations are scripted as shown above: First all variables are defined with intuitive names and then passed to specific CIP functions to calculate results or create graphical illustrations. All variables remain valid until the next global Clear command. Note that Mathematica allows the definition of pure functions which may be used like normal variables. If a specific function definition is to be passed to a CIP method a pure function is commonly used. The CIP methods internally use pure functions for distinct function value evaluations. Pure functions are a powerful functional programming feature of the Mathematica computing platform to simplify many operations in an elegant and efficient manner.

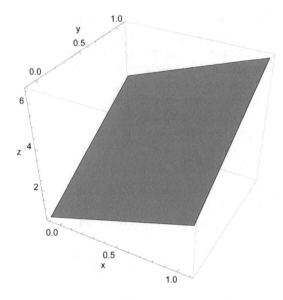

- ... **exactly one optimum**, e.g. a 2D quadratic parabola, a 3D parabolic surface (illustrated below) or a parabolic hyper surface in many dimensions.

```
pureFunction=Function[{x,y},x^2+y^2];
xRange={-2.0,2.0};
yRange={-2.0,2.0};
CIP`Graphics`Plot3dFunction[pureFunction,xRange,yRange,labels]
```

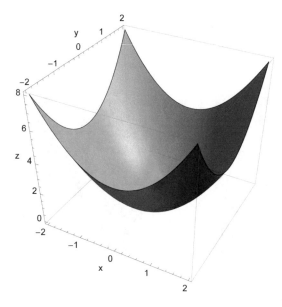

- ... **multiple up to an infinite number of optima** like a 2D sine function, a curved
  3D surface (illustrated below) or a curved hyper surface in multiple dimensions.

```
pureFunction=Function[
 {x,y},1.9*(1.35+Exp[x]*Sin[13.0*(x-0.6)^2]*Exp[-y]*Sin[7.0*y])];
xRange={-0.1,1.1};
yRange={-0.1,1.1};
CIP`Graphics`Plot3dFunction[pureFunction,xRange,yRange,labels]
```

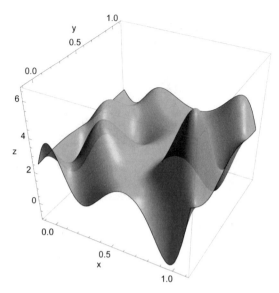

The sketched categorization holds for functions with one argument

$$y = f(x)$$

as well as functions with multiple arguments

$$y = f(x_1, x_2, ..., x_M) = f(\underline{x}) \text{ with } \underline{x} = (x_1, x_2, ..., x_M)$$

i.e. from 2D curves $f(x)$ up to M-dimensional hyper surfaces $f(x_1, x_2, ..., x_M)$. If no optimum exists there is obviously nothing to optimize. For a curve or hyper surface that contains exactly one optimum the optimization problem is usually successfully solvable by analytical methods which are able to calculate the optimum position directly. It is the last category of non-linear functions with multiple optima that cause severe problems - and unfortunately the overwhelming majority of practical applications belong to this drama: The following sections try to reveal some of its tragedy and ways to hold forth a hope again.

1.2.1 Calculus

```
Clear["Global`*"];
<<CIP`Graphics`
```

The standard analytical procedure to determine optima is known from calculus: An example function of the form $y = f(x)$ with one argument x may contain one minimum and one maximum:

```
function=1.0+1.0*x+0.4*x^2-0.1*x^3;
pureFunction=Function[argument,function/.x -> argument];
argumentRange={-2.0,5.0};
functionValueRange={0.0,6.0};
labels={"x","y","Function with one minimum and one maximum"};
CIP`Graphics`Plot2dFunction[pureFunction,argumentRange,
 functionValueRange,labels]
```

Note that the function is defined twice for different purposes: First as a normal symbolic function and in addition as a pure function. The normal function is used in subsequent calculations, the pure function as an argument of the CIP method Plot2dFunction.

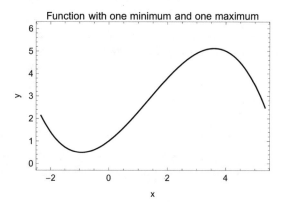

To calculate the positions of the optima the first derivative

```
firstDerivative=D[function,x]
```

$1. + 0.8x - 0.3x^2$

D is Mathematica's operator for partial differentiation to a specified variable which is x in this case.

and their (two) roots are determined:

```
roots=Solve[firstDerivative==0,x]
```

$\{\{x \to -0.927443\}, \{x \to 3.59411\}\}$

Solve is Mathematica's command to solve (systems of) equations. The Solve command returns a list in curly brackets with two rules (also in curly brackets) for setting the x value to solve the equation in question, i.e.

assigning -0.927443 or 3.59411 to x solves the equation. Also note that the number of digits of the result values
is a standard output only: A higher precision could be obtained on demand and is used for internal calculations
(usually the machine precision supported by the hardware).

Then the second derivative

```
secondDerivative=D[function,{x,2}]
```

$0.8 - 0.6x$

D may be told to calculate higher derivatives, i.e. the second derivative in this case.

is used to analyze the type of the two detected optima:

```
secondDerivative/.roots[[1]]
```

1.35647

roots[[1]] denotes the first expression of the roots list above, i.e. the rule $\{x \to$ *-0.927443}: This means that*
the value -0.927443 is to be assigned to x. The /. notation applies this rule to the secondDerivative expres-
sion before, i.e. the x in secondDerivative gets the value -0.927443 and then secondDerivative is numerically
evaluated to 1.35647. These Mathematica specific notations seem to be a bit puzzling at first but they become
convenient and powerful with increased usage.

A value larger zero indicates a minimum at the first optimum position and

```
secondDerivative/.roots[[2]]
```

-1.35647

a value smaller zero a maximum at the second optimum position. The determined
minimum and maximum points

```
minimumPoint={x/.roots[[1]],function/.roots[[1]]};
maximumPoint={x/.roots[[2]],function/.roots[[2]]};
```

may be displayed for visual validation:

```
points2D={minimumPoint,maximumPoint};
CIP`Graphics`Plot2dPointsAboveFunction[points2D,pureFunction,labels,
  GraphicsOptionArgumentRange2D -> argumentRange,
  GraphicsOptionFunctionValueRange2D -> functionValueRange]
```

Method signatures may contain variables and options. Options are set with an arrow as shown in the
Plot2dPointsAboveFunction method above. In contrast to variables the options must not be specified: Then
their default values are used.

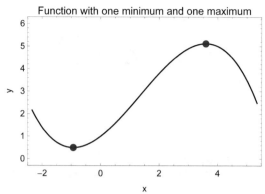

Unfortunately this analytical procedure fails in general. Lets take a somewhat more difficult function with multiple (or more precise: an infinite number of) optima:

```
function=1.0-Cos[x]/(1.0+0.01*x^2);
pureFunction=Function[argument,function/.x -> argument];
argumentRange={-10.0,10.0};
functionValueRange={-0.2,2.2};
labels={"x","y","Function with multiple optima"};
CIP`Graphics`Plot2dFunction[pureFunction,argumentRange,
   functionValueRange,labels]
```

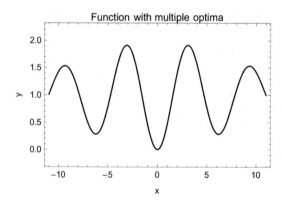

The first derivative may still be obtained

```
firstDerivative=D[function,x]
```

$$\frac{0.02x\text{Cos}[x]}{\left(1.+0.01x^2\right)^2} + \frac{\text{Sin}[x]}{1.+0.01x^2}$$

but the determination of the roots fails

```
roots=Solve[firstDerivative==0,x]
```

The equations appear to involve the variables to be solved for in an essentially non-algebraic way.

$$\text{Solve}\left[\frac{0.02x\text{Cos}[x]}{\left(1.+0.01x^2\right)^2} + \frac{\text{Sin}[x]}{1.+0.01x^2} == 0, x\right]$$

since this non-linear equation can no longer be solved by analytical means. This problem becomes even worse with functions that contain multiple arguments

$$y = f(x_1, x_2, ..., x_M) = f(\underline{x})$$

i.e. with M-dimensional curved hyper surfaces. The necessary condition for an optimum of a M-dimensional hyper surface y is that all partial derivatives become zero:

$$\frac{\partial f(x_1, x_2, ..., x_M)}{\partial x_i} = 0 \; ; \; i = 1, ..., M$$

Whereas the partial derivatives may be successfully evaluated in most cases the resulting system of M (usually non-linear) equations may again not be solvable by analytical means in general. So the calculus-based analytical optimization is restricted to only simple non-linear special cases (linear functions are out of question since they do not contain optima at all). Since these special cases are usually taught extensively at schools and universities (they are ideal for examinations) there is the ongoing impression that the calculus-based solution of optimization problems also achieves success in practice. But the opposite is true: The overwhelming majority of scientific optimization problems is far too difficult for a successful calculus-based treatment. That is one reason why digital computers revolutionized science: With their exponentially growing calculation speed (known as Moore's law which - successfully - predicts a doubling of calculation speed every 18 months) they opened up the perspective for iterative search-based approaches to at least approximate optima in these more difficult and practically relevant cases - a procedure that is simply not feasible with pencil and paper in a man's lifetime.

1.2.2 Iterative optimization

```
Clear["Global`*"];
<<CIP`Graphics`
```

In general the optima of curves and hyper surfaces may only be approximated by iterative step-by-step search procedures - but without any guarantee of success! There are two basic types of iterative optimization strategies:

- **Local optimization:** Beginning at a start position the iterative search method tries to find at least a local optimum (which may not necessarily be the next neighbored optimum to the start position). This local optimum is in general different from the global optimum, i.e. the lowest minimum or the highest maximum of the function.
- **Global optimization:** The iterative search method tries to find the global optimum inside an a priori defined search space.

Global iterative optimization is usually far more computational demanding than local optimization and therefore slower. Both optimization strategies may fail due to two sources of problems:

- **Function related problems:** The function itself to optimize may not contain any optima (e.g. a straight line or a hyperplane) or may otherwise be ill-shaped.
- **Iterative search related problems:** The search algorithm may encounter numerical problems (like division by zero) or simply not find an optimum of required precision within the allowed maximum number of iterations. Whereas in the latter case an increase of the number of iterations should help this solution would fail if the search algorithm is trapped in oscillations around the optimum. Problems are often caused by an inappropriate start position or search space, e.g. if the search algorithm relies on second derivative information but the curvature of the function to be optimized is effectively zero in the search region.

As an example for an unfavorable start position for a minimum detection consider the following situation:

```
function=1.0/x^12-1/x^6;
pureFunction=Function[argument,function/.x -> argument];
xStart=6.0;
startPointForOptimization={xStart,pureFunction[xStart]};
points2D={startPointForOptimization};
argumentRange={0.5,7.0};
functionValueRange={-0.3,0.2};
labels={"x","y","Where to go for the minimum?"};
CIP`Graphics`Plot2dPointsAboveFunction[points2D,pureFunction,labels,
 GraphicsOptionArgumentRange2D -> argumentRange,
 GraphicsOptionFunctionValueRange2D -> functionValueRange]
```

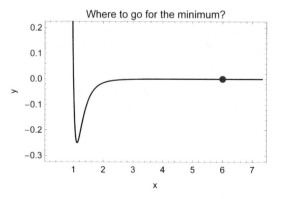

The start position (point) is fairly outside the interesting region that contains the minimum: Its slope (first derivative)

```
D[function, x] /. x -> xStart
```

0.0000214326

and its curvature (second derivative)

```
D[function, {x, 2}] /. x -> xStart
```

−0.0000250037

are nearly zero with the function value itself being nearly constant. In this situation it is difficult for any iterative algorithm to devise a path to the minimum and it is likely for the search algorithm to simply run aground without converging to the minimum.

In practice it is often hard to recognize what went wrong if an optimization failure occurs. And although there are numerous parameters to tune local and global optimization methods for specific optimization problems that does not guarantee to always solve these issues in general. And it becomes clear that any a priori knowledge about the location of an optimum from theoretical considerations or practical experience may play a crucial role. Throughout the later chapters a number of standard problems are discussed and strategies for their circumvention are described.

1.2.3 Iterative local optimization

```
Clear["Global`*"];
```

<<CIP 'Graphics'

Iterative local optimization (or just minimization since maximizing a function f is identical to minimizing $-f$ or f^{-1}) is in principle a simple issue: From a given start position just move downhill as fast as possible by appropriate steps until a local minimum is reached within a desired precision. Thus local optimization methods differ only in the amount of functional information they evaluate to set their step sizes along their chosen downhill directions (see [Press 2007] for details). The evaluation part determines the computational costs of each iteration whereas the directional part determines the convergence speed towards a local minimum where both parts often oppose each other: The more functional information is evaluated the slower a single iteration is performed but the number of iterative steps may be reduced due to more appropriate step sizes and directions.

- Some methods do **only use function value evaluations** at different positions to recognize more or less intelligent downhill paths with adaptive step sizes, e.g. the Simplex method.
- More advanced methods use (first derivative) **slope/gradient information** in addition to function values which allows steepest descent orientations: The so called Gradient method and the more elaborate Conjugate-Gradient and Quasi-Newton methods belong to this type of minimization techniques: The latter two families of methods can find the (one and global) minimum of a M-dimensional parabolic hyper surface with at most M steps (note that this statement just describes a characteristic feature of these algorithms since the optimum of a parabolic hyper surface may simply be calculated with second derivative information by analytical means).
- Also (second derivative) **curvature information** of the function to be minimized may be utilized for a faster convergence near a local minimum as implemented by the so called Newton methods (which were already invented by the grand old father of modern science). If a parabolic hyper surface is under investigation a Newton step leads directly to the minimum, i.e. the Newton method converges to this minimum in one single step (in fact each Newton step assumes a hyper surface to be parabolic and thus calculates the position of its supposed minimum analytically. This assumption is the more accurate the nearer the minimum is located. Since a Newton method has to evaluate an awful lot of functional information for each iterative step which takes its time it is only effective in the proximity of a minimum).

For special types of functions to be minimized like a sum of squares specific combination methods like Levenberg-Marquardt are helpful that try to switch between gradient steps (far from a minimum) and Newton steps (near a minimum) in an effective manner. And besides these general iterative local minimization techniques there are numerous specific solutions for specific optimization tasks that try to take advantage of their specific characteristics. But note that in general there is nothing like the best iterative local optimization method: Being the most effective and therefore fastest method for one minimization problem does not mean to be necessarily superior for another. As a rule of thumb Conjugate-Gradient and Quasi-Newton methods have shown to exert a good compromise between computational costs (function and first derivatives evaluations) and local minimum convergence speed for many practical minimization problems. For the already used multiple optima function

```
function=1.0-Cos[x]/(1.0+0.01*x^2);
pureFunction=Function[argument,function/.x -> argument];
argumentRange={-10.0,10.0};
functionValueRange={-0.2,2.2};

startPosition=8.0;
startPoint={startPosition,function/.x -> startPosition};
points2D={startPoint};

labels={"x","y","Function with multiple optima"};
CIP'Graphics'Plot2dPointsAboveFunction[points2D,pureFunction,labels,
  GraphicsOptionArgumentRange2D -> argumentRange,
  GraphicsOptionFunctionValueRange2D -> functionValueRange]
```

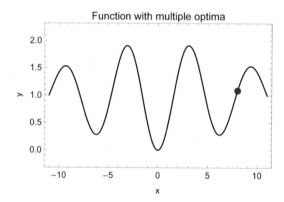

a local minimum may be found from the specified start position (indicated point) with Mathematica's FindMinimum command that provides a unified access to different local iterative search methods (FindMinimum uses a variant of the Quasi-Newton methods by default, see comments on [FindMinimum/FindMaximum] in the references):

```
localMinimum=FindMinimum[function,{x,startPosition}]
```

$\{0.28015, \{x \rightarrow 6.19389\}\}$

FindMinimum returns a list with the function value at the detected local minimum and the rule(s) for the argument value(s) at this minimum

Start point and approximated minimum may be visualized (the arrow indicates the minimization path):

```
minimumPoint={x/.localMinimum[[2]],localMinimum[[1]]};
points2D={startPoint,minimumPoint};
labels={"x","y","Local minimization"};
arrowGraphics=Graphics[{Thick,Red,{Arrowheads[Medium],
 Arrow[{startPoint,minimumPoint}]}}];
functionGraphics=CIP`Graphics`Plot2dPointsAboveFunction[points2D,
 pureFunction,labels,
 GraphicsOptionArgumentRange2D -> argumentRange,
 GraphicsOptionFunctionValueRange2D -> functionValueRange];
Show[functionGraphics,arrowGraphics]
```

Mathematica's Show command allows the overlay of different graphics which are automatically aligned.

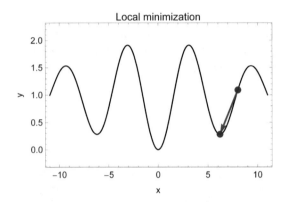

From a different start position a different minimum is found

```
startPosition=2.0;
localMinimum=FindMinimum[function,{x,startPosition}]
```

$\{0.,\{x \to 9.64816 \times 10^{-12}\}\}$

again illustrated as before:

```
startPoint={startPosition,function/.x -> startPosition};
minimumPoint={x/.localMinimum[[2]],localMinimum[[1]]};
points2D={startPoint,minimumPoint};
arrowGraphics=Graphics[{Thick,Red,{Arrowheads[Medium],
 Arrow[{startPoint,minimumPoint}]}}];
functionGraphics=CIP`Graphics`Plot2dPointsAboveFunction[points2D,
 pureFunction,labels,
 GraphicsOptionArgumentRange2D -> argumentRange,
 GraphicsOptionFunctionValueRange2D -> functionValueRange];
Show[functionGraphics,arrowGraphics]
```

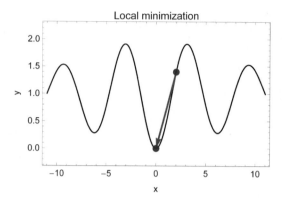

In the last case the approximated minimum is accidentally the global minimum since the start position was near this global optimum. But in general local optimization leads to local optima only.

1.2.4 Iterative global optimization

```
Clear["Global`*"];
<<CIP`Graphics`
```

An optimization of a function usually targets the global optimum of the scientifically relevant argument space. An iterative local search may find the global optimum but is usually only trapped in a local optimum near its start position as demonstrated above. Global optimization strategies try to circumvent this problem by sampling a whole a priori defined search space: They need a set of min/max values for each argument $x_1, x_2, ..., x_M$ of the function $f(x_1, x_2, ..., x_M)$ to be globally optimized where it is assumed that the global optimum lies within the search space that is spanned by these M min/max intervals $[x_{1,min}, x_{1,max}]$ to $[x_{M,min}, x_{M,max}]$. The most straightforward method to achieve this goal seams to be a systematic grid search where the function values are evaluated at equally spaced grid points inside the a priori defined argument search space and then compared to each other to detect the optimum. This grid search procedure is illustrated for an approximation of the global maximum of the curved surface $f(x, y)$ already sketched above

```
function=1.9*(1.35+Exp[x]*Sin[13.0*(x-0.6)^2]*Exp[-y]* Sin[7.0*y]);
pureFunction=
 Function[{argument1,argument2},
  function/.{x -> argument1,y -> argument2}];
```

with a search space of the arguments x and y to be their [0, 1] intervals

```
xMinBorderOfSearchSpace=0.0;
xMaxBorderOfSearchSpace=1.0;
yMinBorderOfSearchSpace=0.0;
yMaxBorderOfSearchSpace=1.0;
```

and 100 equally spaced grid points at $z = 0$ inside this search space (100 grid points means a 10×10 grid, i.e. 10 grid points per dimension):

```
numberOfGridPointsPerDimension=10.0;
gridPoints3D={};
Do[
 Do[
  AppendTo[gridPoints3D,{x,y,0.0}],
   {x,xMinBorderOfSearchSpace,xMaxBorderOfSearchSpace,
   (xMaxBorderOfSearchSpace-xMinBorderOfSearchSpace)/
   (numberOfGridPointsPerDimension-1.0)}
 ],
 {y,yMinBorderOfSearchSpace,yMaxBorderOfSearchSpace,
 (yMaxBorderOfSearchSpace-yMinBorderOfSearchSpace)/
 (numberOfGridPointsPerDimension-1.0)}
];
```

The grid points are calculated with nested Do loops in the xy plane.

This setup can be illustrated as follows (with the grid points located at $z = 0$):

```
xRange={-0.1,1.1};
yRange={-0.1,1.1};
labels={"x","y","z"};
viewPoint3D={3.5,-2.4,1.8};
CIP`Graphics`Plot3dPointsWithFunction[gridPoints3D,pureFunction,
 labels,
 GraphicsOptionArgument1Range3D -> xRange,
 GraphicsOptionArgument2Range3D -> yRange,
 GraphicsOptionViewPoint3D -> viewPoint3D]
```

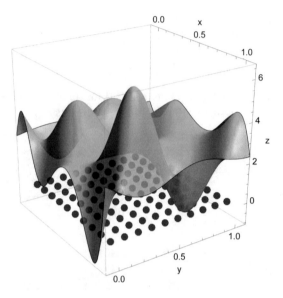

The function values at these grid points are then evaluated and compared

```
winnerGridPoint3D={};
maximumFunctionValue=-Infinity;
Do[
 functionValue=pureFunction[gridPoints3D[[i, 1]],
  gridPoints3D[[i, 2]]];
 If[functionValue>maximumFunctionValue,
  maximumFunctionValue=functionValue;
  winnerGridPoint3D={gridPoints3D[[i, 1]],gridPoints3D[[i, 2]],
   maximumFunctionValue}
 ],
 {i,Length[gridPoints3D]}
];
```

to evaluate the winner grid point

```
winnerGridPoint3D
```

$\{1., 0.222222, 6.17551\}$

that corresponds to the maximum detected function value

```
maximumFunctionValue
```

6.17551

which may be visually validated (with the winner grid point raised to its function value indicated by the arrow and all other grid points still located at $z = 0$):

```
Do[
  If[gridPoints3D[[i,1]] == winnerGridPoint3D[[1]] &&
    gridPoints3D[[i,2]] == winnerGridPoint3D[[2]],
    gridPoints3D[[i]] = winnerGridPoint3D
  ],
  {i,Length[gridPoints3D]}
];

arrowStartPoint={winnerGridPoint3D[[1]],winnerGridPoint3D[[2]],0.0};
arrowGraphics3D=Graphics3D[{Thick,Red,{Arrowheads[Medium],
  Arrow[{arrowStartPoint,winnerGridPoint3D}]}}];
plotStyle3D=Directive[Green,Specularity[White,40],Opacity[0.4]];
functionGraphics3D=CIP`Graphics`Plot3dPointsWithFunction[
  gridPoints3D,pureFunction,labels,
  GraphicsOptionArgument1Range3D -> xRange,
  GraphicsOptionArgument2Range3D -> yRange,
  GraphicsOptionViewPoint3D -> viewPoint3D,
  GraphicsOptionPlotStyle3D -> plotStyle3D];
Show[functionGraphics3D,arrowGraphics3D]
```

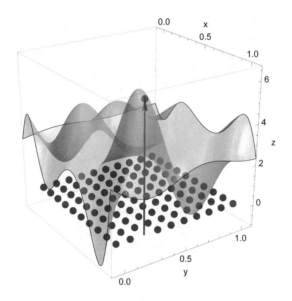

The winner grid point of the global grid search does only approximate the global optimum with an error corresponding to the defined grid spacing. To refine the approximate grid search maximum it may be used as a start point for a following local search since the grid search maximum should be near the global maximum which means that the local search can be expected to converge to the global maximum (but note that there is no guarantee for this proximity and the following convergence in general). Thus the approximate grid search maximum is passed to Mathematica's FindMaximum command (the sister of the FindMinimum command sketched above which utilizes the same algorithms) as a start point for the post-processing local search

```
globalMaximum=FindMaximum[function,{{x,winnerGridPoint3D[[1]]},
 {y,winnerGridPoint3D[[2]]}}]
```

$\{6.54443, \{x \to 0.959215, y \to 0.204128\}\}$

to determine the global maximum with sufficient precision. The improvement obtained by the local refinement process may be inspected (the arrow indicates the maximization path from the winner grid point to the maximum point detected by the post-processing local search in a zoomed view)

```
globalMaximumPoint3D={x/.globalMaximum[[2,1]],
 y/.globalMaximum[[2,2]],globalMaximum[[1]]};
xRange={0.90,1.005};
yRange={0.145,0.26};
arrowGraphics3D=Graphics3D[{Thick,Red,{Arrowheads[Medium],
 Arrow[{winnerGridPoint3D,globalMaximumPoint3D}]}}];
points3D={winnerGridPoint3D,globalMaximumPoint3D};
functionGraphics3D=CIP`Graphics`Plot3dPointsWithFunction[points3D,
 pureFunction,labels,
 GraphicsOptionArgument1Range3D -> xRange,
 GraphicsOptionArgument2Range3D -> yRange,
 GraphicsOptionViewPoint3D -> viewPoint3D,
 GraphicsOptionPlotStyle3D -> plotStyle3D];
Show[functionGraphics3D,arrowGraphics3D]
```

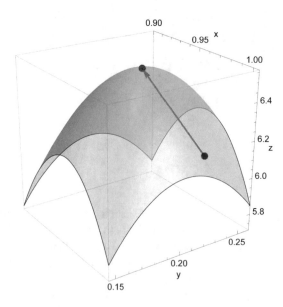

and finally the detected global maximum (point in diagram below) may be visually validated:

```
xRange={-0.1,1.1};
yRange={-0.1,1.1};
points3D={globalMaximumPoint3D};
CIP'Graphics'Plot3dPointsWithFunction[points3D,pureFunction,labels,
  GraphicsOptionArgument1Range3D -> xRange,
  GraphicsOptionArgument2Range3D -> yRange,
  GraphicsOptionViewPoint3D -> viewPoint3D]
```

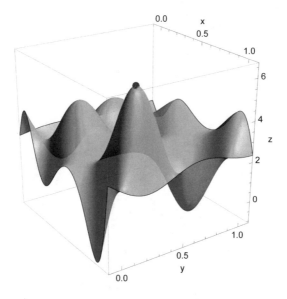

Although a grid search seams to be a rational approach to global optimization it is only an acceptable choice for low-dimensional grids, i.e. global optimization problems with only a small number of function arguments as the example above. This is due to the fact that the number of grid points to evaluate explodes (i.e. grows exponentially) with an increasing number of arguments: The number of grid point is equal to N^M with N to be number of grid points per argument and M the number of arguments. For 12 arguments $x_1, x_2, ..., x_{12}$ with only 10 grid points per argument the grid would already contain one trillion (10^{12}) points so with an increasing number of arguments the necessary function value evaluations at the grid points would become quickly far too slow to be explored in a man's lifetime. As an alternative the number of argument values in the search space to be tested could be confined to a manageable quantity. A rational choice would be randomly selected test points because there is no a priori knowledge about any preferred part of the search space. Note that this random search space exploration would be comparable to a grid search if the number of random test points would equal the number of systematic grid points before (although not looking as tidy). For the current example 20 random test points could be chosen instead of the grid with 100 points:

```
SeedRandom[1];
randomPoints3D=
 Table[
  {RandomReal[{xMinBorderOfSearchSpace,xMaxBorderOfSearchSpace}],
   RandomReal[{yMinBorderOfSearchSpace,yMaxBorderOfSearchSpace}],
   0.0},
  {20}
 ];
CIP`Graphics`Plot3dPointsWithFunction[randomPoints3D,pureFunction,
 labels,
 GraphicsOptionArgument1Range3D -> xRange,
 GraphicsOptionArgument2Range3D -> yRange,
 GraphicsOptionViewPoint3D -> viewPoint3D]
```

The generation of random points can be made deterministic (i.e. always the same sequence of random points is generated) by setting a distinct seed value which is done by the SeedRandom[1] command.

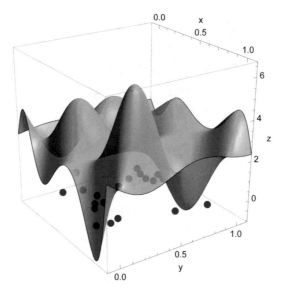

The winner random point is evaluated

```
winnerRandomPoint3D={};
maximumFunctionValue=-Infinity;
Do[
 functionValue=pureFunction[randomPoints3D[[i, 1]],
  randomPoints3D[[i, 2]]];
 If[functionValue>maximumFunctionValue,
  maximumFunctionValue=functionValue;
  winnerRandomPoint3D={randomPoints3D[[i, 1]],randomPoints3D[[i, 2]],
   maximumFunctionValue}
 ],
 {i,Length[randomPoints3D]}
];
```

and visualized (with only the winner random point shown raised to its functions value indicated by the arrow):

```
Do[
  If[randomPoints3D[[i,1]] == winnerRandomPoint3D[[1]] &&
    randomPoints3D[[i,2]] == winnerRandomPoint3D[[2]],
    randomPoints3D[[i]] = winnerRandomPoint3D
  ],
  {i,Length[randomPoints3D]}
];

arrowStartPoint={winnerRandomPoint3D[[1]],winnerRandomPoint3D[[2]],
  0.0};
arrowGraphics3D=Graphics3D[{Thick,Red,{Arrowheads[Medium],
  Arrow[{arrowStartPoint,winnerRandomPoint3D}]}}];
plotStyle3D=Directive[Green,Specularity[White,40],Opacity[0.4]];
functionGraphics3D=CIP`Graphics`Plot3dPointsWithFunction[
  randomPoints3D,pureFunction,labels,
  GraphicsOptionArgument1Range3D -> xRange,
  GraphicsOptionArgument2Range3D -> yRange,
  GraphicsOptionViewPoint3D -> viewPoint3D,
  GraphicsOptionPlotStyle3D -> plotStyle3D];
Show[functionGraphics3D,arrowGraphics3D]
```

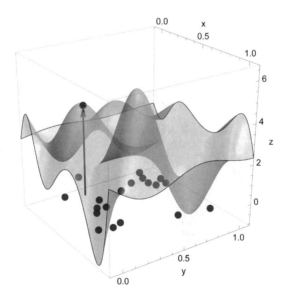

But if this global optimization result

```
winnerRandomPoint3D
```

{0.29287,0.208051,4.49892}

is refined by a post-processing local maximum search starting from the winner random point

```
globalMaximum=FindMaximum[function,
 {{x,winnerRandomPoint3D[[1]]},{y,winnerRandomPoint3D[[2]]}}]
```

$\{4.55146, \{x \to 0.265291, y \to 0.204128\}\}$

only a local maximum is found (point in diagram below) and thus the global maximum is missed:

```
globalMaximumPoint3D={x/.globalMaximum[[2,1]],
 y/.globalMaximum[[2,2]],
 globalMaximum[[1]]};
points3D={globalMaximumPoint3D};
Plot3dPointsWithFunction[points3D,pureFunction,labels,
 GraphicsOptionArgument1Range3D -> xRange,
 GraphicsOptionArgument2Range3D -> yRange,
 GraphicsOptionViewPoint3D -> viewPoint3D]
```

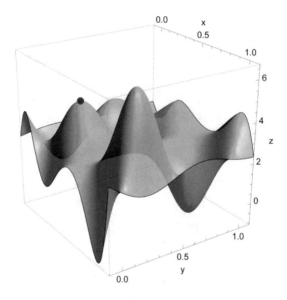

This failure can not be traced to the local optimum search (this worked perfectly from the passed starting position) but must be attributed to an insufficient number of random test points before: If their number is raised the global sampling of the search space would improve and the probability of finding a good test point in the vicinity of the global maximum would increase. But then the same restrictions apply as mentioned with the systematic grid search: With an increasing number of parameters (dimensions) the size of the search space explodes and a random search resembles more and more to be simply looking for a needle in a haystack.

In the face of this desperate situation there was an urgent need for global optimization strategies that are able to tackle difficult search problems in large spaces. As a knight in shining armour a family of so called evolutionary algorithms emerged that rapidly drew a lot of attention. These methods also operate in a basically random manner comparable to a pure random search but in addition they borrow approved refinement strategies from biological evolution to approach the global optimum: These are mutation (random change), crossover or recombination (a kind of random mixing that leads to a directional hopping towards promising search space regions) and selection of the fittest (amplification of the optimal points found so far). The evolution cycles try to speed up the search towards the global optimum by successively composing parts (schemata) of the optimum solution. Mathematica offers an evolutionary-algorithm-based global optimization procedure via the NMinimize and NMaximize commands with the DifferentialEvolution method option (see comments on [NMinimize/NMaximize] for details). The global maximum search

```
globalMaximum=NMaximize[{function,
  {xMinBorderOfSearchSpace<x<xMaxBorderOfSearchSpace,
  yMinBorderOfSearchSpace<y<yMaxBorderOfSearchSpace}},
  {x,y},
  Method -> {"DifferentialEvolution","PostProcess" -> False}]
```

$\{6.54443, \{x \to 0.959215, y \to 0.204128\}\}$

Note the deactivation of the PostProcess in the Method definition: NMaximize automatically applies a local

optimization method to refine the result of a global search - the same was done in the grid and random search

examples above. The deactivation suppresses this refinement to get the pure result of the evolutionary algorithm.

now directly leads to a result of sufficient precision (compare global maximum location above). But it should be noted that evolutionary algorithms in spite of their popularity belong to the methods of last resort: They may be extremely computationally expensive, i.e. time-consuming. Evolutionary algorithms are regarded to be very effective since they imitate the successful biological evolution. This widespread view neglects the fact that natural evolution needed eons to develop life - and living organisms are by no means optimum solutions. If the evolutionary algorithm is applied to the multiple-optima function already demonstrated above

```
function=1.0-Cos[x]/(1.0+0.01*x^2);
pureFunction=Function[argument,function/.x -> argument];
```

with an appropriate search space (not too small, not too large)

```
xMinBorderOfSearchSpace=-10.0;
xMaxBorderOfSearchSpace=15.0;
```

the global minimum (point in diagram below) inside the search space (marked as a background in diagram below)

```
globalMinimum=NMinimize[{function,
  xMinBorderOfSearchSpace<x<xMaxBorderOfSearchSpace},x,
  Method -> {"DifferentialEvolution","PostProcess" -> False}]
```

$\{5.16341 \times 10^{-10}, \{x \to -0.0000318188\}\}$

is also approximated successfully:

```
minimumPoint={x/.globalMinimum[[2]],globalMinimum[[1]]};
points2D={minimumPoint};
argumentRange={-12.0,17.0};
functionValueRange={-0.2,2.2};
labels={"x","y","Global minimization"};
functionGraphics=CIP`Graphics`Plot2dPointsAboveFunction[points2D,
  pureFunction,labels,
  GraphicsOptionArgumentRange2D -> argumentRange,
  GraphicsOptionFunctionValueRange2D -> functionValueRange];
searchSpaceGraphics=Graphics[{RGBColor[0,1,0,0.2],
  Rectangle[{xMinBorderOfSearchSpace,functionValueRange[[1]]},
  {xMaxBorderOfSearchSpace,functionValueRange[[2]]}]}];
Show[functionGraphics,searchSpaceGraphics]
```

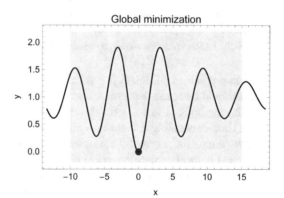

But note: If the search space is inadequately chosen (i.e. the global minimum is outside the interval)

```
xMinBorderOfSearchSpace=50.0;
xMaxBorderOfSearchSpace=60.0;
globalMinimum=NMinimize[{function,
  xMinBorderOfSearchSpace<x<xMaxBorderOfSearchSpace},x,
  Method -> {"DifferentialEvolution","PostProcess" -> False}]
```

$\{0.9619, \{x \rightarrow 50.2272\}\}$

or the search space is simply to large

```
xMinBorderOfSearchSpace=-100000.0;
xMaxBorderOfSearchSpace=100000.0;
globalMinimum=NMinimize[{function,
  xMinBorderOfSearchSpace<x<xMaxBorderOfSearchSpace},x,
  Method -> {"DifferentialEvolution","PostProcess" -> False}]
```

$\{0.805681, \{x \rightarrow 19.2638\}\}$

the global minimum may not be found within the default maximum number of iterations.

1.2.5 Constrained iterative optimization

```
Clear["Global`*"];
<<CIP`Graphics`
```

With the global optimization examples of the previous section the field of constrained optimization was already touched since the a priori defined search space was a constraint of the search (but in fact it was not intended to constrain the optimization procedure: Defining a search space was just a precondition for the global optimization methods to work at all). In general optimization tasks are called unconstrained if they are free from any additional restrictions. If the optimization is subject to one or several constraints the field of constrained optimization is entered. If the function under investigation is not only to be globally minimized but the x value is restricted to lie in an defined interval

```
function=1.0-Cos[x]/(1.0+0.01*x^2);
pureFunction=Function[argument,function/.x -> argument];
xMinConstraint=2.0;
xMaxConstraint=11.0;
constraint=xMinConstraint<x<xMaxConstraint;
constrainedGlobalMinimum=NMinimize[{function,constraint},x,
  Method -> {"DifferentialEvolution","PostProcess" -> False}]
```

$\{0.28015, \{x \rightarrow 6.19386\}\}$

the constrained global minimum (point in diagram below) may differ from the unconstrained one (the constraint is marked as a background in diagram below):

```
constrainedMinimumPoint={x/.constrainedGlobalMinimum[[2]],
  constrainedGlobalMinimum[[1]]};
```

```
points2D={constrainedMinimumPoint};
argumentRange={-12.0,17.0};
functionValueRange={-0.2,2.2};
labels={"x","y","Constrained global minimization"};
functionGraphics=CIP 'Graphics 'Plot2dPointsAboveFunction[points2D,
 pureFunction,labels,
 GraphicsOptionArgumentRange2D -> argumentRange,
 GraphicsOptionFunctionValueRange2D -> functionValueRange];
constraintGraphics=Graphics[{RGBColor[1,0,0,0.1],
 Rectangle[{xMinConstraint,functionValueRange[[1]]},
 {xMaxConstraint,functionValueRange[[2]]}]}];
Show[functionGraphics,constraintGraphics]
```

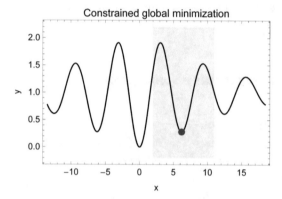

But not only may the unconstrained and constrained global optimum differ: The constrained global optimum may in general not be an optimum of the unconstrained optimization problem at all: This can be illustrated with the following example taken from the Mathematica tutorials. The 3D surface

```
function=
 -1.0/((x+1.0)^2+(y+2.0)^2+1)-2.0/((x-1.0)^2+(y-1.0)^2+1)+2.0;
pureFunction=Function[{argument1,argument2},
 function/.{x -> argument1,y -> argument2}];
xRange={-3.0,3.0};
yRange={-3.0,3.0};
labels={"x","y","z"};
CIP 'Graphics 'Plot3dFunction[pureFunction,xRange,yRange,labels]
```

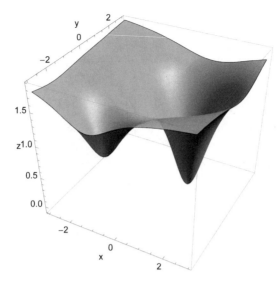

contains two optima: A local and a global minimum. Depending on the start position of the iterative local minimum search method initiated via the FindMinimum command

```
startPosition={-2.5,-1.5};
localMinimum=FindMinimum[function,{{x,startPosition[[1]]},
 {y,startPosition[[2]]}}]
```

$\{0.855748, \{x \to -0.978937, y \to -1.96841\}\}$

the minimization process approximates the local minimum

```
startPoint={startPosition[[1]],startPosition[[2]],
 function/.{x -> startPosition[[1]],y -> startPosition[[2]]}};
minimumPoint={x/.localMinimum[[2,1]],y/.localMinimum[[2,2]],
 localMinimum[[1]]};points3D={startPoint,minimumPoint};
arrowGraphics=Graphics3D[{Thick,Red,{Arrowheads[Medium],
 Arrow[{startPoint,minimumPoint}]}}];
plotStyle3D=Directive[Green,Specularity[White,40],Opacity[0.4]];
functionGraphics=CIP`Graphics`Plot3dPointsWithFunction[points3D,
 pureFunction, labels,
 GraphicsOptionArgument1Range3D -> xRange,
 GraphicsOptionArgument2Range3D -> yRange,
 GraphicsOptionPlotStyle3D -> plotStyle3D];
Show[functionGraphics,arrowGraphics]
```

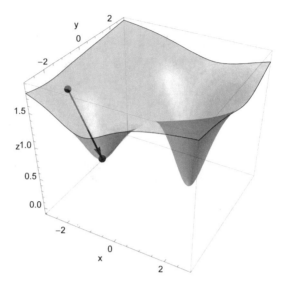

or (with another start point)

```
startPosition={-0.5,2.5};
localMinimum=FindMinimum[function,{{x,startPosition[[1]]},
  {y,startPosition[[2]]}}]
```

$\{-0.071599,\{x \to 0.994861, y \to 0.992292\}\}$

arrives at the global minimum:

```
startPoint={startPosition[[1]],startPosition[[2]],
  function/.{x -> startPosition[[1]],y -> startPosition[[2]]}};
minimumPoint={x/.localMinimum[[2,1]],y/.localMinimum[[2,2]],
  localMinimum[[1]]};points3D={startPoint,minimumPoint};
arrowGraphics=Graphics3D[{Thick,Red,{Arrowheads[Medium],
  Arrow[{startPoint,minimumPoint}]}}];
functionGraphics=CIP`Graphics`Plot3dPointsWithFunction[points3D,
  pureFunction,labels,
  GraphicsOptionArgument1Range3D -> xRange,
  GraphicsOptionArgument2Range3D -> yRange,
  GraphicsOptionPlotStyle3D -> plotStyle3D];
Show[functionGraphics,arrowGraphics]
```

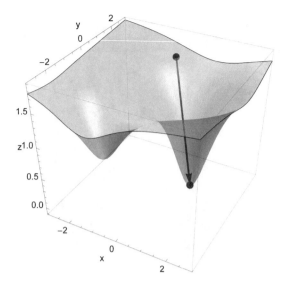

If now the constraint is imposed that

$$x^2 + y^2 > 4.0$$

(the constraint removes a circular argument area around the origin (0,0) of the xy plane) the constrained local minimization algorithm behind the FindMinimum command is activated (see comments on [FindMinimum/FindMaximum] for details). The constrained local minimization process from the first start position

```
startPosition={-2.5,-1.5};
constraint=x^2+y^2>4.0;
localMinimum=FindMinimum[{function,constraint},
  {{x,startPosition[[1]]},{y,startPosition[[2]]}}]
```

$\{0.855748, \{x \rightarrow -0.978937, y \rightarrow -1.96841\}\}$

still results in the local minimum of the unconstrained surface

```
startPoint={startPosition[[1]],startPosition[[2]],
  function/.{x -> startPosition[[1]],y -> startPosition[[2]]}};
minimumPoint={x/.localMinimum[[2,1]],y/.localMinimum[[2,2]],
  localMinimum[[1]]};points3D={startPoint,minimumPoint};
regionFunction=Function[{argument1,argument2},
  constraint/.{x -> argument1,y -> argument2}];
arrowGraphics=Graphics3D[{Thick,Red,{Arrowheads[Medium],
  Arrow[{startPoint,minimumPoint}]}}];
functionGraphics=Plot3dPointsWithFunction[points3D,pureFunction,
  labels,
  GraphicsOptionArgument1Range3D -> xRange,
  GraphicsOptionArgument2Range3D -> yRange,
  GraphicsOptionPlotStyle3D -> plotStyle3D,
```

```
GraphicsOptionRegionFunction -> regionFunction];
Show[functionGraphics, arrowGraphics]
```

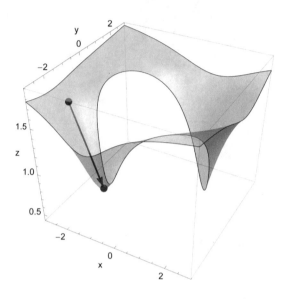

but the second start position

```
startPosition={-0.5,2.5};
localMinimum=FindMinimum[{function,constraint},
  {{x,startPosition[[1]]},{y,startPosition[[2]]}}]
```

$\{0.456856, \{x \to 1.41609, y \to 1.41234\}\}$

leads to a new global minimum since the one of the unconstrained surface is excluded by the constraint:

```
startPoint={startPosition[[1]],startPosition[[2]],
  function/.{x -> startPosition[[1]],y -> startPosition[[2]]}};
minimumPoint={x/.localMinimum[[2,1]],y/.localMinimum[[2,2]],
  localMinimum[[1]]};points3D={startPoint,minimumPoint};
arrowGraphics=Graphics3D[{Thick,Red,{Arrowheads[Medium],
  Arrow[{startPoint,minimumPoint}]}}];
functionGraphics=Plot3dPointsWithFunction[points3D,pureFunction,
  labels,
  GraphicsOptionArgument1Range3D -> xRange,
  GraphicsOptionArgument2Range3D -> yRange,
  GraphicsOptionPlotStyle3D -> plotStyle3D,
  GraphicsOptionRegionFunction -> regionFunction];
Show[functionGraphics, arrowGraphics]
```

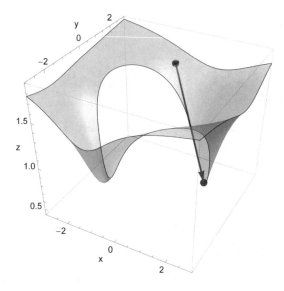

An evolutionary-algorithm-based constrained global search in the displayed argument ranges via NMinimize directly approximates the constrained global minimum

```
Off[NMinimize::cvmit]
localMinimum=NMinimize[{function,constraint},
  {{x,xRange[[1]],xRange[[2]]},{y,yRange[[1]],yRange[[2]]}},
  Method -> {"DifferentialEvolution","PostProcess" -> False}]
```

$\{0.456829, \{x \to 1.41637, y \to 1.41203\}\}$

The Off[NMinimize::cvmit] command suppresses an internal message from NMinimize. Internal messages are usually helpful to understand problems and they advise to interpret results with caution. In this particular case the suppression eases readability.

with sufficient precision (compare above).

In general it holds that the more dimensional the non-linear curved hyper surface is and the more constraints are imposed the more difficult it is to approximate a local or even the global optimum with sufficient precision. The specific optimization problems that are related to the road from curve fitting to machine learning will be discussed in the later chapters where they apply.

1.3 Model functions

Since model functions play an important role throughout the book a basic categorization is helpful. A good starting point is the most prominent model function: The straight line.

1.3.1 Linear model functions with one argument

```
Clear["Global`*"];
<<CIP`Graphics`
```

The well-known functional form of the straight line is

$$y = f(x) = a_1 + a_2 x$$

```
pureFunction=Function[x,1.0+2*x];
argumentRange={0.0,5.0};
functionValueRange={0.0,12.0};
labels={"x","y","Straight line"};
CIP`Graphics`Plot2dFunction[pureFunction,argumentRange,
  functionValueRange,labels]
```

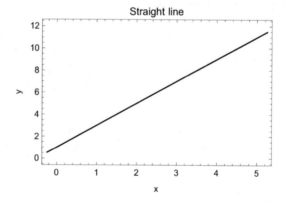

The straight line is linear in two ways: It describes a linear relation between argument x and function value y and is itself linear in its parameters a_1 and a_2, i.e. a_1 and a_2 have exponent 1. A general model function which is linear in its parameters can be defined as follows:

$$y = f(x) = a_1 g_1(x) + a_2 g_2(x) + \ldots + a_L g_L(x) = \sum_{v=1}^{L} a_v g_v(x)$$

This general linear function consists of L parameters a_1 to a_L that are each multiplied by a function $g_v(x)$. The functions $g_v(x)$ depend on x and do only have fixed and known internal parameters. Note that the general linear function does not necessarily describe a linear relation between argument x and function value y: This relation may be highly non-linear, e.g. for a $g_v(x)$ that is equal to e^x. From the point of view of the general linear function the straight line is just a special case with

$$L = 2 \; ; \; g_1(x) = x^0 = 1 \; ; \; g_2(x) = x$$

that leads to

$$y = f(x) = a_1 + a_2 x$$

Another well-known example of this type of linear model functions are polynomials

$$y = f(x) = a_1 + a_2 x + a_3 x^2 + \ldots + a_L x^{L-1} = \sum_{v=1}^{L} a_v x^{v-1}$$

e.g. the quadratic parabola

$$y = f(x) = \sum_{v=1}^{3} a_v x^{v-1} = a_1 + a_2 x + a_3 x^2$$

```
pureFunction=Function[x,11.0-15.0*x+5.0*x^2];
argumentRange={0.0,3.0};
functionValueRange={-1.0,12.0};
labels={"x","y","Quadratic parabola"};
CIP'Graphics'Plot2dFunction[pureFunction,argumentRange,
  functionValueRange,labels]
```

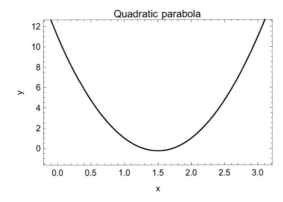

Model functions that are linear in their parameters make up an important special case for curve fitting procedures to experimental data: It can be shown that they lead to optimization problems with only one global optimum which in principle may be calculated with pencil and paper by means of analytic calculation strategies (e.g. see [Hamilton 1964], [Barlow 1989], [Bevington 2002], [Brandt 2002] or [Press 2007]).

Again, note that the term linear model function denotes a function that is linear in its parameters only. It does not necessarily mean a linear dependence of the function value y on the argument x. This subtle difference often causes some misunderstandings in scientific practice as far as non-linear fits are concerned.

1.3.2 Non-linear model functions with one argument

```
Clear["Global`*"];
<<CIP`Graphics`
```

A model function that is not linear in its parameters is called a non-linear model function, e.g.

$$y = f(x) = a_1 e^{a_2 x}$$

To recognize the non-linearity in parameters of the example function a power series expansion is helpful (in this case around $x = 0$ with a display up to the 4th power):

```
Series[Subscript[a, 1]*Exp[Subscript[a, 2]*x],{x,0,4}]
```

$$a_1 + a_1 a_2 x + \frac{1}{2} a_1 a_2^2 x^2 + \frac{1}{6} a_1 a_2^3 x^3 + \frac{1}{24} a_1 a_2^4 x^4 + O[x]^5$$

The cross terms like $a_1 a_2$ or $a_1 a_2^2$ and the higher powers of a_2 like a_2^2, a_2^3, a_2^4 etc. now become directly obvious. A prominent example is the exponential decay model that describes radioactive processes of disintegration or chemical first-order kinetics:

```
pureFunction=Function[x,1.0*Exp[-8.0*x]];
argumentRange={0.0,1.0};
functionValueRange={0.0,1.5};
labels={"x","y","Exponential decay"};
CIP`Graphics`Plot2dFunction[pureFunction,argumentRange,
  functionValueRange,labels]
```

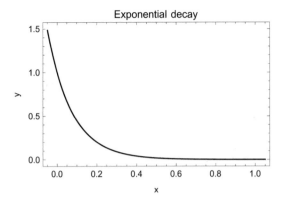

Nature (fortunately) is not linear (otherwise living organisms would not exist) so non-linear model functions play a predominant role in science. But compared to linear models non-linear model functions may cause severe problems in data analysis procedures. They often lead to optimization problems with multiple optima so analytic calculation strategies are no longer applicable in general: Only iterative strategies can be followed that may disastrously fail.

So far only one dimensional model functions with one argument x are discussed. One dimensional model functions play the central part in curve fitting methods where the structural form of the model function is often known but not the values of its parameters (see chapter 2).

1.3.3 Linear model functions with multiple arguments

```
Clear["Global`*"];
<<CIP`Graphics`
```

Model functions with multiple arguments x_1 to x_M may be linear in their parameters and are generally written in the form (that utilizes the general linear function with one argument from above):

$$y = f(x_1, x_2, ..., x_M) = \left(\sum_{v=1}^{L} a_{1v} g_{1v}(x_1)\right) + ... + \left(\sum_{v=1}^{L} a_{Mv} g_{Mv}(x_M)\right)$$
$$y = f(x_1, x_2, ..., x_M) = \sum_{u=1}^{M} \left(\sum_{v=1}^{L} a_{uv} g_{uv}(x_u)\right)$$

The multidimensional analog of the straight line is the hyperplane that is derived from the general linear model function with

$$L = 2$$
$$y = f(x_1, x_2, ..., x_M) = \sum_{u=1}^{M} \left(\sum_{v=1}^{2} a_{uv} g_{uv}(x_u)\right) = \sum_{u=1}^{M} \left(a_{u1} g_{u1}(x_u) + a_{u2} g_{u2}(x_u)\right)$$

$$y = f(x_1, x_2, ..., x_M) = \sum_{u=1}^{M} a_{u1} g_{u1}(x_u) + \sum_{u=1}^{M} a_{u2} g_{u2}(x_u)$$

and

$$a_u = a_{u1} \; ; \; g_{u1}(x_u) = x_u \; ; \; a_{M+1} = \sum_{u=1}^{M} a_{u2} \; ; \; g_{u2}(x_u) = 1$$

that leads to

$$y = f(x_1, x_2, ..., x_M) = \sum_{u=1}^{M} a_u x_u + a_{M+1}$$

A 3D plane with $M = 2$

$$y = f(x_1, x_2) = a_1 x_1 + a_2 x_2 + a_3$$

is visualized below:

```
pureFunction=Function[{x,y},1.0+2.0*x+3.0*y];
xRange={-0.1,1.1};
yRange={-0.1,1.1};
labels={"x","y","z"};
CIP'Graphics'Plot3dFunction[pureFunction,xRange,yRange,labels]
```

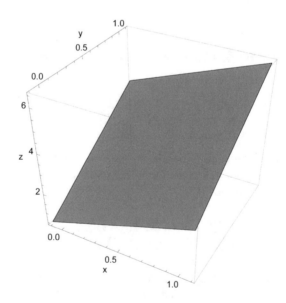

What holds for one dimensional linear model functions still holds for their multidimensional analogs: Model fitting procedures to experimental data lead to optimization problems with one global optimum with analytic calculation strategies for its position (see Multiple Linear and Polynomial Regression in chapter 4).

1.3.4 Non-linear model functions with multiple arguments

```
Clear["Global`*"];
<<CIP`Graphics`
```

Non-linear model functions with multiple arguments x_1 to x_M like

$$y = f(x_1, x_2, ..., x_M) = a_1 \sin(x_1) + \exp\left\{\sum_{u=2}^{M} a_u x_u^2\right\}$$

(where $\exp\{x\}$ denotes e^x) may be viewed as curved hyper surfaces with possible multiple minima and maxima in comparison to linear hyperplanes. The already shown curved 3D surface may again be taken as an example:

```
pureFunction=Function[{x,y},
  1.9*(1.35+Exp[x]*Sin[13.0*(x-0.6)^2]*Exp[-y]* Sin[7.0*y])];
xRange={-0.1,1.1};
yRange={-0.1,1.1};
labels={"x","y","z"};
CIP`Graphics`Plot3dFunction[pureFunction,xRange,yRange,labels]
```

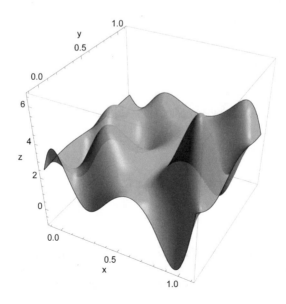

It is these kinds of curved hyper surfaces that answer the most subtle questions about nature but on the other hand they cause the worst data analysis problems. Machine learning methods usually lead to this kind of surfaces to optimize (see chapter 4): They require iterative optimization techniques which in turn need considerable computational power to be applied with success.

1.3.5 Multiple model functions

In a last step multiple model functions may be collected together to generate an output vector y (the answer) for an input vector x (the question)

$$y_1 = f_1(x_1, x_2, ..., x_M)$$
$$y_2 = f_2(x_1, x_2, ..., x_M)$$
$$...$$
$$y_N = f_N(x_1, x_2, ..., x_M)$$

which may be written in an abbreviated vector notation:

$$y = f(x)$$

Note that the output vector y and the function vector f are of dimension N whereas the input vector x is of (maybe different) dimension M. Model function collections of this kind play the crucial role in machine learning methods where function collections are constructed to describe experimental data in multiple dimensions (see chapter 4).

1.3.6 Summary

The Holy Grail of the sciences to calculate nature with

$$output = f(input)$$

may now be written in mathematical detail:

$$y = f(x)$$

Questions about nature are asked with adequately defined input vectors x that are submitted to model functions f to give the answer in form of an adequately defined output vector y. This is a rather general scheme: Nearly everything can be adequately coded in input/output vectors, e.g. molecules, pharmacological effects, material's properties etc. The details of this kind of coding may be subtle and difficult and are the realm of specific areas of science like cheminformatics or bioinformatics. The proper coding is an essential precondition to any data analysis: If the interesting parts of the world are not adequately coded then any association of them by model functions must inevitably fail.

1.4 Data structures

Data structures describe the organization of data that will be used for curve fitting, clustering and machine learning throughout this book. In general algorithms deserve adequate data structures and vice versa. The interplay of algorithms and data structures is at heart of computer science.

1.4.1 Data for curve fitting

```
Clear["Global`*"];
```

For curve fitting methods a xy-error data structure is used. This data structure consists of xy-error data triples (x_i, y_i, σ_i) with an argument value x_i, a corresponding dependent value y_i and the statistical error σ_i of the y_i value. In Mathematica data are stored in lists which are defined by curly brackets. The whole xy-error data structure is a single list with nested sublists that represent the single xy-error data triples. Here is an example of a xy-error data structure with 3 xy-error data triples:

```
xyErrorData={
  {Subscript[x,1],Subscript[y,1],Subscript[\[Sigma],1]},
  {Subscript[x,2],Subscript[y,2],Subscript[\[Sigma],2]},
  {Subscript[x,3],Subscript[y,3],Subscript[\[Sigma],3]}
  }
```

$$\{\{x_1,y_1,\sigma_1\},\{x_2,y_2,\sigma_2\},\{x_3,y_3,\sigma_3\}\}$$

1.4.2 Data for machine learning

```
Clear["Global`*"];
<<CIP`DataTransformation`
```

```
<<CIP`Utility`
```

When it comes to machine learning a data set structure is used. A data set is a list of input/output (I/O) pairs, e.g. the following data set with 3 I/O pairs:

```
dataSet={ioPair1,ioPair2,ioPair3};
```

Each I/O pair consists of an input vector (abbreviated input) and an output vector (abbreviated output):

```
ioPair1={input1,output1};
ioPair2={input2,output2};
ioPair3={input3,output3};
```

Each input and each output is a vector with a defined number of components, e.g. each input may consist of 3 components and each output of 2 components

```
input1={Subscript[in,11],Subscript[in,12],Subscript[in,13]};
output1={Subscript[out,11],Subscript[out,12]};
input2={Subscript[in,21],Subscript[in,22],Subscript[in,23]};
output2={Subscript[out,21],Subscript[out,22]};
input3={Subscript[in,31],Subscript[in,32],Subscript[in,33]};
output3={Subscript[out,31],Subscript[out,32]};
```

where the first index indicates the I/O pair and the second index the component. The whole data set combines to:

```
dataSet
```

$$\{\{\{in_{11},in_{12},in_{13}\},\{out_{11},out_{12}\}\},\{\{in_{21},in_{22},in_{23}\},\{out_{21},out_{22}\}\},\{\{in_{31},in_{32},in_{33}\},\{out_{31},out_{32}\}\}\}$$

Data sets do not contain statistical errors since the machine learning methods discussed in this book are not statistically based and therefore do not take errors into account. But for a proper assessment of a machine learning result it is helpful to know the data errors. The inputs of a data set can be isolated with

```
inputs=CIP`Utility`GetInputsOfDataSet[dataSet]
```

$$\{\{in_{11},in_{12},in_{13}\},\{in_{21},in_{22},in_{23}\},\{in_{31},in_{32},in_{33}\}\}$$

and the outputs accordingly:

```
inputs=CIP`Utility`GetOutputsOfDataSet[dataSet]
```

$\{\{out_{11}, out_{12}\}, \{out_{21}, out_{22}\}, \{out_{31}, out_{32}\}\}$

A data set that contains outputs with more than one output component like the one sketched above may be split in multiple data sets, i.e. a list of data sets

```
dataSetList=
  CIP`DataTransformation`TransformDataSetToMultipleDataSet[dataSet];
```

where each split data set now contains a single output component:

```
dataSetList[[1]]
```

$\{\{\{in_{11}, in_{12}, in_{13}\}, \{out_{11}\}\}, \{\{in_{21}, in_{22}, in_{23}\}, \{out_{21}\}\}, \{\{in_{31}, in_{32}, in_{33}\}, \{out_{31}\}\}\}$

```
dataSetList[[2]]
```

$\{\{\{in_{11}, in_{12}, in_{13}\}, \{out_{12}\}\}, \{\{in_{21}, in_{22}, in_{23}\}, \{out_{22}\}\}, \{\{in_{31}, in_{32}, in_{33}\}, \{out_{32}\}\}\}$

This splitting is used for several machine learning methods and graphical illustrations. As a 3D data set those data sets are denoted that contain inputs with two components and outputs with one component: They may be illustrated by three dimensional graphics in contrast to data sets with higher dimensional inputs or outputs.

1.4.3 Inputs for clustering

```
Clear["Global`*"];
```

The inputs of a data set are defined as the list of inputs of all I/O pairs:

```
input1={Subscript[in,11],Subscript[in,12],Subscript[in,13]};
input2={Subscript[in,21],Subscript[in,22],Subscript[in,23]};
input3={Subscript[in,31],Subscript[in,32],Subscript[in,33]};
inputs={input1,input2,input3}
```

$\{\{in_{11}, in_{12}, in_{13}\}, \{in_{21}, in_{22}, in_{23}\}, \{in_{31}, in_{32}, in_{33}\}\}$

The inputs data structure may be used for clustering tasks.

1.4.4 Inspection, cleaning and splitting of data

```
Clear["Global`*"];
<<CIP`Utility`
<<CIP`ExperimentalData`
<<CIP`DataTransformation`
<<CIP`Graphics`
<<CIP`CalculatedData`
```

The Computational Intelligence Packages (CIP) provide several functions for inspection, cleaning and splitting of data. As an example the adhesive kinetics regression data set and the iris flower classification data set (provided by the CIP ExperimentalData package, see Appendix A) are inspected. The adhesive kinetics data set

```
regressionDataSet=CIP`ExperimentalData`GetAdhesiveKineticsDataSet[];
CIP`Graphics`ShowDataSetInfo[{"IoPairs","InputComponents",
  "OutputComponents"},regressionDataSet]
```

Number of IO pairs = 73

Number of input components = 3

Number of output components = 1

consists of 73 I/O pairs. Each input vector is of dimension 3, the output vector is of dimension 1. The iris flower data set

```
classificationDataSet=
  CIP`ExperimentalData`GetIrisFlowerClassificationDataSet[];
CIP`Graphics`ShowDataSetInfo[{"IoPairs","InputComponents",
  "OutputComponents","ClassCount"},classificationDataSet]
```

Number of IO pairs = 150

Number of input components = 4

Number of output components = 3

Class 1 with 50 members

Class 2 with 50 members

Class 3 with 50 members

consists of 150 I/O pairs with input vectors of dimension 4 and output vectors of dimension 3 (coding for 3 classes, see coding of classes for classification tasks in a following section) where 50 I/O pairs belong to each class. The pure iris flower inputs

```
inputs=CIP`ExperimentalData`GetIrisFlowerInputs[];
CIP`Graphics`ShowInputsInfo[{"InputVectors",
  "InputComponents"},inputs]
```

Number of input vectors = 150

Number of input components = 4

consist of 150 input vectors of dimension 4. Alternatively the iris flower inputs may be directly obtained from the full classification data set:

```
inputsFromDataSet=
  CIP 'Utility 'GetInputsOfDataSet[
   CIP 'ExperimentalData 'GetIrisFlowerClassificationDataSet[]];
CIP 'Graphics 'ShowInputsInfo[{"InputVectors",
  "InputComponents"},inputsFromDataSet]
```

Number of input vectors = 150

Number of input components = 4

Data set cleaning operations are often necessary to avoid fatal (or even worse: subtle) errors during data analysis. If a data set is well formed

```
input1={10.,11.,12.,13.}; output1={101.,102.};
ioPair1={input1,output1};
input2={20.,21.,22.,23.}; output2={201.,202.};
ioPair2={input2,output2};
input3={30.,31.,32.,33.}; output3={301.,302.};
ioPair3={input3,output3};
dataSet={ioPair1,ioPair2,ioPair3};
MatrixForm[CIP 'DataTransformation 'CleanDataSet[dataSet]]
```

$$\begin{pmatrix} \{10.,11.,12.,13.\} & \{101.,102.\} \\ \{20.,21.,22.,23.\} & \{201.,202.\} \\ \{30.,31.,32.,33.\} & \{301.,302.\} \end{pmatrix}$$

a CleanDataSet operation does not affect its structure or values. If components are integer instead of real numbers

```
input1={10,11,12,13}; output1={101.,102.};
ioPair1={input1,output1};
input2={20.,21.,22.,23.}; output2={201.,202.};
ioPair2={input2,output2};
input3={30.,31.,32.,33.}; output3={301.,302.};
ioPair3={input3,output3};
dataSet={ioPair1,ioPair2,ioPair3};
MatrixForm[CIP 'DataTransformation 'CleanDataSet[dataSet]]
```

$$\begin{pmatrix} \{10.,11.,12.,13.\} & \{101.,102.\} \\ \{20.,21.,22.,23.\} & \{201.,202.\} \\ \{30.,31.,32.,33.\} & \{301.,302.\} \end{pmatrix}$$

they are converted to real numbers. If an input or output component value is not defined, i.e. is NaN (Not a Number),

```
input1={10.,NaN,12.,13.}; output1={101.,102.};
```

```
ioPair1={input1,output1};
input2={20.,21.,22.,23.}; output2={201.,202.};
ioPair2={input2,output2};
input3={30.,31.,32.,33.}; output3={301.,302.};
ioPair3={input3,output3};
dataSet={ioPair1,ioPair2,ioPair3};
MatrixForm[CIP`DataTransformation`CleanDataSet[dataSet]]
```

$$\begin{pmatrix} \{10.,12.,13.\} & \{101.,102.\} \\ \{20.,22.,23.\} & \{201.,202.\} \\ \{30.,32.,33.\} & \{301.,302.\} \end{pmatrix}$$

the whole corresponding column is removed. If only the corresponding I/O pair is to be removed the RemoveNonNumberIoPairs operation should be applied:

```
input1={10.,NaN,12.,13.}; output1={101.,102.};
ioPair1={input1,output1};
input2={20.,21.,22.,23.}; output2={201.,202.};
ioPair2={input2,output2};
input3={30.,31.,32.,33.}; output3={301.,302.};
ioPair3={input3,output3};
dataSet={ioPair1,ioPair2,ioPair3};
MatrixForm[CIP`DataTransformation`RemoveNonNumberIoPairs[dataSet]]
```

$$\begin{pmatrix} \{20.,21.,22.,23.\} & \{201.,202.\} \\ \{30.,31.,32.,33.\} & \{301.,302.\} \end{pmatrix}$$

If an input component is identical in all I/O pairs

```
input1={10.,11.,12.,13.}; output1={101.,102.};
ioPair1={input1,output1};
input2={20.,11.,22.,23.}; output2={201.,202.};
ioPair2={input2,output2};
input3={30.,11.,32.,33.}; output3={301.,302.};
ioPair3={input3,output3};
dataSet={ioPair1,ioPair2,ioPair3};
MatrixForm[CIP`DataTransformation`CleanDataSet[dataSet]]
```

$$\begin{pmatrix} \{10.,12.,13.\} & \{101.,102.\} \\ \{20.,22.,23.\} & \{201.,202.\} \\ \{30.,32.,33.\} & \{301.,302.\} \end{pmatrix}$$

the whole corresponding column is removed (the removal of redundant information is favorable for clustering or machine learning operations). The WPBC classification data set (see Appendix A)

```
classificationDataSet=
 CIP`ExperimentalData`GetWPBCClassificationDataSet[];
CIP`Graphics`ShowDataSetInfo[{"IoPairs","InputComponents",
 "OutputComponents","ClassCount"},classificationDataSet]
```

Number of IO pairs = 198

Number of input components = 32

Number of output components = 2

Class 1 with 151 members

Class 2 with 47 members

has missing (NaN) values which have to be removed prior to use (in this case by removal of the corresponding I/O pairs

```
cleanedClassificationDataSet=
  CIP`DataTransformation`RemoveNonNumberIoPairs[classificationDataSet];
CIP`Graphics`ShowDataSetInfo[{"IoPairs","InputComponents",
  "OutputComponents","ClassCount"},cleanedClassificationDataSet]
```

Number of IO pairs = 194

Number of input components = 32

Number of output components = 2

Class 1 with 148 members

Class 2 with 46 members

to obtain a cleaned data set with a reduced number of valid I/O pairs). Large data sets may be split in (two) smaller data sets along one input component: As an example a 3D data set (generated around a 3D function)

```
pureOriginalFunction=
  Function[{x, y},
    1.9*(1.35+Exp[x]*Sin[13.0*(x-0.6)^2]*Exp[-y]*Sin[7.0*y])];
xRange={0.0, 1.5};
yRange={0.0, 1.5};
numberOfDataPerDimension=100;
standardDeviationRange={0.1, 0.1};
dataSet3D=
  CIP`CalculatedData`Get3dFunctionBasedDataSet[pureOriginalFunction,
    xRange,yRange,numberOfDataPerDimension,standardDeviationRange];
labels={"x","y","z"}; pointSize=0.005;
plotStyle3D=Directive[Green,Specularity[White,40],Opacity[0.4]];
CIP`Graphics`Plot3dDataSetWithFunction[
  dataSet3D,pureOriginalFunction,labels,
  GraphicsOptionPointSize -> pointSize,
  GraphicsOptionPlotStyle3D -> plotStyle3D
  ]
```

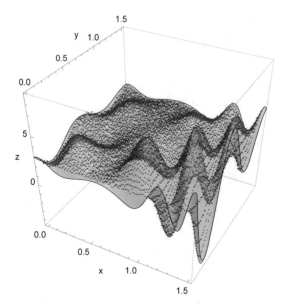

with 10.000 I/O pairs, 2 input components and 1 output component

```
CIP 'Graphics 'ShowDataSetInfo[{"IoPairs", "InputComponents",
  "OutputComponents"},dataSet3D]
```

Number of IO pairs = 10000

Number of input components = 2

Number of output components = 1

is split along input component 1 (which means a split along the *x* axis)

```
component=1;
splitInfo=
  CIP 'DataTransformation 'SplitDataSet[dataSet3D,component];
dataSetSplitInfo1=splitInfo[[1]];
dataSetSplitInfo2=splitInfo[[2]];
splitDataSet3D1=dataSetSplitInfo1[[1]];
splitDataSet3D2=dataSetSplitInfo2[[1]];
```

to obtain two smaller subsets with 5.000 I/O pairs each:

```
CIP 'Graphics 'Plot3dDataSetWithFunction[
  splitDataSet3D1,pureOriginalFunction,labels,
  GraphicsOptionPointSize -> pointSize,
  GraphicsOptionPlotStyle3D -> plotStyle3D
  ]
```

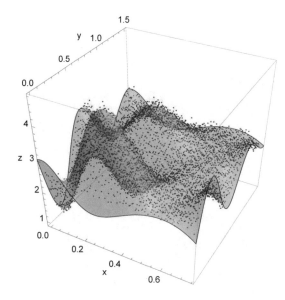

```
CIP`Graphics`ShowDataSetInfo[{"IoPairs","InputComponents",
  "OutputComponents"},splitDataSet3D1]
```

Number of IO pairs = 5000

Number of input components = 2

Number of output components = 1

```
CIP`Graphics`Plot3dDataSetWithFunction[
  splitDataSet3D2,pureOriginalFunction,labels,
  GraphicsOptionPointSize -> pointSize,
  GraphicsOptionPlotStyle3D -> plotStyle3D
  ]
```

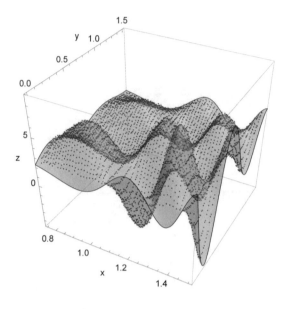

```
CIP`Graphics`ShowDataSetInfo[{"IoPairs","InputComponents",
  "OutputComponents"},splitDataSet3D2]
```

Number of IO pairs = 5000

Number of input components = 2

Number of output components = 1

1.5 Scaling of data

In principal data values should be confined to an order of magnitude around 1 by use of appropriate units, i.e. have values like 0.58, 1.47 or 3.61 but not 10255.24 or 0.00046. This not only makes them easier to comprehend it is also a virtue for numeric computing since computers do only calculate with a finite number of digits: Values that differ orders of magnitude may lead to severe calculation errors due to numerical problems.

Curve fitting methods use xy-error data without any scaling throughout this book. So all data should be reasonably scaled in advance. In contrast all clustering and machine learning methods scale the data as part of their algorithms, e.g. all minimum and maximum values of all single components of inputs and outputs are determined and then a linear transformation for each component from its [min, max] interval to interval [0, 1] is performed (in fact interval [0.05, 0.95] is used in CIP to allow very cautious extrapolations). If outputs of fitted models are calculated the inverse linear transformation is performed. So there is no need for a data preprocessing as far as clustering or machine learning is concerned.

An often neglected subtlety of data transformation may be noticed: If the x and y values of xy-error data are transformed it is essential to also transform the errors by correct error propagation. This is especially important for non-linear transformations. Since the neglect of errors belongs to the most frequent mistakes in practical data analysis its consequences are outlined in the curve fitting chapter 2.

1.6 Data errors

Experimental data are biased by errors in principal. There are three sources of errors that may be distinguished in practice:

- **Gross errors:** These kind of errors are introduced by experimental mistakes or simply bad work. They are completely avoidable by proper performance. Nevertheless gross errors are abundant in available experimental data. Usually it is tried to identify affected data as outliers but this may be very difficult especially for high dimensional data sets.
- **Systematic errors:** They introduce systematic shifts to data. Their cause may be found in subtle calibration problems or specific data preprocessing procedures (a scientific quantity is rarely measured directly). An example from the area of chemical spectra analysis is given in Appendix A and chapter 2. Systematic errors are often difficult to be detected, their avoidance needs a deep understanding and careful inspection of the measurement process.
- **Statistical errors:** These errors originate from the nature of the specific measurement process and can not be avoided in principal. They may only be reduced by replacing a measurement process with an improved one. So statistical errors are an intrinsic property of every measurement and therefore every measured datum must be attributed with its statistical error.

On the contrary simulated data are artificially constructed to only contain defined statistical errors without any systematic or gross errors. So they may play an important role for a proper assessment of a data analysis method.

Somewhere in between experimental and simulated data are calculated data, e.g. data that were produced with a fundamental theory of nature. These data do not contain errors in the statistical sense but are somehow biased by the usually approximate calculation method. In practice calculated data are treated like simulated data.

Experimental data for curve fitting tasks used to statistically estimate parameters for model functions $y = f(x)$ must provide their corresponding errors. This necessity is embodied in the xy-error triple data structure for curve fitting. Each xy-error data triple consists of an argument value x_i, a corresponding dependent value y_i and the statistical error σ_i of the y_i value: (x_i, y_i, σ_i). The errors of the x_i values are usually not taken into account, i.e. all x_i values are considered to be error-free since their errors propagate to corresponding bigger errors σ_i of the dependent y_i values. The errors σ_i are mandatory for the statistical assessment of curve fitting methods though often neglected. The statistical error must be reported or at least be estimated since every measurement is biased: There are no infinite precise measurements possible in this universe. Before starting any data analysis procedure there should always be a clear understanding of all related errors of all quantities. Machine learning methods on the contrary do in general not take data errors explicitly into account since they lack a thorough statistical basis due to the missing model function. But also for the assessment of their results the knowledge of at least the approximate size of the data's errors is helpful. If a machine learning method describes experimental data better than expected from their errors the learning procedure failed: A simple so called overtrained look-up table for the training data was constructed without any power of predictability (see chapter 4).

1.7 Regression versus classification tasks

```
Clear["Global`*"];
<<CIP`DataTransformation`
```

If a machine learning technique is set up to perform a regression task it should build model functions $\underline{f}$ that map input vectors $\underline{x}$ onto output vectors $\underline{y}$

$$\underline{y} = \underline{f}(\underline{x})$$

where the output vectors $\underline{y}$ consists of continuous components with each having a specific scientific meaning.

A classification setup is somewhat different: The machine learning method is trained to assign an input vector $\underline{x}$ to a specific class i. To achieve this goal in terms of the general regression formulation above there must be an adequate coding of the output $\underline{y}$. Throughout this book the following coding is chosen: The number of components of the output vector $\underline{y}$ is set equal to the number of desired classes. Each component y_k of output vector $\underline{y}$ codes one class. As an example the coding of 3 classes leads to the following output vectors:

$$\text{Class 1}: \underline{y} = \begin{pmatrix} 1.0 \\ 0.0 \\ 0.0 \end{pmatrix} ; \text{Class 2}: \underline{y} = \begin{pmatrix} 0.0 \\ 1.0 \\ 0.0 \end{pmatrix} ; \text{Class 3}: \underline{y} = \begin{pmatrix} 0.0 \\ 0.0 \\ 1.0 \end{pmatrix}$$

If this coding is chosen a regression task may be performed with these output vectors. A corresponding data set is called a classification data sets to indicate the coding of its output vectors. To assign an input vector $\underline{x}$ to a specific class the maximum component of the output vector $\underline{y}$ is determined: The attributed class then corresponds to the position of the maximum component in the output vector $\underline{y}$: If a trained machine learning method calculates the output vector

$$\underline{y} = \begin{pmatrix} 0.2 \\ 0.5 \\ 0.3 \end{pmatrix}$$

for an input vector $\underline{x}$ then this input vector is assigned to class 2 since component 2 (0.5) is the maximum component of output vector $\underline{y}$. Note that it is not necessary for a correct classification that the machine learning method achieves a high precision mapping onto the desired output vectors $\underline{y}$ for each class. It is sufficient when the class detection component is the maximum component. The value of 0.5 of the previous example differs considerably from the desired output component value of 1.0 but is absolutely sufficient for a correct classification in this case. Therefore in general a classification task is somewhat less demanding than a regression task for a machine learning method. If a regression task fails it may be at least feasible to classify the data onto different regions of interest.

The I/O pairs of classification data sets like

```
input1={Subscript[in,11],Subscript[in,12],Subscript[in,13]};
output1={0.0,1.0};
ioPair1={input1,output1};
input2={Subscript[in,21],Subscript[in,22],Subscript[in,23]};
output2={0.0,1.0};
ioPair2={input2,output2};
input3={Subscript[in,31],Subscript[in,32],Subscript[in,33]};
output3={0.0,1.0};
ioPair3={input3,output3};
input4={Subscript[in,41],Subscript[in,42],Subscript[in,43]};
output4={1.0,0.0};
ioPair4={input4,output4};
classificationDataSet={ioPair1,ioPair2,ioPair3,ioPair4};
MatrixForm[classificationDataSet]
```

$$\begin{pmatrix} \{in_{11},in_{12},in_{13}\} & \{0.,1.\} \\ \{in_{21},in_{22},in_{23}\} & \{0.,1.\} \\ \{in_{31},in_{32},in_{33}\} & \{0.,1.\} \\ \{in_{41},in_{42},in_{43}\} & \{1.,0.\} \end{pmatrix}$$

may be sorted ascending according to their class memberships

```
sortResult=CIP`DataTransformation`SortClassificationDataSet[
  classificationDataSet];
sortedClassificationDataSet=sortResult[[1]];
MatrixForm[sortedClassificationDataSet]
```

$$\begin{pmatrix} \{in_{41},in_{42},in_{43}\} & \{1.,0.\} \\ \{in_{11},in_{12},in_{13}\} & \{0.,1.\} \\ \{in_{21},in_{22},in_{23}\} & \{0.,1.\} \\ \{in_{31},in_{32},in_{33}\} & \{0.,1.\} \end{pmatrix}$$

with information about the inputs to class relations:

```
classIndexMinMaxList=sortResult[[2]]
```

$\{\{1,1\},\{2,4\}\}$

classIndexMinMaxList contains two elements, i.e. there are two classes: Class 1 with min/max elements {1, 1}
and class 2 with min/max elements {2, 4}. In other words: Class 1 contains one input (with index 1), class 2
contains three inputs (with indices 2, 3 and 4).

Classification task are often connected to pattern recognition: The input vector x codes a pattern (e.g. a MRI created digital image) that is mapped onto a specific class with a specific meaning (e.g. tumor tissue): So the pattern may be recognized. Machine learning methods provide strong pattern recognition abilities in principal (see chapter 4).

1.8 The structure of CIP calculations

The structure of calculations with the Computational Intelligence Packages (CIP) is largely unified: With **Get** methods data are retrieved or simulated (with the CIP ExperimentalData and CalculatedData package) that are then submitted to a **Fit** method (of the CIP CurveFit, Cluster, MLR, MPR, SVM or Perceptron package). The result of the latter is a comprehensive **info data structure** (curveFitInfo, clusterInfo, mlrInfo, mprInfo, svmInfo or perceptronInfo) that can be passed to **Show** methods for evaluation purposes like inspection of the goodness of fit or to **Calculate** methods for model related calculations. The straight forward and intuitive scheme **Get-Fit-Show-Calculate** may easily be remembered and is used throughout the book. Other CIP packages perform auxiliary tasks like the Graphics package that provides standardized 2D and 3D diagrams.

CIP methods use a lot of default settings which are unfortunately necessary for the algorithms to work but the important intricacies may be changed by options which are outlined in detail throughout the book. This will be crucial for success in data analysis applications since the default settings are not generally applicable: They are adequate in one case and lead to a disastrous failure in another.

Note that CIP is open-source and thus available in source code: You may inspect every detail of the implemented methods and even change or improve them. More details about CIP are provided in Appendix A.

The Mathematica program code used throughout the book is initialized at the beginning of each section with

```
Clear["Global`*"];
```

which deletes all prior definitions. This has to be taken into account if code is extracted since it works top-down only.

1.9 A note on reproducibility

It is a main goal of an interactive cookbook to be properly reproduced on a reader's computer system. Deviations in computational speed are inevitable due to hardware differences so that all reported calculation periods do only have a relative meaning (especially when it comes to parallelized calculations). In addition differences concerning the calculation results may be detected for different versions of the Mathematica computing platform: They may range from small numerical deviations in comparison with the reported values up to completely different outcomes of calculations in the case that small numerical deviations leads to qualitatively different decisions under the hood. These changes may be attributed to the constant improvement of the complex basic algorithms of the Mathematica system which are heavily utilized by CIP, e.g. an earlier convergence problem may no longer show up after the implementation of an additional safeguard in a newer version. These possible differences should always be taken into account by comparing the reported book's results with the (emphatically wanted) personal hands-on recalculations by the reader.

Chapter 2
Curve Fitting

```
Clear["Global`*"];
<<CIP`CalculatedData`
<<CIP`Graphics`
<<CIP`CurveFit`
```

Two-dimensional curve fitting starts with experimental xy-error data (points in diagram below) which consist of data triples (x_i, y_i, σ_i) with an argument value x_i, a corresponding dependent value y_i and the (not illustrated) statistical error σ_i of the y_i value (again note that xy-error data are generated by experimental setups which specify a x_i value and measure a corresponding y_i value for that fixed x_i value where the errors of all x_i values are not taken into account, i.e. all x_i values are considered to be error-free since their errors propagate to corresponding bigger errors σ_i of the dependent y_i values):

```
pureModelFunction=Function[x,1.0+1.0*x+0.4*x^2-0.1*x^3];
argumentRange={-2.0,5.0};
numberOfData=100;
standardDeviationRange={0.5,0.5};
xyErrorData=CIP`CalculatedData`GetXyErrorData[pureModelFunction,
  argumentRange,numberOfData,standardDeviationRange];
labels={"x","y","Data"};
CIP`Graphics`PlotXyErrorData[xyErrorData,labels]
```

© Springer International Publishing Switzerland 2016
A. Zielesny, *From Curve Fitting to Machine Learning*, Intelligent Systems
Reference Library 109, DOI 10.1007/978-3-319-32545-3_2

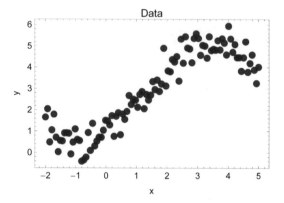

Curve fitting tries to adjust a smooth and balancing model function $f(x)$ (solid line in diagram below)

```
modelFunction=a1+a2*x+a3*x^2-a4*x^3;
argumentOfModelFunction=x;
parametersOfModelFunction={a1,a2,a3,a4};
curveFitInfo=CIP`CurveFit`FitModelFunction[xyErrorData,modelFunction,
  argumentOfModelFunction,parametersOfModelFunction];
labels={"x","y","Curve fitting"};
CIP`CurveFit`ShowFitResult[{"FunctionPlot"},xyErrorData,curveFitInfo,
  CurveFitOptionLabels -> labels];
```

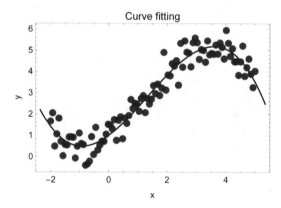

that describes the data adequately (all details will be outlined in a minute). In more mathematical terms curve fitting is a data analysis procedure which tries to construct a linear or non-linear model function

$$y = f(x)$$

from experimental xy-error data. Besides the rare case that the model function $f(x)$ is completely known (then there is nothing to be fitted: The quantity of interest may be directly calculated in this holy grail situation) three different scenarios may be distinguished:

- **Scenario 1:** The structural form of the model function $f(x)$ is theoretically or empirically known but not the values of its parameters, e.g. the structural form is known to be a straight line but the values of its parameters (i.e. of slope and intercept) are unknown.
- **Scenario 2:** The structural form of the model function $f(x)$ is unknown but it may be somehow guessed.
- **Scenario 3:** The structural form of the model function $f(x)$ is unknown and there is no idea what it is about.

Scenario 1 demands a how-to procedure to estimate the unknown parameters of the structurally known model function in an optimum way whereas scenario 2 needs a construction strategy that combines trial and error as well as good guesses in addition (in two dimensions a good guess is quite often feasible in contrast to higher dimensional machine learning problems). For scenario 3 at least some criteria may be derived that allow the construction of something that is smooth and balancing. For scenario 1 (and scenario 2 after the good guess) the estimation of optimum values for the unknown parameters of the model function is the essential step to achieve a good fit. If the statistical distribution of the experimental xy-error data is known this may be performed on a solely statistical ground which then defines the criterion of optimization (see [Hamilton 1964], [Barlow 1989], [Bevington 2002] or [Brandt 2002]). For all further discussions a Gaussian (normal) distribution of experimental errors is always assumed which is the most common case in practice (thanks to the central limit theorem). In addition each data triple (x_i, y_i, σ_i) of the xy-error data is assumed to be statistically independent of each other, i.e. the values of a data triple are by no means influenced by the values of other data triples (which leads to a so called maximum likelihood estimation). Note that this latter assumption is a serious and hard to achieve precondition since a lot of natural (and social) phenomena are subtly correlated to each other. So special care has to be taken for experimental setups to achieve true independence.

If the model function could be successfully fitted to the data it may be used twofold: For interpolation purposes to calculate function values within the experimental argument range $[x_{min}, x_{max}]$ as well as for extrapolation purposes to calculate function values outside this argument range. The latter is possible since the structural form of the model function is a priori known. This is a clear difference to mere data smoothing or machine learning methods that have no initial idea about the model function: Their constructed model functions can not be used for extrapolation purposes in principle (multiple linear regression will be an exception but this method is usually not accounted to fall into the machine learning reign).

When a model function is to be guessed (scenario 2) some general considerations should be taken into account. First the number of parameters should be considerably smaller than the number of data (of course this should apply to scenario 1 too): Otherwise a simple look-up table would be easier to create. The number of parameters should be as small as possible or in other words: The model function with fewer parameters that describes the data satisfactorily is preferred to the model function with more parameters. This is a well-known utilization of Occam's razor - one of the philosophical principles of scientific practice that states that the explanation of any phenomenon should make as few assumptions as possible.

In the case that there is no idea of the functional form of a model function (scenario 3) a convincing data smoothing procedure is outlined that uses smoothing cubic splines. It should be clear that this smoothing model function can not be used for extrapolation purposes as mentioned before.

Chapter 2 starts with an outline of necessary basics: The criteria and quantities for curve fitting and data smoothing are intuitively derived with arguments of plausibility only (section 2.1). To tackle scenario 1 quantities and diagrams to assess the goodness of fit are illustrated by means of a perfect straight line fit to simulated data (section 2.2). The empirical construction of a model function for real experimental data on the basis of trial and error in combination with educated guesses is outlined to illustrate scenario 2 as a next step. The extrapolation problem is addressed in particular (section 2.3). Problems and pitfalls of curve fitting tasks are discussed in detail afterwards: They are at heart of this chapter since they are often the hurdle that prevents practitioners from successful data analysis. Fitting non-linear model functions requires adequate start values for all parameters that allow the fitting procedure to succeed: Problems and search strategies are sketched. The extraction of a model function from experimental data may be challenging up to ambiguous which is discussed for difficult curve fitting problems. Model functions themselves may be inappropriately constructed that may lead to fatal pitfalls. A more subtle kind of inappropriateness of a model function is exemplified by an effort to extract information from data that they simply do not contain (section 2.4). The estimation of parameters' errors, possible corrections and the influence of confidence levels are demonstrated afterwards. Parameters' errors are influenced by the precision of data as well as their number: An iterative method for the estimation of the necessary number of data to achieve a desired parameters' precision is suggested. Experimental data of relatively low precision may lead to large parameter errors for specific model functions: This prevents support or rejection of underlying theoretical considerations. In this context there is a strong temptation for educated cheating which means putting up unjustified statements that seem to be advised by the data analysis procedure - an illustrative example is shown. The discussion of the influence of experimental errors on the fitted optimum parameters' values and the related possible problems of data transformations complement this topic (section 2.5). It is often necessary to enhance theoretical model functions by empirical parameters to successfully describe experimental data. An example is discussed that also makes use of the dangerous removal of outliers (section 2.6). Mere data smoothing without any knowledge of a model function (scenario 3) is demonstrated to create a smooth and balancing description of data (section 2.7). Finally the whole chapter is summarized with a few cookbook recipes for successful curve fitting and data smoothing (section 2.8).

2.1 Basics

A curve fitting procedure may be derived with mathematical statistics (see [Hamilton 1964], [Barlow 1989], [Bevington 2002], [Brandt 2002] and [Press 2007]) whereas this section follows an intuitive approach that only uses arguments of plausibility - but of course comes to the same results: How is a model function to be fit? How may data satisfactorily be smoothed?

2.1.1 Fitting data

At first sight it is obvious that a good fit should minimize the so called residuals, i.e. the deviations between experimental values y_i and their corresponding calculated function values $f(x_i)$:

$$y_i - f(x_i) \longrightarrow \text{minimize!}$$

Since positive and negative residuals should be treated equally they may be squared to get rid of the sign:

$$(y_i - f(x_i))^2 \longrightarrow \text{minimize!}$$

The absolute value or a higher even power of the residuals could be taken as well with respect to plausibility but this would lead to other statistics so the square is taken for statistically independent and normally distributed deviations (behind the scenes: The square stems from the square in the exponential term of a normal distribution where the minimum postulation leads to maximum likelihood). The sum of squared residuals of all K xy-error data triples

$$\sum_{i=1}^{K} (y_i - f(x_i))^2 \longrightarrow \text{minimize!}$$

may be calculated as a quantity to be minimized for a good fit. But so far the errors σ_i of the experimental values y_i are neglected. The smaller a single error σ_i the more precise its corresponding experimental value y_i. If each residual is divided by its corresponding error an individual weight is attributed: The resulting fraction

$$\frac{y_i - f(x_i)}{\sigma_i}$$

is the bigger the smaller the error σ_i is. With the weighted sum of squares

$$\sum_{i=1}^{K} \left(\frac{y_i - f(x_i)}{\sigma_i} \right)^2 \longrightarrow \text{minimize!}$$

a plausible minimization quantity is finally achieved: It becomes smaller the smaller the residuals are, i.e. the better the model function $f(x)$ describes the data. Each single residual is weighted with its error σ_i: The smaller an error σ_i the more the corresponding residual $(y_i - f(x_i))$ is taken into account (i.e. the more it contributes to the sum). This minimization process is known in statistics as the method of least squares and the weighted sum of squares is called χ^2 ("chi-square"):

$$\chi^2 = \sum_{i=1}^{K} \left(\frac{y_i - f(x_i)}{\sigma_i} \right)^2$$

If the L parameters a_1 to a_L of the model function f are explicitly written

$$\chi^2(a_1, ..., a_L) = \sum_{i=1}^{K} \left(\frac{y_i - f(x_i, a_1, ..., a_L)}{\sigma_i} \right)^2$$

it becomes obvious that the quantity χ^2 is a function of these parameters: The parameters a_1 to a_L of the model function f are the variables of the quantity χ^2 which is to be minimized. Thus minimization of $\chi^2(a_1, ..., a_L)$ means finding values for the parameters a_1 to a_L so that the value of $\chi^2(a_1, ..., a_L)$ becomes a global minimum in parameters' value regions that have scientific meaning. The values of the parameters a_1 to a_L at the global minimum of $\chi^2(a_1, ..., a_L)$ are then called the optimum estimates for the true parameter values in a statistical sense. Note that the functional form of f is assumed to be true as a precondition of all statistical procedures: Only parameter values can be estimated but not the structural form of the function f itself.

In summary a linear or non-linear curve fitting procedure is a mere global minimization of the quantity $\chi^2(a_1, ..., a_L)$. The global minimum may be calculated analytically in the case that the model function f is linear in its parameters: Then $\chi^2(a_1, ..., a_L)$ is a parabolic hyper surface and possesses exactly one minimum. But it may only be approximated with an iterative search strategy in the case that f is non-linear in its parameters (compare chapter 1 and [FitModelFunction] in the references). In the latter case the quantity $\chi^2(a_1, ..., a_L)$ may contain multiple minima and the minimization procedure may fail (e.g. get stuck in a local minimum, exceed the defined maximum number of iterations etc.). Failure will be explicitly explored and discussed in subsequent sections.

2.1.2 Useful quantities

There are a number of related statistical quantities that will prove to be useful for further discussions. If the model function describes the data well the residuals $(y_i - f(x_i))$ should be comparable in size to the errors σ_i on average (otherwise the errors σ_i would not be true errors but systematically too large or too small on average). This means that the fractions

$$\frac{y_i - f(x_i)}{\sigma_i} \approx 1$$

should be close to 1 on average. So the sum of squares evaluates approximately to

$$\chi^2 = \sum_{i=1}^{K} \left(\frac{y_i - f(x_i)}{\sigma_i} \right)^2 \approx \sum_{i=1}^{K} (1)^2 = \sum_{i=1}^{K} 1 = 1 + 1 + ... + 1 = K$$

With this result in mind a statistical quantity named χ^2_{red} ("reduced chi-square") can be defined as

$$\chi^2_{red}(a_1, ..., a_L) = \frac{\chi^2(a_1, ..., a_L)}{K - L} = \frac{1}{K - L} \sum_{i=1}^{K} \left(\frac{y_i - f(x_i, a_1, ..., a_L)}{\sigma_i} \right)^2 \approx 1 \text{ for } K \gg L$$

which evaluates to a value close to 1 for a good fit since the number of data K should be considerably larger than the number of parameters of the model function L, i.e. $K \gg L$. The denominator $(K - L)$ is called degrees of freedom since the parameter values are deduced from the data. The residuals of a fit may be condensed into the single statistical quantity σ_{fit} called the standard deviation of the fit. If all errors σ_i are identical (i.e. equal to σ) the standard deviation of the fit is defined as

$$\sigma_{fit} = \sqrt{\frac{1}{K - L} \sum_{i=1}^{K} (y_i - f(x_i, a_1, ..., a_L))^2} \text{ for } \sigma_i = \sigma \; ; \; i = 1, ..., K$$

In general with individual errors σ_i the standard deviation of the fit is expressed as

$$\sigma_{fit} = \sqrt{\frac{1}{K - L} \sum_{i=1}^{K} \left(\frac{y_i - f(x_i, a_1, ..., a_L)}{\sigma_i} \right)^2} \Big/ \sqrt{\frac{1}{K} \sum_{i=1}^{K} \frac{1}{\sigma_i^2}}$$

where the latter equation reduces to the previous one in the case of equal σ_i. The statistical standard deviation of the fit is similar to a purely empirical quantity called the root mean squared error (RMSE). In this context the RMSE of a fit is simply defined as

$$\text{RMSE} = \sqrt{\frac{1}{K} \sum_{i=1}^{K} (y_i - f(x_i))^2}$$

A RMSE may readily be generalized to machine learning applications for problems in multiple dimensions. The quantities $\chi^2_{red}(a_1, ..., a_L)$, σ_{fit} and RMSE respectively may be used to assess the goodness of a fit. As far as the data's errors are concerned a situation quite often encountered in practice is the following: The precise statistical errors σ_i of the y_i values are unknown, but weights w_i for the y_i values are available. The relation of the weights w_i and their corresponding statistical errors σ_i can be written as

$$\sigma_i = \frac{\alpha}{w_i}$$

where the factor α is used to calculate a statistical error from its corresponding weight. Weights are defined to be the heavier the bigger they are: Statistical errors lead to higher weights the smaller they are. Therefore weights and errors are inversely proportional by the constant factor α. If only weights are known the factor α is unknown. But a reasonable estimate of α may be obtained from the χ^2_{red} value (which should be close to 1 as mentioned before):

$$\chi^2_{red}(a_1, ..., a_L) = \frac{\chi^2(a_1, ..., a_L)}{K - L} = \frac{1}{K - L} \sum_{i=1}^{K} \left(\frac{y_i - f(x_i, a_1, ..., a_L)}{\sigma_i} \right)^2 \approx 1$$

$$\frac{1}{K - L} \sum_{i=1}^{K} \left(\frac{y_i - f(x_i, a_1, ..., a_L)}{\frac{\alpha}{w_i}} \right)^2 = \frac{1}{\alpha^2(K - L)} \sum_{i=1}^{K} w_i^2 (y_i - f(x_i, a_1, ..., a_L))^2 \approx 1$$

$$\alpha \approx \sqrt{\frac{1}{(K - L)} \sum_{i=1}^{K} w_i^2 (y_i - f(x_i, a_1, ..., a_L))^2}$$

In practice it is common to correct the errors σ_i of the xy-error data with this method: The original errors $\sigma_{\text{original},i}$ of the xy-error data are assumed to be weights only

$$w_i = \frac{1}{\sigma_{\text{original},i}}$$

and are transformed after the fit into the corrected errors $\sigma_{\text{corrected},i}$:

$$\sigma_{\text{corrected},i} = \frac{\alpha}{w_i} = \frac{\alpha}{\frac{1}{\sigma_{\text{original},i}}} = \alpha \sigma_{\text{original},i}$$

These corrected errors $\sigma_{\text{corrected},i}$ are then used for the derivation of further statistical quantities related to the fit - above all the estimation of errors $\sigma_{a_1}, ..., \sigma_{a_L}$ of the model function's parameters $a_1, ..., a_L$.

2.1.3 Smoothing data

When it comes to mere data smoothing statistics is no longer helpful. As already pointed out statistics is not able to guess a model function in principal - it may only estimate optimum values of a structurally known model function's parameters and related quantities with a bunch of statistical preconditions (like independent and normally distributed data). Therefore data smoothing comprises a set of techniques that somehow construct a smooth and balancing interpolating model function from experimental xy-error data (extrapolation is of course not possible). There is no objective way to smooth data so there is nothing like the best smoothing technique. With data smoothing we are back to the jungle where everything is allowed that leads to a satisfactory result. Among the numerous techniques for smoothing experimental xy-error data the smoothing cubic splines method is sketched in the following (see [Reinsch 1967] and [Reinsch 1971]). This method seems to produce satisfactory results accepted by experimental scientists in general - but this technique is by no means better or superior to others. Data smoothing with cubic splines, i.e. cubic polynomials

$$y = f(x) = a_1 + a_2 x + a_3 x^2 + a_4 x^3,$$

uses the already sketched χ^2_{red} value

$$\chi^2_{\mathrm{red}} = \frac{\chi^2}{K} = \frac{1}{K} \sum_{i=1}^{K} \left(\frac{y_i - f(x_i)}{\sigma_i} \right)^2$$

as a reasonable first control parameter: Good data smoothing should lead to a smoothing model function with

$$\chi^2_{\mathrm{red}} \approx 1$$

For convenience the same notation is used for data smoothing as for statistically fitting model functions. But data smoothing is not statistically based. Quantities like χ^2_{red} have no longer any statistical meaning but are simply used as helpful quantities for the smoothing task. Therefore χ^2_{red} is set to $\frac{\chi^2}{K}$ since there are no statistical degrees of freedom for data smoothing. The same applies to quantities like σ_{fit}: They also use the number of data K instead of the degrees of freedom within this context. The cubic splines are constructed from data point to data point, i.e. for K data points $(K-1)$ cubic splines have to be used. These cubic splines must be adjusted to achieve the initially defined χ^2_{red} value together with the constraint of a criterion of smoothness: The resulting smoothing model function (composed of the piecewise cubic splines) should posses the smallest overall curvature possible to achieve the predefined χ^2_{red} value. Since the curvature is measured by the second derivative $f''(x)$ of the model function the integral of the square of the second derivative over the argument interval $[x_1, x_L]$ is to be minimized

$$\int_{x_1}^{x_L} \left(\frac{d^2 f(x)}{dx^2} \right)^2 dx \longrightarrow \text{minimize!}$$

where the xy-error data are assumed to be sorted ascending according to their argument values x_i. The square of the second derivative is used for equal treatment of positive and negative curvature. Both criteria are of course contradicting each other: The smaller the χ^2_{red} value the larger the curvature integral value and vice versa. With this mutual interplay a satisfactory smooth and balancing model function may be constructed.

2.2 Evaluating the goodness of fit

```
Clear["Global`*"];
<<CIP`CalculatedData`
<<CIP`Graphics`
<<CIP`CurveFit`
```

A simple example is used to demonstrate the curve fitting procedure and the evaluation of the goodness of fit. One thousand (x_i, y_i, σ_i) data triples

```
numberOfData=1000;
```

are simulated around the straight line $y = f(x) = 1 + 2x$

```
pureOriginalFunction=Function[x,1.0+2.0*x];
```

in the argument range [2, 5]

```
argumentRange={2.0,5.0};
```

with a relative error of 5% of the function value (since the straight line is constantly increasing a minimum argument value of 2.0 leads to a minimum function value of 5.0: A relative error of 5% for 5.0 is an absolute value of 0.25. A maximum argument value of 5.0 corresponds to a function value of 11.0 with a 5% relative error of 0.55)

```
errorType="Relative";
standardDeviationRange={0.05,0.05};
```

using the CIP CalculatedData package. All data are normally distributed around their function values, the relative error denotes the standard deviation of the normal distribution used for the data generation (i.e. for a y value of 5 a standard deviation of 0.25 is used, for a y value of 11 a standard deviation of 0.55 respectively):

```
xyErrorData=CIP`CalculatedData`GetXyErrorData[pureOriginalFunction,
  argumentRange,numberOfData,standardDeviationRange,
  CalculatedDataOptionErrorType -> errorType];
```

Here is a plot of the mere simulated data:

```
labels={"x","y","Simulated data"};
pointSize=0.01;
CIP`Graphics`PlotXyErrorData[xyErrorData,labels,
  GraphicsOptionPointSize -> pointSize]
```

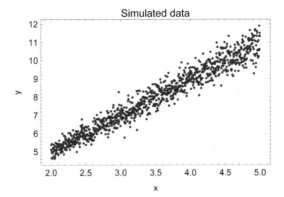

Curve fitting procedures are performed with the CIP CurveFit package. For a fit the model function itself, the argument and the parameters of the model function must be defined

```
modelFunction=a1+a2*x;
argumentOfModelFunction=x;
parametersOfModelFunction={a1,a2};
```

and submitted together with the xy-error data to the FitModelFunction method to produce a result captured in a curveFitInfo data structure (see [FitModelFunction] for algorithmic details):

```
curveFitInfo=CIP`CurveFit`FitModelFunction[xyErrorData,
  modelFunction,argumentOfModelFunction,parametersOfModelFunction];
```

If no error messages are thrown the fit procedure was successful and results can be inspected. The function plot with the fitted straight line and the simulated data painted above provides a first impression:

```
labels={"x","y","Simulated data and model function"};
CIP`CurveFit`ShowFitResult[{"FunctionPlot"},xyErrorData,
  curveFitInfo,
  GraphicsOptionPointSize -> pointSize,
  CurveFitOptionLabels -> labels];
```

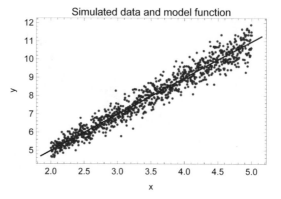

The fit looks perfect which is also affirmed by inspection of the residuals, i.e. the deviations between the data and the model function: The residuals plot for the relative residuals

```
CIP'CurveFit'ShowFitResult[{"RelativeResidualsPlot"},xyErrorData,
  curveFitInfo,GraphicsOptionPointSize -> pointSize];
```

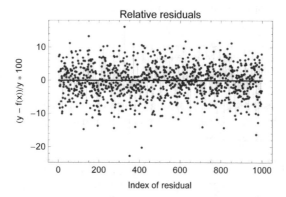

exhibits statistically distributed residuals predominantly in the expected value range of ± 5% without any systematic deviation patterns. Residuals plots are probably the most important goodness-of-fit visualizations: If they look good the fit in general is good (but compare comments on educated cheating below). Note that the index of an residual corresponds to the x value of its data triple: Residual with index 1 corresponds to the data triple with the smallest x-value, the residual with the highest index to the data triple with the maximum x-value. The standard deviation of the fit σ_{fit}

```
CIP `CurveFit `ShowFitResult[{"SDFit"},xyErrorData,curveFitInfo];
```

Standard deviation of fit = 3.699×10^{-1}

lies well within the range of (absolute) simulated errors from 0.25 to 0.55. The value of χ^2_{red}

```
CIP `CurveFit `ShowFitResult[{"ReducedChiSquare"},xyErrorData,
    curveFitInfo];
```

Reduced chi-square of fit = 9.955×10^{-1}

is close to 1 as expected. The fitted model function is:

```
CIP `CurveFit `ShowFitResult[{"ModelFunction"},xyErrorData,
    curveFitInfo];
```

Fitted model function:

$1.01901 + 1.98941x$

Note that the estimated optimum parameter values are not identical to the true parameter value of 1.0 and 2.0 used for the data generation. The errors of the simulated data are propagated to corresponding errors of the estimated optimum parameter's values so the latter are also not exact but biased by errors:

```
CIP `CurveFit `ShowFitResult[{"ParameterErrors"},xyErrorData,
    curveFitInfo];
```

	Value	Standard error	Confidence region
Parameter a1 =	1.01901	0.0454137	{0.973574, 1.06445}
Parameter a2 =	1.98941	0.014094	{1.9753, 2.00351}

The estimated optimum value of parameter a_1 is 1.02, its standard error is 0.05: So the parameter value lies with a standard statistical probability of 68.3% in the confidence region 1.02 ± 0.05, i.e. interval [0.97, 1.07]. Within linear statistics an awful lot of additional statistical quantities could be deduced. Within the scope of this book the discussion is restricted to basic quantities that play the most important role for evaluation and analysis purposes and those quantities and diagrams that may readily be generalized to machine learning applications for problems with more dimensions. The empirical root mean squared error RMSE should also lie within the range of (absolute) simulated errors from 0.25 to 0.55

```
CIP `CurveFit `ShowFitResult[{"RMSE"},xyErrorData,curveFitInfo];
```

Root mean squared error (RMSE) = 4.044×10^{-1}

and is similar to σ_{fit} as expected. The mean, median, standard deviation and maximum values of the (absolute) relative residuals do correspond perfectly to the simulated errors:

```
CIP `CurveFit`ShowFitResult[{"RelativeResidualsStatistics"},
  xyErrorData,curveFitInfo];
```

Definition of 'Residual (percent)': 100*(Data - Model)/Data

Out 1 : Residual (percent): Mean/Median/Maximum Value = $4.01 / 3.28 / 2.27 \times 10^{1}$

Out 1 means output component 1: In two-dimensional curve fitting there is only one output component whereas machine learning problems with more dimensions may contain several output components. Another frequently used diagram is the model-versus-data plot: The output (function value) of the model function is plotted against the corresponding data value:

```
CIP `CurveFit`ShowFitResult[{"ModelVsDataPlot",
  "CorrelationCoefficient"},xyErrorData,curveFitInfo,
  GraphicsOptionPointSize -> pointSize];
```

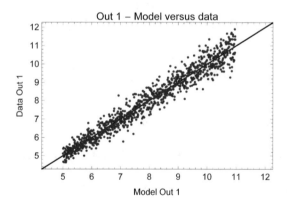

Out 1 : Correlation coefficient = 0.973565

A statistical (Pearson) correlation coefficient was calculated in addition that condenses the agreement between data and output values into a single quantity (where a value closer to one means a desired high correlation between both quantities and a value closer to zero an unwanted low correlation which thus motivates a closer look at the used model function with respect to its appropriateness). In an alternative diagram all model function values are sorted in ascending order and are jointly plotted with the corresponding data values above:

```
CIP `CurveFit `ShowFitResult[{"SortedModelVsDataPlot"},xyErrorData,
  curveFitInfo];
```

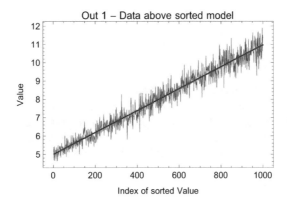

The data line above should statistically/randomly crawl around the model line below (the model function outputs) as shown in this perfect example. If the statistical distribution of relative residuals is approximated by the frequency of relative residuals within a number of interval bins (default: 20 bins) that cover the whole range of relative residual values a normal distribution around zero is approximated as expected

```
CIP `CurveFit `ShowFitResult[
  {"RelativeResidualsDistribution"},xyErrorData,
  curveFitInfo];
```

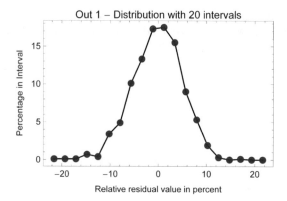

since a normal distribution was used to generate the data. The width of the approximated Gaussian bell curve corresponds perfectly to the 5% value of the standard deviation used for the data generation above. All plots reveal an excellent and very convincing model function fit. In the next section it is shown how the sketched quantities and diagrams may be utilized to construct a model function for real experimental data.

2.3 How to guess a model function

```
Clear["Global`*"];
<<CIP`ExperimentalData`
<<CIP`Graphics`
<<CIP`CurveFit`
```

As a practical example a model function for the temperature dependence of the viscosity of water is to be constructed. The viscosity of a liquid is a dynamic property which is the result of the specific molecular interactions describable in the reign of quantum theory. But the dynamics of these interactions is too complex to be calculated ab-initio on the grounds of today's science. Moreover water is not a simple liquid in chemical terms although it is so well-known from everyday life: The water molecules form specific dynamic supramolecular structures due to their ability to create hydrogen bonds - specific weak quantum-mechanical interactions that also hold our DNA strands together. The experimental data are provided by the CIP ExperimentalData package (see Appendix A for reference):

```
xyErrorData=CIP`ExperimentalData`GetWaterViscosityXyErrorData[];
```

They describe the temperature dependence of the viscosity η of water in the temperature range from 293.15 to 323.15 K (20 to 50 degree Celsius) with a very small estimated experimental error of $0.0001\ (10^{-4})$ cP (**centi-P**oise is the scientific unit of viscosity) as is illustrated by the mere data plot:

```
labels={"T [K]","\[Eta] [cP]","Viscosity of water"};
CIP`Graphics`PlotXyErrorData[xyErrorData,labels]
```

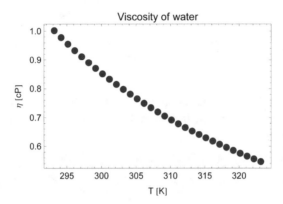

The dependence of the viscosity on the temperature is distinct but not dramatically non-linear as may be shown by an initial straight-line fit:

$$\eta = f(T) = a_1 + a_2 T$$

```
modelFunction=a1+a2*T;
argumentOfModelFunction=T;
parametersOfModelFunction={a1,a2};
curveFitInfo=CIP`CurveFit`FitModelFunction[xyErrorData,
 modelFunction,argumentOfModelFunction,parametersOfModelFunction];
labels={"T [K]","\[Eta] [cP]","Data above model function"};
CIP`CurveFit`ShowFitResult[{"FunctionPlot"},xyErrorData,
 curveFitInfo,CurveFitOptionLabels -> labels];
```

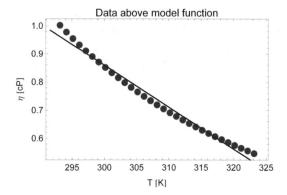

But the residuals (i.e. the deviations between data and linear model) are orders of magnitude larger than the experimental errors and they reveal a distinct systematic deviation pattern:

```
CIP 'CurveFit 'ShowFitResult[{"AbsoluteResidualsPlot",
  "CorrelationCoefficient"},xyErrorData,curveFitInfo];
```

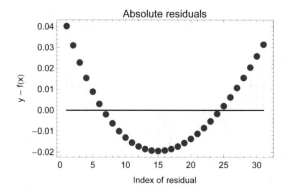

Out 1 : Correlation coefficient = 0.991451

Thus a linear straight line is only a poor model for the data. Note that the popular correlation coefficient is very close to 1 which indicates a high correlation between data and machine output: This is often cited by practitioners as a convincing goodness of fit criterion but it is a number which has to be judged with caution (see discussion below). An improvement may be attempted by introduction of a third parameter to build a (non-linear) quadratic parabola

$$\eta = f(T) = a_1 + a_2 T + a_3 T^2$$

as a model function:

```
modelFunction=a1+a2*T+a3*T^2;
parametersOfModelFunction={a1,a2,a3};
curveFitInfo=CIP'CurveFit'FitModelFunction[xyErrorData,
 modelFunction,argumentOfModelFunction,parametersOfModelFunction];
CIP'CurveFit'ShowFitResult[{"FunctionPlot"},xyErrorData,
 curveFitInfo,CurveFitOptionLabels -> labels];
```

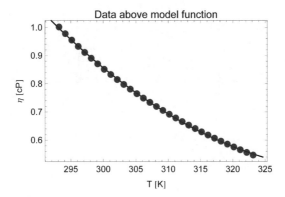

The function plot looks better and the residuals are reduced by an order of magnitude but are still beyond acceptability:

```
CIP'CurveFit'ShowFitResult[{"AbsoluteResidualsPlot",
 "CorrelationCoefficient"},xyErrorData,curveFitInfo];
```

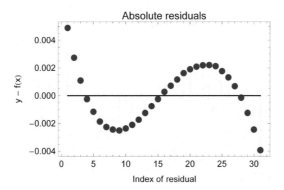

Out 1 : Correlation coefficient = 0.999888

In principal the degree of the fit polynomial could be raised with additional parameters to improve the fit but this strategy is generally a poor one: The high-order polynomials tend to oscillate and may not be predictive for extrapolation or even interpolation purposes (also compare below). Since the viscosity is decreasing with increasing temperature a two-parameter inversely proportional approach seems to be a plausible alternative trial:

$$\eta = f(T) = a_1 + \frac{a_2}{T}$$

```
modelFunction=a1+a2/T;
parametersOfModelFunction={a1,a2};
curveFitInfo=CIP`CurveFit`FitModelFunction[xyErrorData,
 modelFunction,argumentOfModelFunction,parametersOfModelFunction];
CIP`CurveFit`ShowFitResult[{"FunctionPlot"},xyErrorData,
 curveFitInfo,CurveFitOptionLabels -> labels];
```

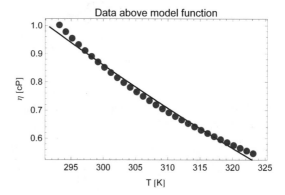

Unfortunately there is no real improvement but it might be a good idea to shift the data along the T axis with a third parameter in addition:

$$\eta = f(T) = a_1 + \frac{a_2}{a_3 - T}$$

```
modelFunction=a1+a2/(a3-T);
parametersOfModelFunction={a1,a2,a3};
startParameters={{a1,1.0},{a2,-10.0},{a3,250.0}};
curveFitInfo=CIP`CurveFit`FitModelFunction[xyErrorData,
 modelFunction,argumentOfModelFunction,parametersOfModelFunction,
 CurveFitOptionStartParameters -> startParameters];
CIP`CurveFit`ShowFitResult[{"FunctionPlot"},xyErrorData,
 curveFitInfo,CurveFitOptionLabels -> labels];
```

Note that start parameters had to be introduced to perform a successful fit: This necessity is addressed in the next sections to ease the current discussion.

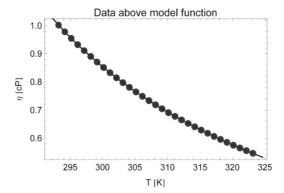

For the first time the function plot seems to be convincing. The residuals plot shows a dramatic improvement

```
CIP'CurveFit'ShowFitResult[{"AbsoluteResidualsPlot",
 "CorrelationCoefficient"},xyErrorData,curveFitInfo];
```

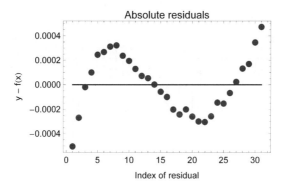

Out 1 : Correlation coefficient = 0.999998

with residuals in the order of the experimental error. But an unlovely systematic deviation pattern can still be detected: This indicates that the true functional form is still missed. As another alternative a two-parameter decaying exponential function may be tried

$$\eta = f(T) = a_1 \exp\{a_2 T\}$$

```
modelFunction=a1*Exp[a2*T];
parametersOfModelFunction={a1,a2};
```

```
startParameters={{a1,0.1},{a2,-0.001}};
curveFitInfo=CIP'CurveFit'FitModelFunction[xyErrorData,
 modelFunction,argumentOfModelFunction,parametersOfModelFunction,
 CurveFitOptionStartParameters -> startParameters];
CIP'CurveFit'ShowFitResult[{"FunctionPlot"},xyErrorData,
 curveFitInfo,CurveFitOptionLabels -> labels];
```

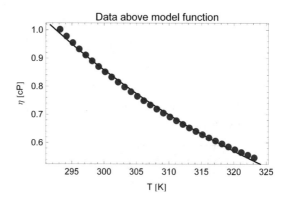

again with a poor result

```
CIP'CurveFit'ShowFitResult[{"AbsoluteResidualsPlot",
 "CorrelationCoefficient"},xyErrorData,curveFitInfo];
```

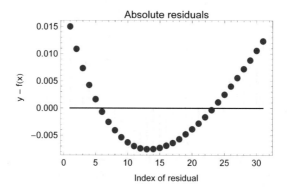

Out 1 : Correlation coefficient = 0.998755

so the use of an inverse argument in the exponential may be a choice

$$\eta = f(T) = a_1 \exp\left\{\frac{a_2}{T}\right\}$$

```
modelFunction=a1*Exp[a2/T];
parametersOfModelFunction={a1,a2};
startParameters={{a1,0.1},{a2,1000.0}};
curveFitInfo=CIP`CurveFit`FitModelFunction[xyErrorData,
 modelFunction,argumentOfModelFunction,parametersOfModelFunction,
 CurveFitOptionStartParameters -> startParameters];
CIP`CurveFit`ShowFitResult[{"FunctionPlot"},xyErrorData,
 curveFitInfo,CurveFitOptionLabels -> labels];
```

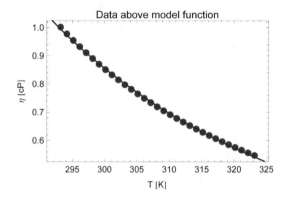

that actually offers a better outcome:

```
CIP`CurveFit`ShowFitResult[{"AbsoluteResidualsPlot",
 "CorrelationCoefficient"},xyErrorData,curveFitInfo];
```

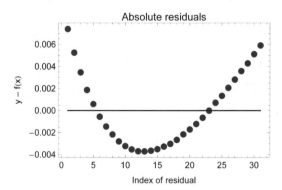

Out 1 : Correlation coefficient = 0.999704

The introduction of the exponential function with a reciprocal argument produced the best two-parameter fit so far (this is also a historical result obtained by Andrade, see [Andrade 1934]: Note that the fitted model function can be linearized by a logarithmic transformation - the only feasible solution for non-linear problems in the precomputing era). Since shifting along the T axis with a third parameter was successful earlier it is tried again with the new functional form:

$$\eta = f(T) = a_1 \exp\left\{\frac{a_2}{a_3 - T}\right\}$$

```
modelFunction=a1*Exp[a2/(a3-T)];
parametersOfModelFunction={a1,a2,a3};
startParameters={{a1,0.1},{a2,-500.0},{a3,150.0}};
curveFitInfo=CIP`CurveFit`FitModelFunction[xyErrorData,
  modelFunction,argumentOfModelFunction,parametersOfModelFunction,
  CurveFitOptionStartParameters -> startParameters];
CIP`CurveFit`ShowFitResult[{"FunctionPlot"},xyErrorData,
  curveFitInfo,CurveFitOptionLabels -> labels];
```

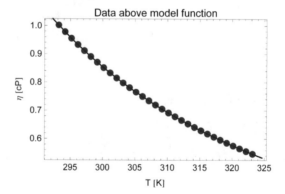

From visual inspection the fit looks perfect and the residuals plot

```
CIP`CurveFit`ShowFitResult[{"AbsoluteResidualsPlot",
  "CorrelationCoefficient"},xyErrorData,curveFitInfo];
```

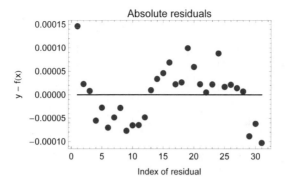

Out 1 : Correlation coefficient = 1.

reveals residuals that satisfactorily correspond to the experimental error of 0.0001 cP (this model function was historically found by Vogel after laborious linearization work, see [Vogel 1921]). Since a systematic pattern of deviations is still obvious two nearby improvements are finally tested which do not increase the number of parameters. First the initial factor is divided by T to try a combination with the inversely proportional approach tested earlier

$$\eta = f(T) = \frac{a_1}{T} \exp\left\{\frac{a_2}{a_3 - T}\right\}$$

```
modelFunction=a1/T*Exp[a2/(a3-T)];
parametersOfModelFunction={a1,a2,a3};
startParameters={{a1,0.1},{a2,-500.0},{a3,150.0}};
curveFitInfo=CIP`CurveFit`FitModelFunction[xyErrorData,
  modelFunction,argumentOfModelFunction,parametersOfModelFunction,
  CurveFitOptionStartParameters -> startParameters];
CIP`CurveFit`ShowFitResult[{"AbsoluteResidualsPlot",
  "CorrelationCoefficient"},xyErrorData,curveFitInfo];
```

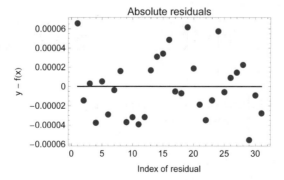

Out 1 : Correlation coefficient = 1.

```
ShowFitResult[{"SDFit"},xyErrorData,curveFitInfo];
```

Standard deviation of fit = 3.321×10^{-5}

and in addition the shift along the T axis is generalized:

$$\eta = f(T) = \frac{a_1}{a_3 - T} \exp\left\{\frac{a_2}{a_3 - T}\right\}$$

```
modelFunction=a1/(a3-T)*Exp[a2/(a3-T)];
parametersOfModelFunction={a1,a2,a3};
startParameters={{a1,0.1},{a2,-500.0},{a3,150.0}};
curveFitInfo=CIP`CurveFit`FitModelFunction[xyErrorData,
 modelFunction,argumentOfModelFunction,parametersOfModelFunction,
 CurveFitOptionStartParameters -> startParameters];
CIP`CurveFit`ShowFitResult[{"AbsoluteResidualsPlot",
 "CorrelationCoefficient"},xyErrorData,curveFitInfo];
```

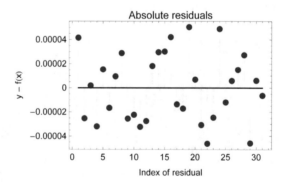

Out 1 : Correlation coefficient = 1.

```
CIP`CurveFit`ShowFitResult[{"SDFit"},xyErrorData,curveFitInfo];
```

Standard deviation of fit = 2.937×10^{-5}

The latter model function produces an absolutely convincing result: Systematic deviation patterns are vanished and the residuals are even below the estimated experimental error. This is also revealed by the χ^2_{red} value of

```
CIP`CurveFit`ShowFitResult[{"ReducedChiSquare"},xyErrorData,
 curveFitInfo];
```

Reduced chi-square of fit = 8.625×10^{-2}

which is considerably below 1 (this finding will be discussed in a subsequent chapter in combination with parameter errors). The optimum description of the data achieved by empirical construction may now be stated:

```
CIP`CurveFit`ShowFitResult[{"ModelFunction"},xyErrorData,
  curveFitInfo];
```

Fitted model function:

$$19.3098e^{-\frac{200.831}{179.802-T}} \frac{200.831}{179.802-T}$$

For a satisfactory description of the data a three-parameter model function is necessary compared to the two-parameter results (this view is also supported by the historical trend from Andrade to Vogel, see above). The final model function may be used for interpolation and extrapolation purposes. If the correlation coefficient is again inspected for the sketched trial and error model generation procedure its relative value corresponds to the true goodness of fit of each model (the better the model the closer is the correlation coefficient to one). But note that its absolute value is always very near to 1 so care has to be taken if a guessed model function is only cited with its correlation coefficient without any further information (which quite often occurs in practice) since this does not necessarily mean a good fit. For the water-viscosity data it is finally possible to precisely show what is meant by reasonable extrapolation with the following plot:

```
pureFunction=Function[x,CIP`CurveFit`CalculateFunctionValue[x,
  curveFitInfo]];
argumentRange={263.0,383.0};
plotRange={0.0,3.0};
plotStyle={{Thickness[0.005],Black}};
labels={"T [K]","\[Eta] [cP]","Extrapolation problems"};
extrapolationGraphics=
  CIP`Graphics`PlotXyErrorDataAboveFunctions[xyErrorData,
  {pureFunction},argumentRange,plotRange,plotStyle,labels];
intervalGraphics=Graphics[{RGBColor[0,1,0,0.2],
  Rectangle[{273.15,0.0},{373.15,3.0}]}];
Show[extrapolationGraphics,intervalGraphics]
```

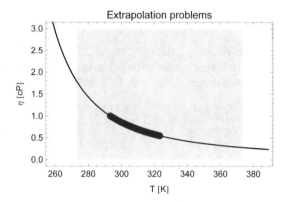

The model function does not know that liquid water undergoes phase transitions if the temperature is lowered or raised beyond the illustrated background region in the diagram: Below 273.15 K water is solid matter (ice) with a practically infinite viscosity, above 373.15 K water is gaseous (vapor) with a dramatically reduced viscosity. To calculate a viscosity at 260 K is possible

```
argument=260;
CIP`CurveFit`CalculateFunctionValue[argument,curveFitInfo]
```

2.94556

but this value is not of this world. Whereas extrapolations around the data argument range may be helpful and sufficiently precise any large-scale extrapolations should always be regarded with suspicion. In summary it should be noted that the outlined construction strategy is very common for an educated guess of a model function. A combination of experience with mere trial and error is very often successful in two-dimensional curve fitting.

2.4 Problems and pitfalls

Linear as well as non-linear curve fitting was shown to be an optimization task (again note that the terms linear and non-linear denote the linearity or non-linearity of the model function with regard to its parameters a_1 to a_L, not the linear or non-linear dependence of the function value y on the argument value x): The global minimum of the $\chi^2(a_1,...,a_L)$ surface is to be found. As discussed in chapter 1 minimization procedures may fail. Failure leads to wrong estimates for the parameters' values and the parameters' errors or even a crash (i.e. an internal termination) of the whole fitting procedure.

Linear curve fitting implies the minimization of a parabolic $\chi^2(a_1,...,a_L)$ hyper surface that contains only one global minimum which can be calculated directly by analytical means (see [Hamilton 1964], [Bevington 2002] or [Brandt 2002] for details). But this calculation involves a matrix inversion which can be a numerically ill-conditioned operation, i.e. problems may occur because computers can only calculate with a finite number of digits. These numerical problems can be tackled with state-of-the-art algorithms so failure usually happens in consequence of the implementation of deficient algorithms with missing safeguards against numerical instabilities. Since CIP is based on Mathematica which provides state-of-the-art algorithms linear curve fitting almost always works without problems. But there should be some awareness if alternative software applications are used as black boxes for linear curve fitting to avoid unnoticed pitfalls: There is a lot of dangerous stuff around - may it be commercial or free.

The situation with non-linear curve fitting is fundamentally different: Since $\chi^2(a_1, ..., a_L)$ may be an arbitrarily difficult and complex curved hyper surface for a non-linear model function it may possess a plethora of minima. There is no way to directly calculate the global minimum by analytical means in principle. The $\chi^2(a_1, ..., a_L)$ hyper surface can only be searched by iterative local minimization procedures that start at user-defined parameters' values and explore their surroundings (compare chapter 1 and [FitModelFunction] in the references). In addition to these principal issues the numerical problems sketched for linear curve fitting may be encountered as well or even in a more serious manner. So in practice there may be an evil mixture of problems - some that can be avoided by state-of-the-art software and others that can only be attributed to the nature of the fitting problem and may be tackled by specific strategies. Some practical problems together with possible solution strategies are outlined in the following.

2.4.1 Parameters' start values

```
Clear["Global`*"];
<<CIP`CalculatedData`
<<CIP`Graphics`
<<CIP`CurveFit`
```

The necessity of adequate parameters' start values may be illustrated by an example. Fifty xy-error data triples

```
numberOfData=50;
```

are simulated around the non-linear Gaussian-peak shaped function

$$y = f(x) = \tfrac{1}{2}x + 3\exp\left\{-(x-4)^2\right\}$$

```
pureOriginalFunction=Function[x,0.5*x+3.0*Exp[-(x-4.0)^2]];
```

in the argument range [1.0, 7.0]

```
argumentRange={1.0,7.0};
```

with an absolute standard deviation of 0.5

```
standardDeviationRange={0.5,0.5};
xyErrorData=CIP`CalculatedData`GetXyErrorData[pureOriginalFunction,
 argumentRange,numberOfData,standardDeviationRange];
```

and finally plotted for visual inspection:

```
labels={"x","y","Data above original function"};
CIP`Graphics`PlotXyErrorDataAboveFunction[xyErrorData,
 pureOriginalFunction,labels]
```

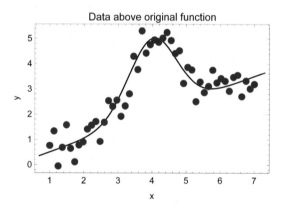

If the data are fitted with the corresponding model function with three parameters
and the CIP default settings

```
modelFunction=a1*x+a2*Exp[-(x-a3)^2];
argumentOfModelFunction=x;
parametersOfModelFunction={a1,a2,a3};
curveFitInfo=CIP`CurveFit`FitModelFunction[xyErrorData,
 modelFunction,argumentOfModelFunction,parametersOfModelFunction];
CIP`CurveFit`ShowFitResult[{"FunctionPlot","ModelFunction"},
 xyErrorData,curveFitInfo];
```

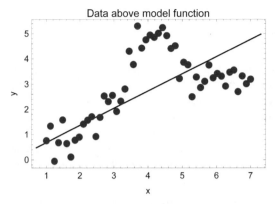

Fitted model function:
$5.53945e^{-(-25.4806+x)^2} + 0.681816x$

the achieved result is simply wrong. What happened? Internally the FitModel-Function method generates random parameters' start values for the local minimization procedure - and these start values are simply inadequate in this case (but they may work perfectly in other fitting procedures). So start values for the parameters must be provided by hand. Since the true parameter values are 0.5, 3.0 and 4.0 (see above) everything works fine if parameters' start values are specified near the solution:

```
startParameters={{a1,0.4},{a2,3.1},{a3,3.9}};
curveFitInfo=CIP`CurveFit`FitModelFunction[xyErrorData,
 modelFunction,argumentOfModelFunction,parametersOfModelFunction,
 CurveFitOptionStartParameters -> startParameters];
CIP`CurveFit`ShowFitResult[{"FunctionPlot","ModelFunction"},
 xyErrorData,curveFitInfo];
```

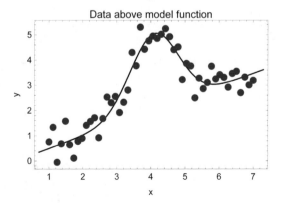

Fitted model function:

$$3.01029e^{-(-4.09318+x)^2} + 0.495279x$$

A perfect fit is the result. It is well-known to practitioners that fitting Gaussian-peak shaped model functions requires a good guess for the parameter value in the exponential term: This start value may be deduced from the mere data in this case: The maximum is around $x = 4$ so use a value around 4 as a start value for parameter a_3.

The worst case occurs if a_3 is chosen to be very unfavorable: Then the whole fitting procedure may crash (i.e. may internally be terminated) as shown in the following example:

```
startParameters={{a1,0.4},{a2,3.1},{a3,-3.0}};
curveFitInfo=CIP 'CurveFit 'FitModelFunction[xyErrorData,
  modelFunction,argumentOfModelFunction,parametersOfModelFunction,
  CurveFitOptionStartParameters -> startParameters];
```

Underflow occurred in computation.

A value was calculated during the minimization process that was smaller than the smallest allowed value of the Mathematica system and therefore an underflow error message (and subsequent error messages) were generated. Note that this behavior can not simply be traced to a bad algorithm: The default Levenberg-Marquardt algorithm used by FitModelFunction for two-dimensional non-linear curve fitting is a state-of-the-art algorithm for this purpose. But it may fail in principle: It can not safeguard every possible calculation. It might be a good idea to simply change the minimization algorithm: An alternative minimization algorithm will usually generate a different outcome. That is why a library of algorithms is most often a severe advantage. But not in this case: If the algorithm is changed from Levenberg-Marquardt to Conjugate-Gradient

```
method={"ConjugateGradient"};
curveFitInfo=CIP 'CurveFit 'FitModelFunction[xyErrorData,
  modelFunction,argumentOfModelFunction,parametersOfModelFunction,
  CurveFitOptionStartParameters -> startParameters,
  CurveFitOptionMethod -> method];
```

The line search decreased the step size to within tolerance specified by AccuracyGoal and PrecisionGoal but was unable to find a sufficient decrease in the norm of the residual.

a similar problem as before occurs in the line search subroutine of this algorithm. Another switch to the mere Gradient algorithm

```
method={"Gradient"};
curveFitInfo=CIP 'CurveFit 'FitModelFunction[xyErrorData,
  modelFunction,argumentOfModelFunction,parametersOfModelFunction,
  CurveFitOptionStartParameters -> startParameters,
  CurveFitOptionMethod -> method];
```

```
CIP`CurveFit`ShowFitResult[{"ModelFunction"},xyErrorData,
  curveFitInfo];
```

Fitted model function:

$$13.0817e^{-(1.36437+x)^2} + 0.681692x$$

does not help either: The minimization procedure seems to have converged (since there are no error messages) but it stopped somewhere over the rainbow: The result is simply wrong. So the only practical solution is to provide good parameters' start values by hand, i.e. by ...

- ... **knowledge:** Parameters may be known to lie within defined intervals by experience or they may have a scientific meaning (i.e. they are theoretically well-defined) so that their values are approximately known in advance. In some cases a start value for a parameter may be deduced by visual inspection as in the example above for parameter a_3 in the exponential term.
- ... **trial and error:** Not a promising strategy but often the only practical possibility: It may be very exhaustive and disappointing but science is often more devoted to mere trial and error than scientists like to tell.

In the next section the trial and error case is tackled with more strategic approaches but these also can not solve the problem in principle.

2.4.2 How to search for parameters' start values

```
Clear["Global`*"];
<<CIP`CalculatedData`
<<CIP`Graphics`
<<CIP`CurveFit`
```

To get good parameters' start values a global search of the parameter space is necessary: A huge task! In chapter 1 different strategies for a global search were discussed like a grid or a random search. CIP implements a purely random search strategy as an option for the GetStartParameters method of the CIP CurveFit package with search type "Random":

```
searchType="Random";
```

For the curve fitting task outlined of the last subsection

```
numberOfData=50;
pureOriginalFunction=Function[x,0.5*x+3.0*Exp[-(x-4.0)^2]];
argumentRange={1.0,7.0};
standardDeviationRange={0.5,0.5};
```

```
xyErrorData=CIP `CalculatedData`GetXyErrorData[pureOriginalFunction,
  argumentRange,numberOfData,standardDeviationRange];
modelFunction=a1*x+a2*Exp[-(x-a3)^2];
argumentOfModelFunction=x;
parametersOfModelFunction={a1,a2,a3};
```

a parameters' search space is defined by the individual intervals of each parameter

```
parameterIntervals={{0.0,10.0},{0.0,10.0},{0.0,10.0}};
```

with a 100 random trial points:

```
numberOfTrialPoints=100;
```

If the GetStartParameters method is called with these settings

```
startParameters=CIP `CurveFit`GetStartParameters[xyErrorData,
  modelFunction,argumentOfModelFunction,parametersOfModelFunction,
  parameterIntervals,CurveFitOptionSearchType -> searchType,
  CurveFitOptionNumberOfTrialPoints -> numberOfTrialPoints]
```

$\{\{a1,0.670859\},\{a2,7.48994\},\{a3,8.29601\}\}$

the resulting parameters' start values correspond to the smallest value of $\chi^2(a_1,a_2,a_3)$ that was detected by random. These start values are now used as an input for the model function fit by setting the CurveFitOptionStartParameters option with the result:

```
curveFitInfo=CIP `CurveFit`FitModelFunction[xyErrorData,
  modelFunction,argumentOfModelFunction,parametersOfModelFunction,
  CurveFitOptionStartParameters -> startParameters];
CIP `CurveFit`ShowFitResult[{{"FunctionPlot","ModelFunction"},
  xyErrorData,curveFitInfo];
```

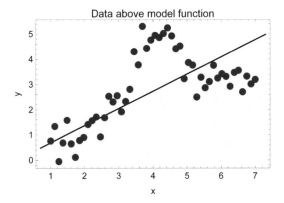

Fitted model function:
$$97.9599e^{-(-17.502+x)^2} + 0.681816x$$

The result is still not correct: 100 trial points do not lead to sufficiently precise parameters' start values since the random grid is too coarsely meshed. Therefore their number is increased tenfold to 1000 and the search is repeated:

```
numberOfTrialPoints=1000;
startParameters=CIP`CurveFit`GetStartParameters[xyErrorData,
 modelFunction,argumentOfModelFunction,parametersOfModelFunction,
 parameterIntervals,CurveFitOptionSearchType -> searchType,
 CurveFitOptionNumberOfTrialPoints -> numberOfTrialPoints]
```

$\{\{a1, 0.679172\}, \{a2, 1.7166\}, \{a3, 4.48335\}\}$

With the new start values

```
curveFitInfo=CIP`CurveFit`FitModelFunction[xyErrorData,
 modelFunction,argumentOfModelFunction,parametersOfModelFunction,
 CurveFitOptionStartParameters -> startParameters];
CIP`CurveFit`ShowFitResult[{"FunctionPlot","ModelFunction"},
 xyErrorData,curveFitInfo];
```

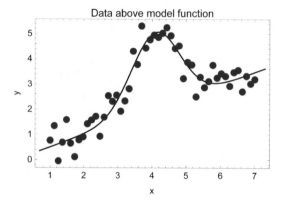

Fitted model function:

$3.01029 e^{-(-4.09318+x)^2} + 0.495279x$

a successful fit is finally obtained: The determined start values were precise enough for the local minimization algorithm to converge to the global minimum of $\chi^2(a_1, a_2, a_3)$. Although this may seem promising it again should be noticed that a random search is a rather limited option in general: Since the parameter space becomes really large with an increasing number of parameters a random search within tolerable periods of time will be likely to fail. The glimmer of hope of chapter 1 in this desperate situation were evolutionary algorithms. Method GetStartParameters uses the differential-evolution algorithm via Mathematica's NMinimize command as its default global search strategy (see [NMinimize/NMaximize] in the references) which also proofs to be successful for the current task:

```
startParameters=CIP 'CurveFit 'GetStartParameters[xyErrorData,
  modelFunction,argumentOfModelFunction,parametersOfModelFunction,
  parameterIntervals]
```

$\{\{a1, 0.460666\}, \{a2, 3.67567\}, \{a3, 4.12436\}\}$

A fit with the obtained start parameters

```
curveFitInfo=CIP 'CurveFit 'FitModelFunction[xyErrorData,
  modelFunction,argumentOfModelFunction,parametersOfModelFunction,
  CurveFitOptionStartParameters -> startParameters];
CIP 'CurveFit 'ShowFitResult[{"FunctionPlot","ModelFunction"},
  xyErrorData,curveFitInfo];
```

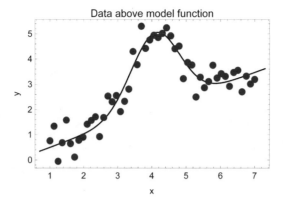

Fitted model function:

$3.01029e^{-(-4.09318+x)^2} + 0.495279x$

shows that the evolutionary search was able to determine start values in the proximity of the global minimum of $\chi^2(a_1, a_2, a_3)$ which were close enough for a successful local refinement.

2.4.3 More difficult curve fitting problems

```
Clear["Global`*"];
<<CIP`CalculatedData`
<<CIP`Graphics`
<<CIP`CurveFit`
```

Extracting the correct model function from experimental data may be arbitrarily difficult up to impossible due to the nature of the curve fitting problem. To demonstrate an example fifty xy-error data triples

```
numberOfData=50;
```

with a very high precision (absolute standard deviation of 0.001)

```
standardDeviationRange={0.001,0.001};
```

are generated in an argument range [1, 8]

```
argumentRange={1.0,8.0};
```

around a model function with two Gaussian peaks in close proximity (around $x = 4$ and $x = 5.5$)

$$y = f(x) = \tfrac{1}{2}x + 3\exp\left\{-(x-4)^2\right\} + 2\exp\left\{-(x-5.5)^2\right\}$$

```
pureOriginalFunction=
 Function[x, 0.5*x+3.0*Exp[-(x-4.0)^2]+2.0*Exp[-(x-5.5)^2]];
```

where the smaller one around $x = 5.5$ appears to be the shoulder of the other:

```
xyErrorData=CIP'CalculatedData'GetXyErrorData[pureOriginalFunction,
 argumentRange,numberOfData,standardDeviationRange];
labels={"x","y","Data above original function"};
CIP'Graphics'PlotXyErrorDataAboveFunction[xyErrorData,
 pureOriginalFunction,labels]
```

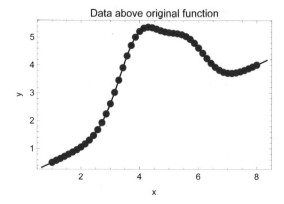

With the previous discussions in mind it should be obvious that a successful curve fitting procedure needs very good start values for the parameters in this case. Again at least start values for the parameters in the exponentials could be obtained by mere visual inspection of the generated data (peaks around $x = 4$ and $x = 5.5$) but a more general strategy is explored that uses the advised start-parameter search on the basis of an evolutionary algorithm with the GetStartParameters method of the CIP CurveFit package. With the 5-parameter model function

```
modelFunction=a1*x+a2*Exp[-(x-a3)^2]+a4*Exp[-(x-a5)^2];
argumentOfModelFunction=x;
parametersOfModelFunction={a1,a2,a3,a4,a5};
```

and a well defined parameters' search space

```
parameterIntervals=
 {{0.0,10.0},{0.0,10.0},{0.0,10.0},{0.0,10.0},{0.0,10.0}};
```

the proposed parameters' start values for the fit procedure are:

```
maximumNumberOfIterations=10;
startParameters=CIP `CurveFit `GetStartParameters[xyErrorData,
 modelFunction,argumentOfModelFunction,parametersOfModelFunction,
 parameterIntervals,
 CurveFitOptionMaximumIterations -> maximumNumberOfIterations]
```

$\{\{a1,0.443407\},\{a2,4.81041\},\{a3,4.83269\},\{a4,0.0454534\},\{a5,4.45465\}\}$

A fit with these parameters' start values

```
curveFitInfo=CIP `CurveFit `FitModelFunction[xyErrorData,
 modelFunction,argumentOfModelFunction,parametersOfModelFunction,
 CurveFitOptionStartParameters -> startParameters];
CIP `CurveFit `ShowFitResult[{"FunctionPlot","ModelFunction"},
 xyErrorData,curveFitInfo];
```

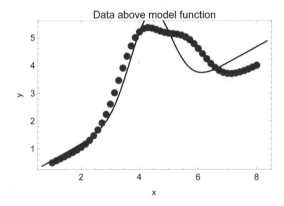

Fitted model function:
$$-0.173247e^{-(-42.6798+x)^2}+3.33414e^{-(-4.38782+x)^2}+0.584755x$$

leads to an unsatisfying result. Obviously the parameters' start values search was not successful - remember that there is no guarantee for an evolutionary strategy to succeed. This failure might be attributed to the applied setting of the internal number of iterations (i.e. the number of generations for evolution) to only 10. In this particular case the parameter space should be explored more thoroughly with an increased number of iterations (note that this number must always be restricted to balance between accuracy and speed):

```
maximumNumberOfIterations=100;
startParameters=CIP`CurveFit`GetStartParameters[xyErrorData,
  modelFunction,argumentOfModelFunction,parametersOfModelFunction,
  parameterIntervals,
  CurveFitOptionMaximumIterations -> maximumNumberOfIterations]
```

$\{\{a1,0.535511\},\{a2,1.39872\},\{a3,5.55438\},\{a4,2.80221\},\{a5,4.13868\}\}$

With the improved parameters' start values the curve fitting procedure

```
curveFitInfo=CIP`CurveFit`FitModelFunction[xyErrorData,
  modelFunction,argumentOfModelFunction,parametersOfModelFunction,
  CurveFitOptionStartParameters -> startParameters];
CIP`CurveFit`ShowFitResult[{"FunctionPlot","ModelFunction"},
  xyErrorData,curveFitInfo];
```

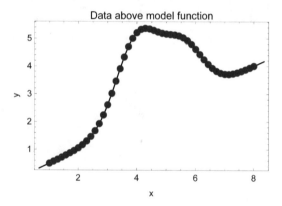

Fitted model function:
$2.00039e^{-(-5.49993+x)^2} + 2.99947e^{-(-4.00006+x)^2} + 0.5x$

is successful: A perfect fit is obtained. The sketched curve fitting problem will certainly become more difficult if the two Gaussian peaks are moved together, e.g.

$$y = f(x) = \tfrac{1}{2}x + 3\exp\left\{-(x-4)^2\right\} + 2\exp\left\{-(x-4.5)^2\right\}$$

```
pureOriginalFunction=
  Function[x, 0.5*x+3.0*Exp[-(x-4.0)^2]+2.0*Exp[-(x-4.5)^2]];
```

where the two peaks now are closely neighbored around $x = 4$ and $x = 4.5$. After xy-error data generation as before a visual inspection shows

```
xyErrorData=CIP 'CalculatedData 'GetXyErrorData[pureOriginalFunction,
  argumentRange,numberOfData,standardDeviationRange];
labels={"x","y","Data above Original Function"};
CIP 'Graphics 'PlotXyErrorDataAboveFunction[xyErrorData,
  pureOriginalFunction,labels]
```

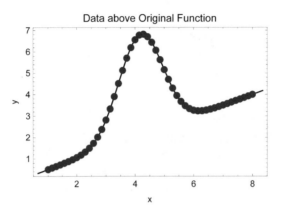

that the shoulder becomes invisible and only one merged peak appears. Note that without an a priori knowledge about the two existing peaks (the data are artificial) only one peak would be anticipated. If again the parameters' start value search is used with an insufficient number of iterations

```
maximumNumberOfIterations=20;
startParameters=CIP 'CurveFit 'GetStartParameters[xyErrorData,
  modelFunction,argumentOfModelFunction,parametersOfModelFunction,
  parameterIntervals,
  CurveFitOptionMaximumIterations -> maximumNumberOfIterations]
```

$\{\{a1,0.554496\},\{a2,4.20733\},\{a3,4.10765\},\{a4,3.44001\},\{a5,9.99968\}\}$

the results are dubious, i.e. values are too close to the search boundaries. A fit with these start values give evidence for this assessment:

```
curveFitInfo=CIP 'CurveFit 'FitModelFunction[xyErrorData,
  modelFunction,argumentOfModelFunction,parametersOfModelFunction,
  CurveFitOptionStartParameters -> startParameters];
CIP 'CurveFit 'ShowFitResult[{"FunctionPlot","ModelFunction",
  "AbsoluteResidualsPlot"},xyErrorData,curveFitInfo];
```

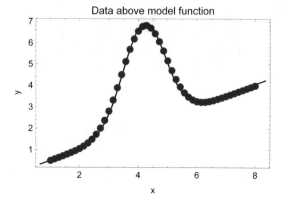

Fitted model function:

$$158.045e^{-(-58.0059+x)^2} + 4.8003e^{-(-4.19422+x)^2} + 0.508776x$$

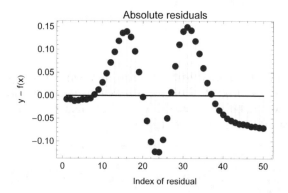

The second Gaussian peak is sent to infinity. This leads to a systematic deviation pattern of the residuals which is a clear indication that something is missed. A further refinement of the parameters' space exploration becomes necessary with an additional increase of the number of iterations:

```
maximumNumberOfIterations=100;
startParameters=CIP `CurveFit `GetStartParameters[xyErrorData,
  modelFunction,argumentOfModelFunction,parametersOfModelFunction,
  parameterIntervals,
  CurveFitOptionMaximumIterations -> maximumNumberOfIterations]
```

$\{\{a1,0.505924\},\{a2,4.76675\},\{a3,4.18593\},\{a4,0.125418\},\{a5,5.46316\}\}$

The following fit

```
curveFitInfo=CIP 'CurveFit 'FitModelFunction[xyErrorData,
  modelFunction,argumentOfModelFunction,parametersOfModelFunction,
  CurveFitOptionStartParameters -> startParameters];
CIP 'CurveFit 'ShowFitResult[{"FunctionPlot","ModelFunction",
  "AbsoluteResidualsPlot"},xyErrorData,curveFitInfo];
```

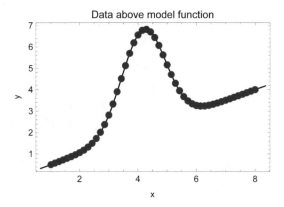

Fitted model function:

$$2.00853e^{-(-4.49914+x)^2} + 2.99136e^{-(-3.99927+x)^2} + 0.500001x$$

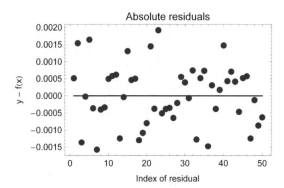

now leads to a satisfactory result. This successful outcome was invoked by a continuously intensified brute-force strategy that led to an enhanced thoroughness of parameters' space exploration. In summary it is a remarkable fact that data analysis is able to reveal invisible peaks that would not be assumed by mere visual inspection but only if they are known to be there. On the other hand subtle interpretation problems will emerge if things become slightly more difficult in the case of less precise data. For an illustration low-precision data are generated with a high absolute standard deviation of 0.6 around the last example function:

```
standardDeviationRange={0.6,0.6};
xyErrorData=CIP `CalculatedData `GetXyErrorData[pureOriginalFunction,
 argumentRange,numberOfData,standardDeviationRange];
labels={"x","y","Data above original function"};
CIP `Graphics `PlotXyErrorDataAboveFunction[xyErrorData,
 pureOriginalFunction,labels]
```

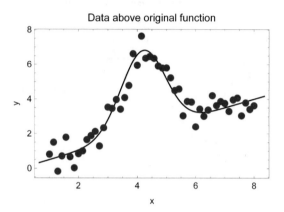

The thorough parameters' start value search with again a large number of evolutionary steps

```
maximumNumberOfIterations=100;
startParameters=CIP `CurveFit `GetStartParameters[xyErrorData,
 modelFunction,argumentOfModelFunction,parametersOfModelFunction,
 parameterIntervals,
 CurveFitOptionMaximumIterations -> maximumNumberOfIterations]
```

$\{\{a1,0.512016\},\{a2,4.55652\},\{a3,4.28594\},\{a4,0.302777\},\{a5,3.01461\}\}$

and a following fit

```
curveFitInfo=CIP `CurveFit `FitModelFunction[xyErrorData,
 modelFunction,argumentOfModelFunction,parametersOfModelFunction,
 CurveFitOptionStartParameters -> startParameters];
CIP `CurveFit `ShowFitResult[{"FunctionPlot","AbsoluteResidualsPlot",
 "SDFit","ModelFunction"},xyErrorData,curveFitInfo];
```

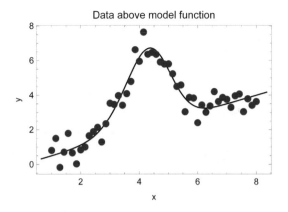

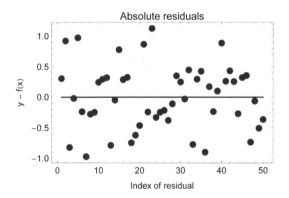

Standard deviation of fit = 5.478×10^{-1}

Fitted model function:

$$4.25758e^{-(-4.37855+x)^2} + 0.725734e^{-(-3.3966+x)^2} + 0.49908x$$

lead to a result of good quality with effectively two different peaks. But the precision of peak detection is no longer satisfying. Moreover this result is no longer convincing if alternatives are taken into consideration. This can be shown with an alternative fit of the corresponding model function with one Gaussian peak which would be assumed by mere visual inspection:

```
modelFunction=a1*x+a2*Exp[-(x-a3)^2];
argumentOfModelFunction=x;
parametersOfModelFunction={a1,a2,a3};
```

With adequate start values

```
parameterIntervals={{0.0,10.0},{0.0,10.0},{0.0,10.0}};
```

```
startParameters=CIP `CurveFit `GetStartParameters[xyErrorData,
  modelFunction,argumentOfModelFunction,parametersOfModelFunction,
  parameterIntervals]
```

$\{\{a1,0.523746\},\{a2,5.85935\},\{a3,4.12508\}\}$

the alternative one-peak fit

```
curveFitInfo=CIP `CurveFit `FitModelFunction[xyErrorData,
  modelFunction,argumentOfModelFunction,parametersOfModelFunction,
  CurveFitOptionStartParameters -> startParameters];
CIP `CurveFit `ShowFitResult[{"FunctionPlot","AbsoluteResidualsPlot",
  "SDFit","ModelFunction"},xyErrorData,curveFitInfo];
```

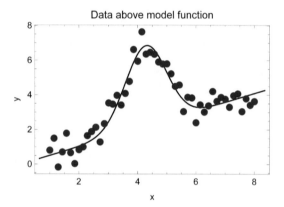

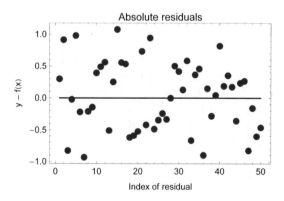

Standard deviation of fit = 5.527×10^{-1}

Fitted model function:

$$4.65303e^{-(-4.27207+x)^2} + 0.511526x$$

leads to a result of comparable quality (the standard deviations of the fits are nearly identical and the residuals patterns are equally good): The latter model function should be preferred according to Occam's razor since it contains less parameters (unless the existence of two peaks is certainly known in advance). Depending on the precision of the data and the nature of the fitting problem severe ambiguities can appear in data analysis. In the last case peaks may be found or may be argued for that can not be supported by the mere data in the light of alternative models. So all data analysis procedures are prone to be misused for the sake of a scientist's mere opinion and not the truth (where the scientist is always assumed to pursue the most noble intentions). As a rule of thumb an adequate distrust is indicated for statements like *it is clearly shown by thorough data analysis that* .. Curve fitting should always be data driven and it should not be tried to get more out of them than possible. The old and latent tendency to overstretch data analysis once led to the famous sentence by John von Neumann: *With four parameters I can fit an elephant, and with five I can make him wiggle his trunk* ([Dyson 2004]).

2.4.4 Inappropriate model functions

```
Clear["Global`*"];
<<CIP`CalculatedData`
<<CIP`Graphics`
<<CIP`CurveFit`
```

Model functions may be unfavorable up to simply wrong. An example of the latter is demonstrated as follows: Fifty fairly precise data

```
numberOfData=50;
```

with an absolute standard deviation of 0.01

```
standardDeviationRange={0.01,0.01};
```

are generated in the argument range [1, 5]

```
argumentRange={1.0,5.0};
```

around function

$$y = f(x) = 2\exp\{-1.5x\}$$

```
pureOriginalFunction=Function[x,2.0*Exp[-0.75*x]];
```

to give

```
xyErrorData=CIP`CalculatedData`GetXyErrorData[pureOriginalFunction,
  argumentRange,numberOfData,standardDeviationRange];
labels={"x","y","Data above original function"};
CIP`Graphics`PlotXyErrorDataAboveFunction[xyErrorData,
  pureOriginalFunction,labels]
```

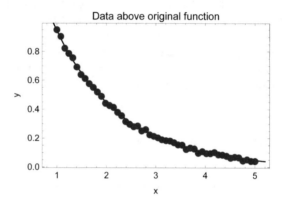

For the fit we use an empirical model function constructed without to much meditation:

$$y = f(x) = a_1 \exp\{-a_2 x + a_3\}$$

```
modelFunction=a1*Exp[-a2*x+a3];
argumentOfModelFunction=x;
parametersOfModelFunction={a1,a2,a3};
```

The result

```
curveFitInfo=CIP`CurveFit`FitModelFunction[xyErrorData,
  modelFunction,argumentOfModelFunction,parametersOfModelFunction];
ShowFitResult[{"FunctionPlot","AbsoluteResidualsPlot",
  "ModelFunction"},xyErrorData,curveFitInfo];
```

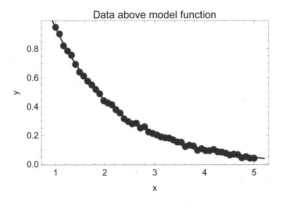

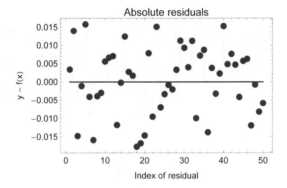

Fitted model function:

$1.91781e^{0.0463659-0.752542x}$

appears to be a perfect fit. So what is wrong with the model function? The answer is that it contains redundant parameters since parameters a_1 and a_3 essentially mean the same: They are both mere prefactors to the exponential term

$$y = f(x) = a_1 \exp\{-a_2 x + a_3\} = a_1 \exp\{a_3\}\exp\{-a_2 x\}$$

and therefore they are arbitrary. Only their product is the true prefactor used to generate the data. All infinite other combinations of values resulting to the same prefactor would be valid as well. Although redundant parameters can always be avoided by proper inspection of the model function they do occur easily if non-mathematicians (i.e. the overwhelming majority of scientists) construct difficult empirical models. Usually the fitting algorithms simply crash if redundant parameters are defined in a model function. It is only due to Mathematica's algorithmic safeguards that lead to an arbitrary but correct result. A more subtle problem occurs if the model function is correct but simply inappropriate to the data since it tries to extract information which is simply not there. This may be shown with fifty fairly precise data with an absolute standard deviation of 0.1

```
standardDeviationRange={0.1,0.1};
```

in the argument range [1, 7]

```
argumentRange={1.0,7.0};
```

around one Gaussian peak:

$$y = f(x) = \tfrac{1}{2}x + 3\exp\{-(x-4)^2\}$$

```
pureOriginalFunction=Function[x,0.5*x+3.0*Exp[-(x-4.0)^2]];
xyErrorData=CIP `CalculatedData `GetXyErrorData[pureOriginalFunction,
   argumentRange,numberOfData,standardDeviationRange];
CIP `Graphics `PlotXyErrorDataAboveFunction[xyErrorData,
   pureOriginalFunction,labels]
```

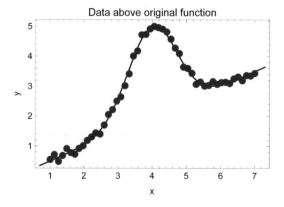

A model function with two Gaussian peaks is prepared

```
modelFunction=a1*x+a2*Exp[-(x-a3)^2]+a4*Exp[-(x-a5)^2];
argumentOfModelFunction=x;
parametersOfModelFunction={a1,a2,a3,a4,a5};
```

which inevitably tries to extract two Gaussian peaks from the data which just contain one peak. After an successful search for parameters' start values

```
parameterIntervals=
 {{0.0,10.0},{0.0,10.0},{0.0,10.0},{0.0,10.0},{0.0,10.0}};
startParameters=GetStartParameters[xyErrorData,modelFunction,
 argumentOfModelFunction,parametersOfModelFunction,
 parameterIntervals]
```

$\{\{a1,0.533351\},\{a2,0.764096\},\{a3,4.36781\},\{a4,2.40032\},\{a5,3.9497\}\}$

the following fit

```
curveFitInfo=CIP`CurveFit`FitModelFunction[xyErrorData,
 modelFunction,argumentOfModelFunction,parametersOfModelFunction,
 CurveFitOptionStartParameters -> startParameters];
CIP`CurveFit`ShowFitResult[{"FunctionPlot","AbsoluteResidualsPlot",
 "ModelFunction"},xyErrorData,curveFitInfo];
```

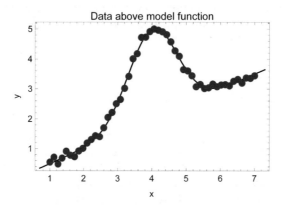

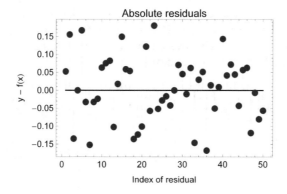

Fitted model function:

$$2.50893e^{-(-4.01878+x)^2} + 0.489941e^{-(-4.01878+x)^2} + 0.499246x$$

shows what happened: The same peak was found twice with arbitrary prefactors that only have meaning as a sum. Again note: If alternative software to CIP/Mathematica is used the fitting algorithms usually crash if a model function is inappropriate as outlined in the latter example.

2.5 Parameters' errors

```
Clear["Global`*"];
<<CIP`ExperimentalData`
<<CIP`CurveFit`
```

The second most important information that may be extracted from a successful curve fitting procedure in accordance with the optimum estimates of the parameters' values are estimates of the parameters' errors.

2.5.1 Correction of parameters' errors

Since the xy-error data are biased by errors these errors propagate to the errors of the estimated parameters' values: The parameters' errors therefore are deduced from the data's errors. This is certainly the best procedure if the data's errors are true experimentally obtained errors, e.g. each y value is measured multiple times and then reported as the statistical mean y_i with the statistical standard deviation of the mean σ_i for an argument value x_i. But often the reported errors σ_i can only be regarded as rough estimates of the true errors. Moreover these estimates are usually overestimated since scientists tend to be cautious: A bigger error is the better error if the error is not known precisely. Then of course the resulting parameters' errors of a model function fit are also overestimated. As an example the water-viscosity data are inspected again (compare above):

```
xyErrorData=CIP`ExperimentalData`GetWaterViscosityXyErrorData[];
modelFunction=a1/(a3-T)*Exp[a2/(a3-T)];
argumentOfModelFunction=T;
parametersOfModelFunction={a1,a2,a3};
startParameters={{a1,0.1},{a2,-500.0},{a3,150.0}};
curveFitInfo=CIP`CurveFit`FitModelFunction[xyErrorData,
  modelFunction,argumentOfModelFunction,parametersOfModelFunction,
  CurveFitOptionStartParameters -> startParameters];
CIP`CurveFit`ShowFitResult[{"ReducedChiSquare","ParameterErrors"},
  xyErrorData,curveFitInfo];
```

Reduced chi-square of fit = 8.625×10^{-2}

	Value	Standard error	Confidence region
Parameter a1 = -19.3098	0.108964	{-19.4208, -19.1989}
Parameter a2 = -200.831	1.86845	{-202.734, -198.929}
Parameter a3 = 179.802	0.445257	{179.348, 180.255}

The χ^2_{red} value of 0.086 indicates that the fitted residuals are fair below the corresponding errors σ_i of the y_i values since χ^2_{red} should be close to 1 for a good fit with good data's errors. Consequently the data's errors should be decreased for a resulting χ^2_{red} value near 1. A correction for the data's errors may be calculated when they are assumed to be only weights of the y_i values and not their true statistical errors (see above). The FitModelFunction method can be told to estimate parameters' errors with the corrected and not the original errors by changing the option CurveFitOptionVarianceEstimator from its default value to "ReducedChiSquare":

```
varianceEstimator="ReducedChiSquare";
curveFitInfo=CIP`CurveFit`FitModelFunction[xyErrorData,
 modelFunction,argumentOfModelFunction,parametersOfModelFunction,
 CurveFitOptionStartParameters -> startParameters,
 CurveFitOptionVarianceEstimator -> varianceEstimator];
CIP`CurveFit`ShowFitResult[{"ReducedChiSquare","ParameterErrors"},
 xyErrorData,curveFitInfo];
```

Reduced chi-square of fit = 8.625×10^{-2}

	Value	Standard error	Confidence region
Parameter a1 = -19.3098	0.0320015	{-19.3424, -19.2772}
Parameter a2 = -200.831	0.548743	{-201.39, -200.272}
Parameter a3 = 179.802	0.130767	{179.669, 179.935}

The parameters' standard errors and their confidence regions are reduced by more than a factor of 3 in comparison to the result before. The outlined error correction is often used as a standard procedure for curve fitting. But in practice it simply depends on the problem and the scientist's mood to use the cautious (higher) error estimates for all subsequent derivations as well.

2.5.2 Confidence levels of parameters' errors

```
Clear["Global`*"];
<<CIP`ExperimentalData`
<<CIP`CurveFit`
```

Another important option that may be modified for the estimation of parameters' errors is their level of confidence which affects the width of their confidence regions. With the default setting of 68.3% the parameters' confidence regions correspond to the standard errors, i.e. a confidence region spans the interval $[a_i - \sigma_{a_i}, a_i + \sigma_{a_i}]$ where σ_{a_i} is the standard error of parameter a_i. In many cases a higher confidence level of e.g. 95% or 99% is required. This may be specified with option CurveFitOptionConfidenceLevel of method FitModelFunction. Here a confidence level of 99.9%

```
confidenceLevelOfParameterErrors=0.999;
```

is used for the water-viscosity fit for with the corrected errors (see previous section):

```
xyErrorData=CIP`ExperimentalData`GetWaterViscosityXyErrorData[];
modelFunction=a1/(a3-T)*Exp[a2/(a3-T)];
argumentOfModelFunction=T;
parametersOfModelFunction={a1,a2,a3};
startParameters={{a1,0.1},{a2,-500.0},{a3,150.0}};
varianceEstimator="ReducedChiSquare";
curveFitInfo=CIP`CurveFit`FitModelFunction[xyErrorData,
  modelFunction,argumentOfModelFunction,parametersOfModelFunction,
  CurveFitOptionStartParameters -> startParameters,
  CurveFitOptionVarianceEstimator -> varianceEstimator,
  CurveFitOptionConfidenceLevel -> confidenceLevelOfParameterErrors];
CIP`CurveFit`ShowFitResult[{"ReducedChiSquare","ParameterErrors"},
  xyErrorData,curveFitInfo];
```

Reduced chi-square of fit = 8.625×10^{-2}

	Value	Standard error	Confidence region
Parameter a1 =	-19.3098	0.0320015	{-19.4274, -19.1922}
Parameter a2 =	-200.831	0.548743	{-202.847, -198.815}
Parameter a3 =	179.802	0.130767	{179.321, 180.282}

Note that the standard errors are not affected since they are related to the standard confidence level of 68.3% but the confidence regions increased considerably: Now it can be assured with a probability of 99.9% that the parameters' values are within the denoted regions.

2.5.3 Estimating the necessary number of data

```
Clear["Global`*"];
<<CIP`CalculatedData`
<<CIP`Graphics`
<<CIP`CurveFit`
```

An practically important issue related to the parameters' errors is the following: A theoretical model function with well-defined parameters is known. A specific measurement process with its intrinsic measurement errors is available. How many experimental data in a defined argument range must be measured to get a reasonable statement about a parameters' value with a specific level of confidence? To get an impression the Gaussian-peak shaped model function is taken again as an example. If a measurement process imposes an absolute error of 0.5 on each measurement the following parameters' errors and confidence regions are obtained for fifty (x_i, y_i, σ_i) data triples in the argument range [1.0, 7.0] with a confidence level of 68.3%:

```
numberOfData=50;
standardDeviationRange={0.5,0.5};
pureOriginalFunction=Function[x,0.5*x+3.0*Exp[-(x-4.0)^2]];
argumentRange={1.0,7.0};
xyErrorData=CIP `CalculatedData`GetXyErrorData[pureOriginalFunction,
  argumentRange,numberOfData,standardDeviationRange];
labels={"x","y","Data above original function"};
CIP `Graphics`PlotXyErrorDataAboveFunction[xyErrorData,
  pureOriginalFunction,labels]
```

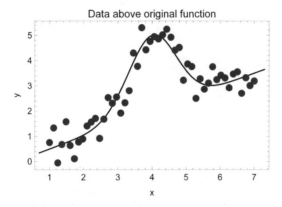

Data above original function

```
modelFunction=a1*x+a2*Exp[-(x-a3)^2];
argumentOfModelFunction=x;
parametersOfModelFunction={a1,a2,a3};
startParameters={{a1,0.4},{a2,2.9},{a3,4.1}};
curveFitInfo=CIP `CurveFit`FitModelFunction[xyErrorData,
  modelFunction,argumentOfModelFunction,parametersOfModelFunction,
  CurveFitOptionStartParameters -> startParameters];
CIP `CurveFit`ShowFitResult[{"ModelFunction","ReducedChiSquare",
  "ParameterErrors"},xyErrorData,curveFitInfo];
```

Fitted model function:

$3.01029e^{-(-4.09318+x)^2} + 0.495279x$

Reduced chi-square of fit = 8.17×10^{-1}

	Value	Standard error	Confidence region
Parameter a1 = 0.495279		0.0205374	{0.474521, 0.516038}
Parameter a2 = 3.01029		0.196361	{2.81182, 3.20877}
Parameter a3 = 4.09318		0.0528048	{4.0398, 4.14655}

If the number of data is increased the parameters' values will become more precise and the parameters' errors and their related confidence regions are reduced:

```
numberOfData=1500;
xyErrorData=CIP 'CalculatedData 'GetXyErrorData[pureOriginalFunction,
  argumentRange,numberOfData,standardDeviationRange];
pointSize=0.01;
CIP 'Graphics 'PlotXyErrorDataAboveFunction[xyErrorData,
  pureOriginalFunction,labels,GraphicsOptionPointSize -> pointSize]
```

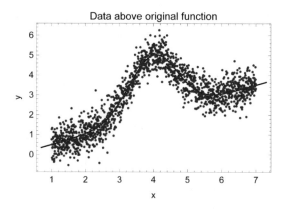

Data above original function

```
curveFitInfo=CIP 'CurveFit 'FitModelFunction[xyErrorData,
  modelFunction,argumentOfModelFunction,parametersOfModelFunction,
  CurveFitOptionStartParameters -> startParameters];
CIP 'CurveFit 'ShowFitResult[{"ModelFunction","ReducedChiSquare",
  "ParameterErrors"},xyErrorData,curveFitInfo];
```

Fitted model function:

$2.97527e^{-(-3.99917+x)^2} + 0.502659x$

Reduced chi-square of fit = 9.819×10^{-1}

	Value	Standard error	Confidence region
Parameter a1 =	0.502659	0.00374083	{0.498917, 0.506401}
Parameter a2 =	2.97527	0.0352989	{2.93996, 3.01058}
Parameter a3 =	3.99917	0.00966191	{3.9895, 4.00883}

As a second alternative another measurement process may be available with a decreased intrinsic error that it imposes on the data (here the absolute error is reduced by a factor of 10 from 0.5 to 0.05):

```
numberOfData=50;
standardDeviationRange={0.05,0.05};
xyErrorData=CIP 'CalculatedData 'GetXyErrorData[pureOriginalFunction,
  argumentRange,numberOfData,standardDeviationRange];
```

```
CIP `Graphics `PlotXyErrorDataAboveFunction[xyErrorData,
 pureOriginalFunction,labels]
```

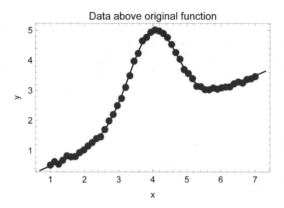

Data above original function

```
curveFitInfo=CIP `CurveFit `FitModelFunction[xyErrorData,
 modelFunction,argumentOfModelFunction,parametersOfModelFunction,
 CurveFitOptionStartParameters -> startParameters];
CIP `CurveFit `ShowFitResult[{"ModelFunction","ReducedChiSquare",
 "ParameterErrors"},xyErrorData,curveFitInfo];
```

Fitted model function:

$2.99923e^{-(-4.00939+x)^2} + 0.499635x$

Reduced chi-square of fit = 8.163×10^{-1}

	Value	Standard error	Confidence region
Parameter a1 =	0.499635	0.0020301	{0.497583, 0.501687}
Parameter a2 =	2.99923	0.01941	{2.97961, 3.01885}
Parameter a3 =	4.00939	0.00529798	{4.00403, 4.01474}

Improved estimates of the parameters' values as well as decreased parameters' errors and smaller confidence regions are the result. Unfortunately the latter possibility of an alternative measurement process with increased precision is only rarely encountered in practice. So the only method of choice is usually to increase the number of data which means more time and more money. To estimate this critical quantity in advance the simulation of the necessary number of experimental data is always helpful and indicated. The CIP CurveFit package provides the Get-NumberOfData method to fulfill this task: This method tries to detect the necessary number of data necessary to achieve a desired width of the confidence region of a specified parameter for a specified confidence level by an iterative process. If a width of the confidence region of 0.01 for parameter a_3 is desired

```
desiredWidthOfConfidenceRegion=0.01;
indexOfParameter=3;
```

the necessary number of data for the latter example would be

```
numberOfData=CIP `CurveFit `GetNumberOfData[
  desiredWidthOfConfidenceRegion,indexOfParameter,
  pureOriginalFunction,argumentRange,standardDeviationRange,
  modelFunction,argumentOfModelFunction,parametersOfModelFunction,
  CurveFitOptionStartParameters -> startParameters]
```

57

For a halved confidence region of 0.005

```
desiredWidthOfConfidenceRegion=0.005;
```

the number of data must be increased to about 221:

```
numberOfData=CIP `CurveFit `GetNumberOfData[
  desiredWidthOfConfidenceRegion,indexOfParameter,
  pureOriginalFunction,argumentRange,standardDeviationRange,
  modelFunction,argumentOfModelFunction,parametersOfModelFunction,
  CurveFitOptionStartParameters -> startParameters]
```

221

Note that there is a strong non-linear relation between the necessary number of data and the width of a confidence region: To half the width of a confidence region in value there is a considerable increase of the number of data necessary in general.

2.5.4 Large parameters' errors and educated cheating

```
Clear["Global `*"];
<<CIP `CalculatedData `
<<CIP `CurveFit `
```

For specific model functions very precise experimental data are necessary to estimate its parameters' values with a sufficient precision. A good example are power laws that play an important role in different areas of science like critical phenomena or the analysis of biological (scale-free) networks. A power law of the form

$$y = f(x) = a_1|x - a_2|^{-a_3}$$

that diverges at $x = a_2$ with a so called critical exponent a_3 will be discussed in the following. Power law fits are often used to prove or reject a specific theoretical prediction whereupon the critical exponent a_3 enjoys the highest attention: Therefore this parameter is to be estimated with an utmost precision. For a power law fit a search for parameters' start values is not necessary in most cases since all parameters are approximately known in advance from theory or visual inspection of the data: The critical exponent a_3 comes from theory, the location of the divergence a_2 may be directly deduced from the data so only the prefactor a_1 is in question. As an example fifty high precision normally distributed data

```
numberOfData=50;
```

will be generated around the power law

$$y = f(x) = 2\,|x - 10|^{-0.63}$$

```
pureOriginalFunction=Function[x,2.0*Abs[x-10.0]^(-0.63)];
```

in the argument range [8.0, 9.9]

```
argumentRange={8.0,9.9};
```

with a relative standard deviation of 0.1%:

```
errorType="Relative";
standardDeviationRange={0.001,0.001};
```

The arguments will be spaced by a logarithmic scale to push more data into the divergence region (as is usually performed by a proper design of experiment):

```
argumentDistance="LogLargeToSmall";
```

The xy-error data are generated with method GetXyErrorData of the CIP CalculatedData package:

```
xyErrorData=CIP`CalculatedData`GetXyErrorData[pureOriginalFunction,
  argumentRange,numberOfData,standardDeviationRange,
  CalculatedDataOptionErrorType -> errorType,
  CalculatedDataOptionDistance -> argumentDistance];
```

The model function to fit is set in accordance

```
modelFunction=a1*Abs[x-a2]^(-a3);
argumentOfModelFunction=x;
parametersOfModelFunction={a1,a2,a3};
```

The necessary start parameters are chosen to be near the true parameters:

```
startParameters={{a1,1.9},{a2,9.99},{a3,-0.6}};
```

A high confidence level of 99.9% is advised for the confidence region of the parameters:

```
confidenceLevelOfParameterErrors=0.999;
```

For these simulated data a perfect fit results:

```
curveFitInfo=CIP`CurveFit`FitModelFunction[xyErrorData,
 modelFunction,argumentOfModelFunction,parametersOfModelFunction,
 CurveFitOptionStartParameters -> startParameters,
 CurveFitOptionConfidenceLevel ->
 confidenceLevelOfParameterErrors];
CIP`CurveFit`ShowFitResult[{"FunctionPlot","RelativeResidualsPlot",
 "SDFit","ReducedChiSquare","ParameterErrors"},xyErrorData,
 curveFitInfo];
```

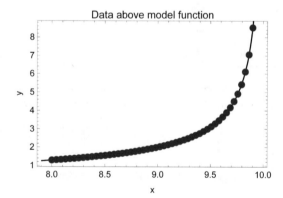

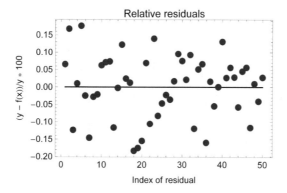

Standard deviation of fit = 1.78×10^{-3}

Reduced chi-square of fit = 8.514×10^{-1}

	Value	Standard error	Confidence region
Parameter	a1 = 2.0006	0.000557139	{1.99864, 2.00255}
Parameter	a2 = 10.0004	0.000295439	{9.99932, 10.0014}
Parameter	a3 = 0.630468	0.000458764	{0.628857, 0.632078}

The critical exponent a_3 is found to be in a small confined interval [0.629, 0.632] around 0.63 with a high probability of 99.9%. If a theoretical model would predict the value of 0.63 this fit would rightly be regarded as a *strong experimental evidence* (by cautious scientists) up to a *convincing experimental proof* (by more enthusiastic ones). Unfortunately experimental data for power law fits are often far less precise. This has a dramatic influence on the confidence region of the critical exponent a_3 as shown in the next example. The relative error of the data is increased by a factor of 100 to 10%:

```
standardDeviationRange={0.1,0.1};
xyErrorData=CIP'CalculatedData'GetXyErrorData[pureOriginalFunction,
 argumentRange,numberOfData,standardDeviationRange,
 CalculatedDataOptionErrorType -> errorType,
 CalculatedDataOptionDistance -> argumentDistance];
```

The corresponding fit

```
curveFitInfo=CIP'CurveFit'FitModelFunction[xyErrorData,
 modelFunction,argumentOfModelFunction,parametersOfModelFunction,
 CurveFitOptionStartParameters -> startParameters,
 CurveFitOptionConfidenceLevel ->
 confidenceLevelOfParameterErrors];
CIP'CurveFit'ShowFitResult[{"FunctionPlot","RelativeResidualsPlot",
 "SDFit","ReducedChiSquare","ParameterErrors"},xyErrorData,
 curveFitInfo];
```

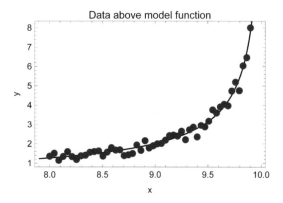

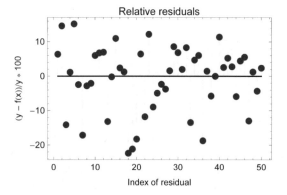

Standard deviation of fit = 1.779×10^{-1}

Reduced chi-square of fit = 8.512×10^{-1}

	Value	Standard error	Confidence region
Parameter a1 =	2.0764	0.0830715	{1.78482, 2.36797}
Parameter a2 =	10.0444	0.043694	{9.89101, 10.1977}
Parameter a3 =	0.68482	0.0580304	{0.481139, 0.888501}

again looks perfect but the confidence region of the critical exponent a_3 is found to be nearly as large $(0.89 - 0.48 = 0.41)$ as the absolute value of the parameter itself (0.68): So its evidence for support or rejection of a specific theoretical prediction almost vanished. The bitter truth is that simply nothing can be deduced from the data - a result that most principal investigators hate since it means wasted time and money. And that's where the educated cheating starts. Let's say the theoretical prediction of the critical exponent a_3 is 0.73 (remember that the data were generated with a true value of 0.63): Simply fix parameter a_3 to 0.73

```
modelFunction=a1*Abs[x-a2]^(-0.73);
parametersOfModelFunction={a1,a2};
```

```
startParameters={{a1,1.9},{a2,9.99}};
```

and fit parameters a_1 and a_2 only:

```
curveFitInfo=CIP `CurveFit `FitModelFunction[xyErrorData,
 modelFunction,argumentOfModelFunction,parametersOfModelFunction,
 CurveFitOptionStartParameters -> startParameters,
 CurveFitOptionConfidenceLevel ->
 confidenceLevelOfParameterErrors];
CIP `CurveFit `ShowFitResult[{"FunctionPlot","RelativeResidualsPlot",
 "SDFit","ReducedChiSquare","ParameterErrors"},xyErrorData,
 curveFitInfo];
```

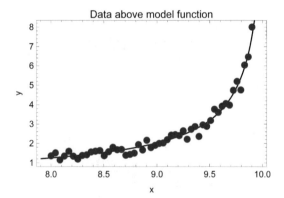

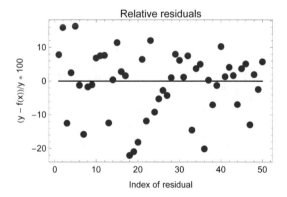

Standard deviation of fit = 1.772×10^{-1}

Reduced chi-square of fit = 8.445×10^{-1}

	Value	Standard error	Confidence region
Parameter $a1$ =	2.13617	0.0487269	{1.96538, 2.30696}
Parameter $a2$ =	10.0775	0.017931	{10.0146, 10.1403}

A very good looking fit is the result with a very convincing residuals plot which may easily be published to be *in perfect agreement with the theoretical prediction of 0.73*. But with about the same evidence it could be argued for a critical exponent a_3 of value 0.53:

```
modelFunction=a1*Abs[x-a2]^(-0.53);
curveFitInfo=CIP`CurveFit`FitModelFunction[xyErrorData,
 modelFunction,argumentOfModelFunction,parametersOfModelFunction,
 CurveFitOptionStartParameters -> startParameters,
 CurveFitOptionConfidenceLevel ->
 confidenceLevelOfParameterErrors];
CIP`CurveFit`ShowFitResult[{"FunctionPlot","RelativeResidualsPlot",
 "SDFit","ReducedChiSquare","ParameterErrors"},xyErrorData,
 curveFitInfo];
```

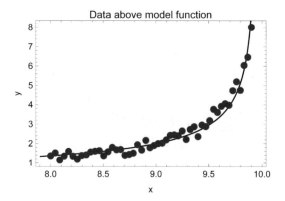

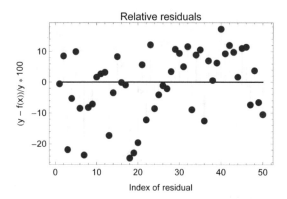

Standard deviation of fit = 2.008×10^{-1}

Reduced chi-square of fit = 1.083

	Value	Standard error	Confidence region
Parameter a1 =	1.95121	0.0335399	{1.83365, 2.06877}
Parameter a2 =	9.95776	0.00900621	{9.92619, 9.98933}

The fit again is convincing and *in perfect agreement with ...* The situation becomes only somewhat better if the number of data is increased. For a fivefold data boost

```
numberOfData=250;
xyErrorData=CIP`CalculatedData`GetXyErrorData[pureOriginalFunction,
 argumentRange,numberOfData,standardDeviationRange,
 CalculatedDataOptionErrorType -> errorType,
 CalculatedDataOptionDistance -> argumentDistance];
```

and a fit with the complete 3-parameter model function

```
modelFunction=a1*Abs[x-a2]^(-a3);
parametersOfModelFunction={a1,a2,a3};
startParameters={{a1,1.9},{a2,9.99},{a3,-0.6}};
curveFitInfo=CIP`CurveFit`FitModelFunction[xyErrorData,
 modelFunction,argumentOfModelFunction,parametersOfModelFunction,
 CurveFitOptionStartParameters -> startParameters,
 CurveFitOptionConfidenceLevel ->
 confidenceLevelOfParameterErrors];
pointSize=0.01;
CIP`CurveFit`ShowFitResult[{"FunctionPlot","RelativeResidualsPlot",
 "SDFit","ReducedChiSquare","ParameterErrors"},xyErrorData,
 curveFitInfo,GraphicsOptionPointSize -> pointSize];
```

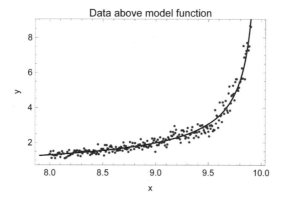

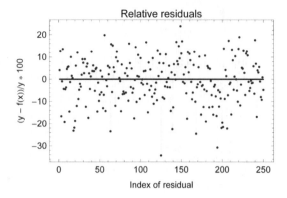

Standard deviation of fit = 1.92×10^{-1}

Reduced chi-square of fit = 9.885×10^{-1}

	Value	Standard error	Confidence region
Parameter a1 =	1.96714	0.0222953	{1.89289, 2.0414}
Parameter a2 =	9.97971	0.0120999	{9.93941, 10.02}
Parameter a3 =	0.602932	0.0193802	{0.538389, 0.667475}

```
numberOfIntervals=10;
CIP`CurveFit`ShowFitResult[
  {"RelativeResidualsDistribution"},xyErrorData,
  curveFitInfo,NumberOfIntervalsOption -> numberOfIntervals];
```

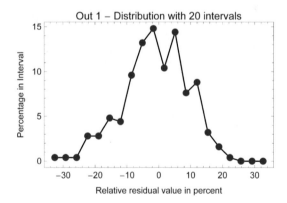

the estimated value of the critical exponent a_3 improves and its confidence region inevitably shrinks. The distribution of the residuals looks like a distorted bell curve. But the evidence for both false theoretical predictions with values 0.73 and 0.53 would still be convincing: Theoretical prediction 0.73

```
modelFunction=a1*Abs[x-a2]^(-0.73);
parametersOfModelFunction={a1,a2};
startParameters={{a1,1.9},{a2,9.99}};
curveFitInfo=CIP `CurveFit `FitModelFunction[xyErrorData,
 modelFunction,argumentOfModelFunction,parametersOfModelFunction,
 CurveFitOptionStartParameters -> startParameters,
 CurveFitOptionConfidenceLevel ->
 confidenceLevelOfParameterErrors];
CIP `CurveFit `ShowFitResult[{"FunctionPlot","RelativeResidualsPlot",
 "SDFit","ReducedChiSquare","ParameterErrors"},xyErrorData,
 curveFitInfo,GraphicsOptionPointSize -> pointSize];
```

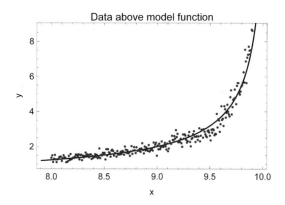

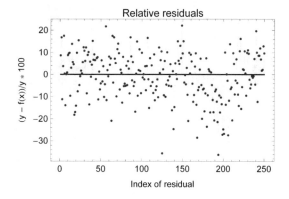

Standard deviation of fit = 2.028×10^{-1}

Reduced chi-square of fit = 1.103

	Value	Standard error	Confidence region
Parameter a1 =	2.10871	0.0216214	{2.03671, 2.18072}
Parameter a2 =	10.067	0.00791856	{10.0406, 10.0934}

```
CIP 'CurveFit 'ShowFitResult[
  {"RelativeResidualsDistribution"},xyErrorData,
  curveFitInfo,NumberOfIntervalsOption -> numberOfIntervals];
```

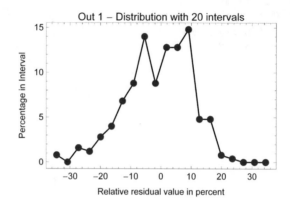

looks approximately as good as the 3-parameter-fit and theoretical prediction 0.53:

```
modelFunction=a1*Abs[x-a2]^(-0.53);
curveFitInfo=CIP 'CurveFit 'FitModelFunction[xyErrorData,
  modelFunction,argumentOfModelFunction,parametersOfModelFunction,
  CurveFitOptionStartParameters -> startParameters,
  CurveFitOptionConfidenceLevel ->
  confidenceLevelOfParameterErrors];
CIP 'CurveFit 'ShowFitResult[{"FunctionPlot","RelativeResidualsPlot",
  "SDFit","ReducedChiSquare","ParameterErrors"},xyErrorData,
  curveFitInfo,GraphicsOptionPointSize -> pointSize];
```

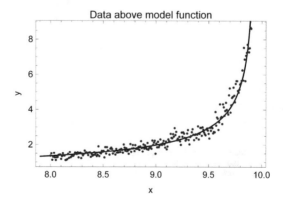

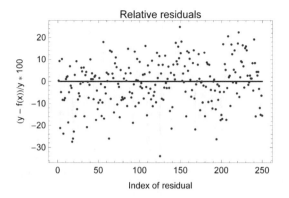

Standard deviation of fit = 1.991×10^{-1}

Reduced chi-square of fit = 1.063

	Value	Standard error	Confidence region
Parameter a1 =	1.92718	0.0147138	{1.87818, 1.97618}
Parameter a2 =	9.94512	0.00372884	{9.93271, 9.95754}

```
CIP`CurveFit`ShowFitResult[
  {"RelativeResidualsDistribution"},xyErrorData,
  curveFitInfo,NumberOfIntervalsOption -> numberOfIntervals];
```

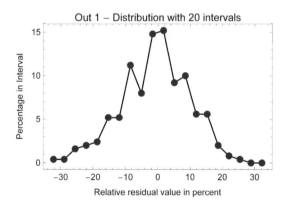

So a lot more experimental data would be needed to really make clear decisions. With the aid of the GetNumberOfData method of the CIP CurveFit package the necessary number of data for a desired width of a parameters' confidence region may be estimated (see the previous section). For a desired confidence region width of 0.04 for parameter a_3

```
desiredWidthOfConfidenceRegion=0.04;
indexOfParameter=3;
```

the number of data must be increased to

```
modelFunction=a1*Abs[x-a2]^(-a3);
argumentOfModelFunction=x;
parametersOfModelFunction={a1,a2,a3};
startParameters={{a1,1.9},{a2,9.99},{a3,-0.6}};
numberOfData=CIP `CurveFit `GetNumberOfData[
  desiredWidthOfConfidenceRegion,indexOfParameter,
  pureOriginalFunction,argumentRange,standardDeviationRange,
  modelFunction,argumentOfModelFunction,parametersOfModelFunction,
  CurveFitOptionStartParameters -> startParameters,
  CurveFitOptionConfidenceLevel ->
  confidenceLevelOfParameterErrors,
  CalculatedDataOptionErrorType -> errorType,
  CalculatedDataOptionDistance -> argumentDistance]
```

3344

The corresponding fit with this estimated number of data

```
xyErrorData=CIP `CalculatedData `GetXyErrorData[pureOriginalFunction,
  argumentRange,numberOfData,standardDeviationRange,
  CalculatedDataOptionErrorType -> errorType,
  CalculatedDataOptionDistance -> argumentDistance];
curveFitInfo=CIP `CurveFit `FitModelFunction[xyErrorData,
  modelFunction,argumentOfModelFunction,parametersOfModelFunction,
  CurveFitOptionStartParameters -> startParameters,
  CurveFitOptionConfidenceLevel ->
  confidenceLevelOfParameterErrors];
CIP `CurveFit `ShowFitResult[{"FunctionPlot","RelativeResidualsPlot",
  "SDFit","ReducedChiSquare","ParameterErrors"},xyErrorData,
  curveFitInfo,GraphicsOptionPointSize -> pointSize];
```

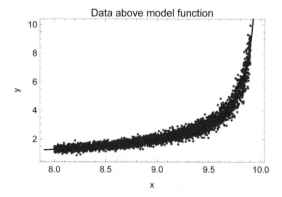

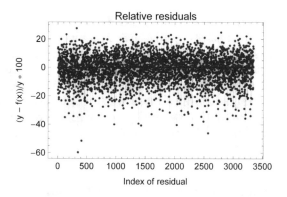

Standard deviation of fit = 1.912×10^{-1}

Reduced chi-square of fit = 9.798×10^{-1}

	Value	Standard error	Confidence region
Parameter a1 =	1.99724	0.00747252	{1.97263, 2.02185}
Parameter a2 =	9.99805	0.00417287	{9.98431, 10.0118}
Parameter a3 =	0.628188	0.00604781	{0.60827, 0.648106}

```
numberOfIntervals=30;
CIP`CurveFit`ShowFitResult[
  {"RelativeResidualsDistribution"},xyErrorData,
  curveFitInfo,NumberOfIntervalsOption -> numberOfIntervals];
```

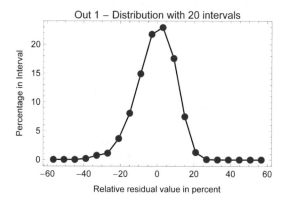

finally allows estimates within the required precision. For many experimental se-tups however the necessary increase of data would be completely out of reach due to restrictions in time and money. Therefore the sketched kind of educated cheat-ing is unfortunately more widespread than it ought to be (and even worse is often combined with an elimination of outliers after the fit: A "very successful strategy" to tune the data). In most cases experimentalists do not even have a bad conscience since the final plots look good. Therefore a clear trend can be detected for experimental data analysis to follow theoretical predictions (this can be superbly shown in the field of critical phenomena where the theoretical predictions changed over the decades and the experimental data analysis with them in close accordance). But it should not be forgotten that cheating simply has nothing to do with science - and in the end someone will detect it regardless how educated it was hidden.

2.5.5 Experimental errors and data transformation

```
Clear["Global`*"];
<<CIP`CalculatedData`
<<CIP`CurveFit`
```

The errors σ_i of the y_i values do not only influence the errors of the parameters of the fitted model function but they also influence the parameters' values themselves. This is often ignored but obvious if it is remembered that curve fitting means minimization of $\chi^2(a_1,...,a_L)$:

$$\chi^2(a_1,...,a_L) = \sum_{i=1}^{K} \left(\frac{y_i - f(x_i,a_1,...,a_L)}{\sigma_i} \right)^2 \longrightarrow \text{minimize!}$$

Since the errors σ_i are part of the sum of squares they contribute to the determination of the minimum location of $\chi^2(a_1,...,a_L)$. Only in the special case that all errors σ_i are equal

$$\sigma_i = \sigma$$

they are a mere factor σ that can be factored out of the sum and therefore does not influence the minimum. The influence of the errors σ_i can be illustrated with the following (artificial) example of twenty simulated data

```
numberOfData=20;
```

around the function

$$y = f(x) = 2e^{-\frac{1}{x}}$$

```
pureOriginalFunction=Function[x,2.0*Exp[-1.0/x]];
```

in the argument range [1.0, 8.0]

```
argumentRange={1.0,8.0};
```

with a relative standard deviation of 5%

```
standardDeviationRange={0.05,0.05};
errorType="Relative";
xyErrorData=CIP`CalculatedData`GetXyErrorData[pureOriginalFunction,
  argumentRange,numberOfData,standardDeviationRange,
  CalculatedDataOptionErrorType -> errorType];
```

that are fitted with corrected estimates of parameters' errors for comparison purposes:

```
modelFunction=a1*Exp[-a2/x];
argumentOfModelFunction=x;
parametersOfModelFunction={a1,a2};
varianceEstimator="ReducedChiSquare";
curveFitInfo=CIP`CurveFit`FitModelFunction[xyErrorData,
  modelFunction,argumentOfModelFunction,parametersOfModelFunction,
  CurveFitOptionVarianceEstimator -> varianceEstimator];
CIP`CurveFit`ShowFitResult[{"FunctionPlot","SDFit",
  "ReducedChiSquare","ParameterErrors"},xyErrorData,curveFitInfo];
```

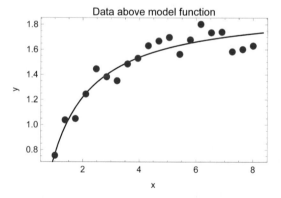

Standard deviation of fit = 7.272×10^{-2}

Reduced chi-square of fit = 1.122

	Value	Standard error	Confidence region
Parameter a1 =	1.93972	0.0396424	{1.89894, 1.98049}
Parameter a2 =	0.92742	0.051963	{0.873972, 0.980868}

If the errors σ_i are asymmetrically enlarged by different factors from 10.0 (tenfold increase) to 1.0 (no change)

```
minFactor=1.0;
maxFactor=10.0;
errorTransformationFactors=Table[i,{i,maxFactor,minFactor,
 -(maxFactor-minFactor)/(Length[xyErrorData]-1)}];
newXyErrorData=Table[{xyErrorData[[i,1]],xyErrorData[[i,2]],
 xyErrorData[[i,3]]*errorTransformationFactors[[i]]},
 {i,Length[xyErrorData]}];
```

the estimated optimum values of the parameters become clearly different:

```
curveFitInfo=CIP'CurveFit'FitModelFunction[newXyErrorData,
 modelFunction,argumentOfModelFunction,parametersOfModelFunction,
 CurveFitOptionVarianceEstimator -> varianceEstimator];
CIP'CurveFit'ShowFitResult[{"FunctionPlot","SDFit",
 "ReducedChiSquare","ParameterErrors"},xyErrorData,curveFitInfo];
```

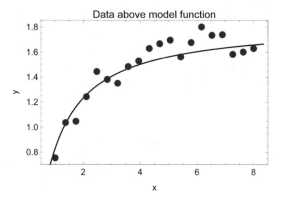

Standard deviation of fit = 8.842×10^{-2}

Reduced chi-square of fit = 1.659

	Value	Standard error	Confidence region
Parameter a1 =	1.83037	0.0368555	{1.79246, 1.86828}
Parameter a2 =	0.787937	0.107727	{0.67713, 0.898743}

The parameter estimates changed by around 5-10% of their absolute values although the x_i and y_i values were not changed at all. Also the values of corrected parameters' errors increased due to the increase of the data's errors.

As far as the popular data transformations are considered the outlined context may play a more or less pronounced role. It is still common in lab data analysis to linearize model functions to a straight line if possible (despite the existence of non-linear curve fitting software). For the model function above linearization may be easily performed by simple application of the natural logarithm

$$y = f(x) = a_1 e^{-\frac{a_2}{x}}$$

$$\ln(y) = \ln\left(a_1 e^{-\frac{a_2}{x}}\right) = \ln a_1 - a_2 \frac{1}{x}$$

which results in a straight line

$$y = f(x) = a_1 + a_2 x$$

with the necessary non-linear data transformations:

$$x_i \to \frac{1}{x_i} \; ; \; y_i \to \ln(y_i)$$

If data are transformed it is often forgotten that the errors σ_i must be transformed too according to standard error propagation:

$$\sigma_i \to \sqrt{\left(\frac{\partial \ln(y_i)}{\partial y_i}\right)^2 \sigma_i^2} = \left(\frac{\partial \ln(y_i)}{\partial y_i}\right)\sigma_i = \frac{\sigma_i}{y_i}$$

Note that the neglect of this error transformation is perhaps the second most frequent mistake in lab data analysis. (The most frequent mistake is the lab journal's report of a mean in combination with the standard deviation of a single measurement and not the correct standard deviation of the mean.) In summary each data triple of the xy-error data must be transformed as follows:

$$(x_i, y_i, \sigma_i) \to \left(\frac{1}{x_i}, \ln(y_i), \frac{\sigma_i}{y_i}\right)$$

Standard error propagation assumes vanishingly small errors since it belongs to linear statistics (with Taylor series expansions up to the first derivative only). Therefore the transformed errors and the original errors do only correspond in an approximate manner. This may have more or less influence on the estimated values of the parameters after linearization depending on the specific fit problem.

2.6 Empirical enhancement of theoretical model functions

```
Clear["Global`*"];
<<CIP`ExperimentalData`
<<CIP`DataTransformation`
<<CIP`Graphics`
<<CIP`CurveFit`
```

Suppose there is a well-defined theoretical model function but the x and y quantities it associates can not be measured directly. Preprocessing steps are necessary to construct the data in question which may introduce systematic errors. An example is outlined in Appendix A that shows the extraction of kinetics data for a chemical reaction (in this case the hydrolysis of acetanhydride) from time dependent infrared (IR) spectra: There are two different methods advised to extract the data: One straight forward method denoted 1 and one more elaborate method denoted 2. The results are provided by the CIP ExperimentalData package. The data produced by method 1 are as follows:

```
xyData=
 CIP 'ExperimentalData 'GetAcetanhydrideKineticsData1[
 ];
errorValue=1.0;
xyErrorData=CIP 'DataTransformation 'AddErrorToXYData[xyData,
 errorValue];
labels={"Time [min]","Absorption",
 "Kinetics of hydrolysis of acetanhydride 1"};
CIP 'Graphics 'PlotXyErrorData[xyErrorData,labels]
```

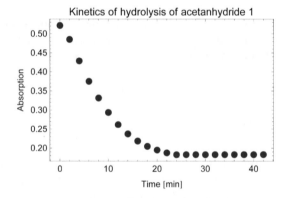

Note that a standard weight of 1.0 was added as an error to the xy data to obtain xy-error data since the preprocessing method did not yield any error estimate. All estimates for parameters' errors thus need a correction deduced from χ^2_{red} (see previous sections).

The hydrolysis of acetanhydride in water is a reaction of (pseudo) first-order which is theoretically described by a simple exponential decay:

$$y = f(x) = a_1 e^{-a_2 x}$$

But a direct fit of this model function

```
modelFunction=a1*Exp[-a2*x];
argumentOfModelFunction=x;
parametersOfModelFunction={a1,a2};
varianceEstimator="ReducedChiSquare";
curveFitInfo=CIP 'CurveFit 'FitModelFunction[xyErrorData,
```

```
modelFunction,argumentOfModelFunction,parametersOfModelFunction,
CurveFitOptionVarianceEstimator -> varianceEstimator];
labels={"Time [min]","Absorption","Data 1 above Model Function"};
CIP`CurveFit`ShowFitResult[{"FunctionPlot"},xyErrorData,
curveFitInfo,CurveFitOptionLabels -> labels];
```

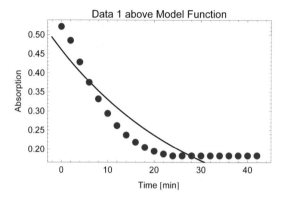
Data 1 above Model Function

fails completely. Due to preprocessing method 1 the data do not direct to a zero absorption with increasing time (acetanhydride vanishes with reaction progress) but to a constant value above zero (a so called background caused by the extraction process, see Appendix A). Therefore the theoretical model function must be enhanced by (at least) an empirical constant background parameter a_3 that takes this deficiency into account:

$$y = f(x) = a_1 e^{-a_2 x} + a_3$$

```
modelFunction=a1*Exp[-a2*x]+a3;
parametersOfModelFunction={a1,a2,a3};
startParameters={{a1,0.4},{a2,0.1},{a3,0.2}};
```

The enhanced fit

```
curveFitInfo=CIP`CurveFit`FitModelFunction[xyErrorData,
modelFunction,argumentOfModelFunction,parametersOfModelFunction,
CurveFitOptionStartParameters -> startParameters,
CurveFitOptionVarianceEstimator -> varianceEstimator];
CIP`CurveFit`ShowFitResult[{"FunctionPlot","AbsoluteResidualsPlot",
"ParameterErrors"},xyErrorData,curveFitInfo,
CurveFitOptionLabels -> labels];
```

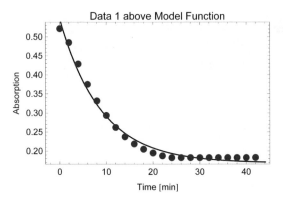

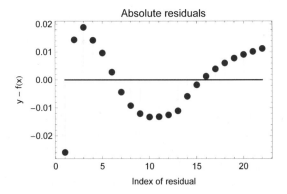

	Value	Standard error	Confidence region
Parameter a1 =	0.378912	0.00979157	{0.368856, 0.388968}
Parameter a2 =	0.112982	0.00687499	{0.105921, 0.120043}
Parameter a3 =	0.168445	0.0051983	{0.163106, 0.173783}

leads to an improved description of the data but reveals a strong systematic deviation pattern of the residuals. In contrast to method 1 the more elaborate preprocessing method 2 tries to estimate the background contribution in advance (see details
in Appendix A):

```
xyData=
 CIP`ExperimentalData`GetAcetanhydrideKineticsData2[
 ];
errorValue=1.0;
xyErrorData=CIP`DataTransformation`AddErrorToXYData[xyData,
 errorValue];
labels={"Time [min]","Absorption",
 "Kinetics of Hydrolysis of Acetanhydride 2"};
CIP`Graphics`PlotXyErrorData[xyErrorData,labels]
```

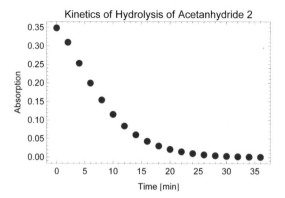

Now the absorption values seem to direct to zero. But a direct fit of the pure theoretical model

```
modelFunction=a1*Exp[-a2*x];
parametersOfModelFunction={a1,a2};
curveFitInfo=CIP`CurveFit`FitModelFunction[xyErrorData,
  modelFunction,argumentOfModelFunction,parametersOfModelFunction,
  CurveFitOptionVarianceEstimator -> varianceEstimator];
labels={"Time [min]","Absorption","Data 2 above Model Function"};
CIP`CurveFit`ShowFitResult[{"FunctionPlot","AbsoluteResidualsPlot",
  "ParameterErrors"},xyErrorData,curveFitInfo,
  CurveFitOptionLabels -> labels];
```

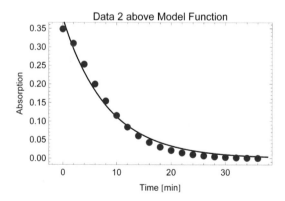

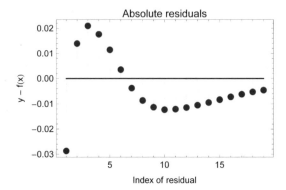

	Value	Standard error	Confidence region
Parameter a1 =	0.377157	0.0104281	{0.366413, 0.387901}
Parameter a2 =	0.121256	0.00534371	{0.11575, 0.126761}

still suggests the use of an additional constant background parameter a_3

```
modelFunction=a1*Exp[-a2*x]+a3;
parametersOfModelFunction={a1,a2,a3};
startParameters={{a1,0.4},{a2,0.1},{a3,0.2}};
curveFitInfo=CIP`CurveFit`FitModelFunction[xyErrorData,
  modelFunction,argumentOfModelFunction,parametersOfModelFunction,
  CurveFitOptionStartParameters -> startParameters,
  CurveFitOptionVarianceEstimator -> varianceEstimator];
CIP`CurveFit`ShowFitResult[{"FunctionPlot","AbsoluteResidualsPlot",
  "ParameterErrors"},xyErrorData,curveFitInfo,
  CurveFitOptionLabels -> labels];
```

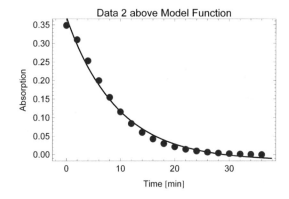

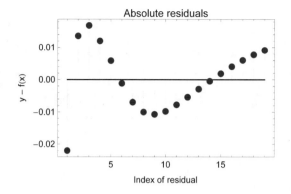

	Value	Standard error	Confidence region
Parameter	a1 = 0.38797	0.00863007	{0.379061, 0.396878}
Parameter	a2 = 0.106077	0.00623914	{0.0996363, 0.112517}
Parameter	a3 = -0.0174831	0.00606562	{-0.0237445, -0.0112218}

which results in a estimated value for a_3 that is at least close to zero (so the background correction of the more elaborate preprocessing method was not in vain). The visual inspection of the data also suggests to treat the first two points as outliers: If they are removed from the xy-error data

```
xyErrorData=Drop[xyErrorData,2];
labels={"Time [min]","Absorption",
 "Kinetics of Hydrolysis of Acetanhydride 2"};
CIP `Graphics `PlotXyErrorData[xyErrorData,labels]
```

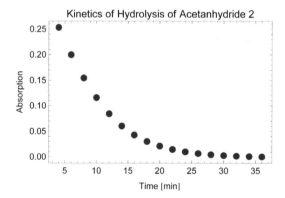

the remaining data do not only look better but lead to an improved fit

```
modelFunction=a1*Exp[-a2*x]+a3;
parametersOfModelFunction={a1,a2,a3};
```

```
startParameters={{a1,0.4},{a2,0.1},{a3,0.2}};
curveFitInfo=CIP`CurveFit`FitModelFunction[xyErrorData,
 modelFunction,argumentOfModelFunction,parametersOfModelFunction,
 CurveFitOptionStartParameters -> startParameters,
 CurveFitOptionVarianceEstimator -> varianceEstimator];
labels={"Time [min]","Absorption","Data 2 above Model Function"};
CIP`CurveFit`ShowFitResult[{"FunctionPlot","AbsoluteResidualsPlot",
 "ParameterErrors"},xyErrorData,curveFitInfo,
 CurveFitOptionLabels -> labels];
```

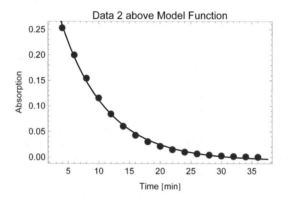

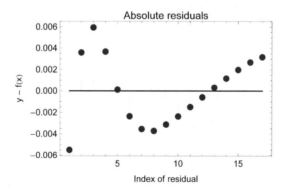

		Value	Standard error	Confidence region
Parameter	a1 =	0.453659	0.00879298	{0.444541, 0.462778}
Parameter	a2 =	0.134134	0.00371886	{0.130277, 0.13799}
Parameter	a3 =	-0.00662982	0.00180625	{-0.00850296, -0.00475667}

with significantly smaller residuals and a further decreased background parameter a_3. There is still a clear systematic deviation pattern of the residuals but this
is probably the best we can get. Always keep in mind that the introduction of new
empirical parameters and the removal of apparent outliers are dangerous procedures
that are an ideal basis for educated cheating: In the end you can obtain (nearly) any

result you like to get and this is not the objective of science which claims to describe the real world out there. But careful and considerate use of these procedures may extract information from data that would otherwise be lost.

2.7 Data smoothing with cubic splines

```
Clear["Global`*"];
<<CIP `CalculatedData`
<<CIP `Graphics`
<<CIP `CurveFit`
<<CIP `ExperimentalData`
<<FunctionApproximations`
```

Data smoothing with cubic splines is controlled by the specified χ^2_{red} value (see above). Depending on the χ^2_{red} value there are two smoothing extremes: A small χ^2_{red} value enforces small residuals and restricts the smoothing function to close proximity of the data points whereas a high χ^2_{red} value allows for larger residuals but forces the curvature of the smoothing function to minimize towards a straight line. This may be demonstrated with fifty simulated xy-error data around the Gaussian-peak shaped function:

```
numberOfData=50;
pureOriginalFunction=Function[x,0.5*x+3.0*Exp[-(x-4.0)^2]];
argumentRange={1.0,7.0};
standardDeviationRange={0.1,0.1};
xyErrorData=CIP `CalculatedData`GetXyErrorData[pureOriginalFunction,
 argumentRange,numberOfData,standardDeviationRange];
labels={"x","y","Data above Original Function"};
CIP `Graphics`PlotXyErrorDataAboveFunction[xyErrorData
 ,pureOriginalFunction,labels]
```

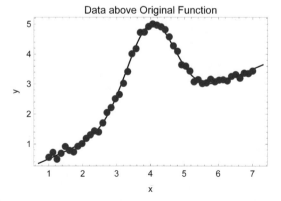

A small χ^2_{red} value of 0.01

```
reducedChiSquare=0.01;
```

leads to mere interpolation between the data without smoothing

```
curveFitInfo=
  CIP `CurveFit `FitCubicSplines[xyErrorData,reducedChiSquare];
  CIP `CurveFit `ShowFitResult[{"FunctionPlot","AbsoluteResidualsPlot",
  "ReducedChiSquare","CorrelationCoefficient"},xyErrorData,
  curveFitInfo];
```

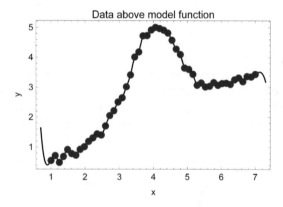

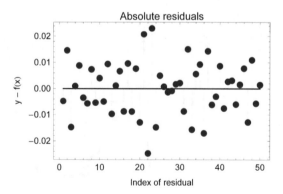

Reduced chi-square of fit = $1. \times 10^{-2}$

Out 1 : Correlation coefficient = 0.999973

and small residuals. Note that the correlation coefficient that indicates the agreement of data and machine output is (almost) one which means a perfect correlation:

Since the data are erroneous this outcome indicates a so called overfitting of the data (which is to be avoided for convincing smoothing). A high χ^2_{red} value of 100

```
reducedChiSquare=100.0;
```

leads to a straight line

```
curveFitInfo=CIP 'CurveFit 'FitCubicSplines[xyErrorData,
  reducedChiSquare];
CIP 'CurveFit 'ShowFitResult[{"FunctionPlot","AbsoluteResidualsPlot",
  "ReducedChiSquare","CorrelationCoefficient"},xyErrorData,
  curveFitInfo];
```

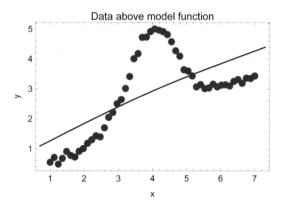

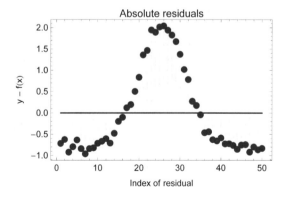

Reduced chi-square of fit = $1. \times 10^2$

Out 1 : Correlation coefficient = 0.683152

without adequate data description and a small correlation coefficient (which is as unfavorable as a perfect correlation for erroneous data). In practice a χ_{red}^2 value is initially chosen that is around 1 to produce a smooth and balancing model function with a convincing residuals plot

```
reducedChiSquare=1.0;
curveFitInfo=CIP`CurveFit`FitCubicSplines[xyErrorData,
  reducedChiSquare];
CIP`CurveFit`ShowFitResult[{"FunctionPlot","AbsoluteResidualsPlot",
  "ReducedChiSquare","CorrelationCoefficient"},xyErrorData,
  curveFitInfo];
```

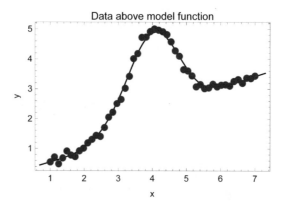

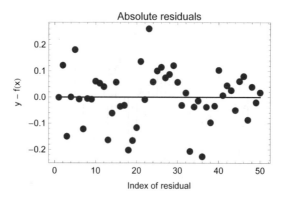

Reduced chi-square of fit = 1.001

Out 1 : Correlation coefficient = 0.997407

and a reasonable high correlation coefficient. Since the smoothing cubic splines procedure tries to minimize the overall curvature over the whole argument range the curved peak region of the current example is comparatively poorly described: A systematic deviation pattern of positive residuals is visible in this middle region of the residuals plot. In this case the χ^2_{red} value should be lowered which enforces smaller residuals to describe the peak region more precisely:

```
reducedChiSquare=0.6;
curveFitInfo=CIP `CurveFit `FitCubicSplines[xyErrorData,
  reducedChiSquare];
CIP `CurveFit `ShowFitResult[{"FunctionPlot","AbsoluteResidualsPlot",
  "ReducedChiSquare","CorrelationCoefficient"},xyErrorData,
  curveFitInfo];
```

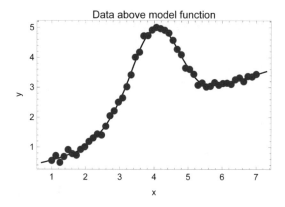

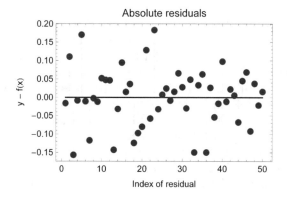

Reduced chi-square of fit = 6. $\times 10^{-1}$

Out 1 : Correlation coefficient = 0.998401

This results in an overall acceptable fit. Note that the correlation coefficient is not too valuable for a goodness-of-smoothing discussion since a higher value does not imply better smoothing due to the increased tendency towards overfitting. The smoothing model function may be finally compared (overlayed) with the original Gaussian-peak shaped function that was used for the simulated data generation

```
pureSmoothingFunction=Function[x,CalculateFunctionValue[x,
  curveFitInfo]];
pureFunctions={pureOriginalFunction,pureSmoothingFunction};
plotRange={0.0,5.0};
plotStyle={{Thickness[0.005],Black},{Thickness[0.005],Blue}};
labels={"x","y","Original + smoothing cubic splines"};
CIP`Graphics`Plot2dFunctions[pureFunctions,argumentRange,plotRange,
  plotStyle,labels]
```

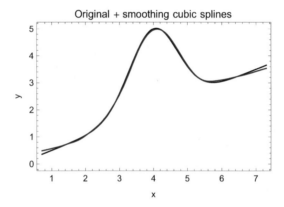

to demonstrate their close proximity and thus a successful model approximation by mere data smoothing. The cubic splines based smoothing model function may be used for interpolating calculations of function values and derivatives. Calculations outside the data's argument range are possible but useless since the cubic splines may have an arbitrary value there: As already mentioned reasonable extrapolations are in principle out of reach if the structural form of the model function is not known. For publishing purposes the internal representation of the smoothing model function is somewhat lengthy: For each (x_i, y_i, σ_i) data triple of the xy-error data a cubic polynomial with 4 parameters is constructed so that the 50 data triples require 200 parameters for the cubic splines. To achieve a more condensed representation the smoothing function may be approximated by a rational function which is constructed by mere trial and error (in this case with a numerator of order 8 and a denominator of order 4):

```
rationalFunction=FunctionApproximations`RationalInterpolation[
  CalculateFunctionValue[x,curveFitInfo],{x,8,4},
  {x,argumentRange[[1]],argumentRange[[2]]}]
```

$$\frac{0.492884-0.61194x+0.646898x^2-0.467755x^3+0.204442x^4-0.0524233x^5+0.00768635x^6-0.000597131x^7+0.0000192755x^8}{1-0.866354x+0.295339x^2-0.0462791x^3+0.0028032x^4}$$

This condensed representation

```
pureRationalFunction=Function[argument,
 rationalFunction/.x -> argument];
pureFunctions={pureOriginalFunction,pureSmoothingFunction,
 pureRationalFunction};
plotStyle={{Thickness[0.005],Black},{Thickness[0.005],Blue},
 {Thickness[0.005],Red}};
labels={"x","y","Original + splines + rational function"};
CIP`Graphics`Plot2dFunctions[pureFunctions,argumentRange,plotRange,
 plotStyle,labels]
```

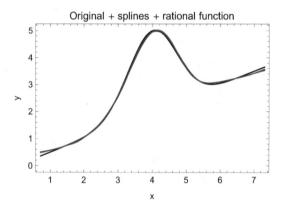

is of sufficient precision as shown by the final overlay. Another application of data smoothing is the representation of calculated data. This sounds absurd since calculated data can be calculated so there seems to be no need for data smoothing. But data calculation may be computationally very expensive in many cases. For example ab-initio quantum-chemical calculations for molecular properties are very time-consuming and therefore require a considerable percentage of the world's overall available computational power. If for example the potential energy surface (PES) of the diatomic molecule hydrogen fluoride is to be described the Schroedinger equation has to be solved for every desired distance between hydrogen and fluoride: Every single calculation may take from seconds up to minutes or hours depending on the level of approximation. The CIP ExperimentalData package contains a set of high precision single point calculations for hydrogen fluoride (see Appendix A):

```
xyErrorData=CIP`ExperimentalData`GetHydrogenFluoridePESXyErrorData[];
labels={"H-F Distance [Angstrom]","Energy [Hartree]",
 "PES of hydrogen fluoride (HF)"};
CIP`Graphics`PlotXyErrorData[xyErrorData,labels]
```

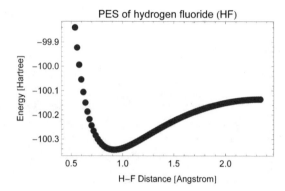

The data are reported with a very small absolute error of 10^{-6}. So a very precise model function that is very near a pure interpolating function is in need. The standard approach with high-degree polynomials

```
modelFunction=
 A0+A1*R+A2*R^2+A3*R^3+A4*R^4+A5*R^5+A6*R^6+A7*R^7+A8*R^8+A9*R^9;
argumentOfModelFunction=R;
parametersOfModelFunction={A0,A1,A2,A3,A4,A5,A6,A7,A8,A9};
curveFitInfo=CIP`CurveFit`FitModelFunction[xyErrorData,
 modelFunction,argumentOfModelFunction,parametersOfModelFunction];
labels={"H-F Distance [Angstrom]","Energy [Hartree]",
 "PES of HF: Polynom fit"};
CIP`CurveFit`ShowFitResult[{"FunctionPlot","AbsoluteResidualsPlot",
 "ReducedChiSquare","AbsoluteResidualsStatistics",
 "CorrelationCoefficient"},xyErrorData,curveFitInfo,
 CurveFitOptionLabels -> labels];
```

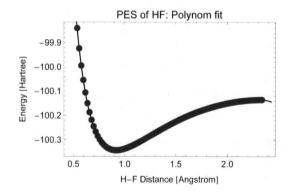

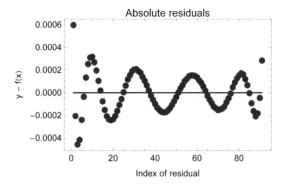

Reduced chi-square of fit = 3.262×10^4

Definition of 'Residual (absolute)': Data - Model

Out 1 : Residual (absolute): Mean/Median/Maximum Value = 1.4×10^{-4} / 1.35×10^{-4} / 5.97×10^{-4}

Out 1 : Correlation coefficient = 0.999998

leads to the well-known systematic oscillations (compare above) around the data that are beyond the required precision. A rational function fit is a little better

```
modelFunction=
  A0+A1*R^-1+A2*R^-2+A3*R^-3+A4*R^-4+A5*R^-5+A6*R^-6+A7*R^-7+
  A8*R^-8+A9*R^-9;
argumentOfModelFunction=R;
parametersOfModelFunction={A0,A1,A2,A3,A4,A5,A6,A7,A8,A9};
curveFitInfo=CIP`CurveFit`FitModelFunction[xyErrorData,
  modelFunction,argumentOfModelFunction,parametersOfModelFunction];
labels={"H-F Distance [Angstrom]","Energy [Hartree]",
  "PES of HF: Rational function fit"};
CIP`CurveFit`ShowFitResult[{"FunctionPlot","AbsoluteResidualsPlot",
  "ReducedChiSquare","AbsoluteResidualsStatistics",
  "CorrelationCoefficient"},xyErrorData,curveFitInfo,
  CurveFitOptionLabels -> labels];
```

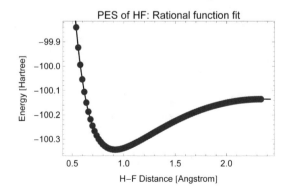

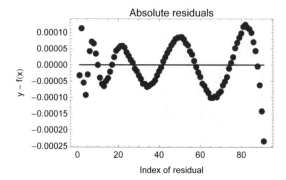

Reduced chi-square of fit = 5.16×10^3

Definition of 'Residual (absolute)': Data - Model

Out 1 : Residual (absolute): Mean/Median/Maximum Value = 5.69×10^{-5} / 5.55×10^{-5} / 2.35×10^{-4}

Out 1 : Correlation coefficient = 1.

but also beyond acceptability. Data smoothing with cubic splines and a χ^2_{red} value of 1 however

```
reducedChiSquare=1.0;
curveFitInfo=CIP`CurveFit`FitCubicSplines[xyErrorData,
 reducedChiSquare];
labels={"H-F Distance [Angstrom]","Energy [Hartree]",
 "PES of HF: Smoothing cubic splines"};
CIP`CurveFit`ShowFitResult[{"FunctionPlot","AbsoluteResidualsPlot",
 "ReducedChiSquare","AbsoluteResidualsStatistics",
 "CorrelationCoefficient"},xyErrorData,curveFitInfo,
 CurveFitOptionLabels -> labels];
```

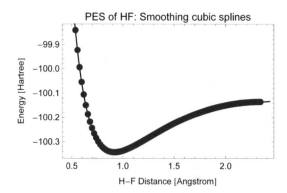

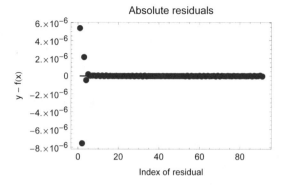

Reduced chi-square of fit = 1.

Definition of 'Residual (absolute)': Data - Model

Out 1 : Residual (absolute): Mean/Median/Maximum Value = 1.92×10^{-7} / 1.99×10^{-8} / 7.57×10^{-6}

Out 1 : Correlation coefficient = 1.

achieves an acceptable interpolation with residuals well within the required order of magnitude: Deviations are only more pronounced in the divergence region at small interatomic distances. Also note that a correlation coefficient of effectively one does not indicate overfitting in this situation since the data errors are very small. A check of the smoothing model function is a calculation of the minimum energy distance between hydrogen and fluoride that is known to be 0.917 Angstrom:

```
argumentRange={0.65,1.3};
functionValueRange={-100.35,-100.2};
CIP`Graphics`Plot2dFunction[Function[arg,CalculateFunctionValue[arg,
  curveFitInfo]],argumentRange,functionValueRange,labels]
```

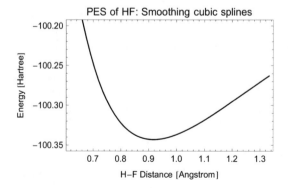

```
FindMinimum[CIP`CurveFit`CalculateFunctionValue[x,curveFitInfo],
   {x,0.5,1.5}]
```

$\{-100.343, \{x \to 0.917413\}\}$

This correct result together with a vanishing derivative value at the minimum (which ought to be 0)

```
CIP`CurveFit`CalculateDerivativeValue[1,0.917,curveFitInfo]
```

-0.000933712

assures an overall satisfactory model function that may be successfully used for interpolation purposes.

2.8 Cookbook recipes for curve fitting

As demonstrated in the previous sections curve fitting can be a challenging task. In this last section some cookbook recipes for curve fitting and data smoothing summarize different aspects outlined above.

- **The data:** Start with a thorough (visual) inspection of the data to avoid the GIGO (garbage-in/garbage-out) effect. Data analysis is not magic, it can not extract information out of nothing. Are the data reasonably scaled and distributed? Are the reported errors convincing? If no errors are available apply the standard weight 1.0 and correct errors with χ^2_{red}. There are additional subtle problems with data that contain outliers, i.e. single data points with extraordinarily large errors. Outliers usually indicate experimental failure. If outliers can be easily detected they should always be removed from the data since they tend to mask themselves in a fitting procedure (they draw the model towards them to become invisible). Data which are known to be prone to contain outliers may deserve a completely different statistical treatment like the so called robust estimation which is beyond this introduction (see [Hampel 1986] or [Rousseeuw 2003] for further reading).
- **The model function:** Is a well-defined model function available? Does the number of data well exceed the number of parameters? Then go on. If no model function is known it might be worth to try to construct one by educated trial and error: This is quite often successful. Avoid model functions with redundant or highly similar parameters. If an educated guess seems to be unfeasible try data smoothing.
- **Linear or non-linear model function:** Is the model function linear in its parameters? Then the fit will work without further considerations. If not: Are parameters' start values approximately known in advance? Then try these values for local minimization. Otherwise an extensive start values search may be advised.

Don't give up too early if things are difficult. The parameters' start values are often the most difficult part of the game.

- **Problems with the fitting procedure:** If the fitting procedure crashes try to use an alternative minimization algorithm. If nothing helps there seems to be a severe problem with the model function or the parameters' start values. Do you use professional curve fitting software? A lot of programs do use (too) simple algorithms without appropriate safeguards that fail needlessly.

- **Goodness of fit:** Are the fitted parameters' values reasonable? Otherwise the minimization procedure sent you somewhere over the rainbow. Is the data plot above the fitted model function convincing, i.e. smooth and balancing? If not the fit failed. Is the residuals plot well within experimental errors and free of systematic deviation patterns? Then probably everything worked well. Other goodness of fit quantities may be used to support your assessment.

- **Parameter errors:** Is the χ^2_{red} value close to 1? If not the reported experimental errors are poor and should be corrected. Do you need high confidence? Then adjust the parameters' confidence level in accordance. Are the parameters' errors too large to make any decisions? Try to avoid strategies of educated cheating unless your career or PhD is in danger (then you should at least provide a convincing residuals plot because this is what most reviewers believe in).

- **Data transformation for linearization:** If possible simply avoid it and use nonlinear curve fitting software. Final diagrams may then be linearized for your audience.

- **Data smoothing:** Adjust the set screws until you like the result with a convincing residuals plot. Then apply the smoothing model for interpolation (but never for extrapolation) purposes.

This chapter sketched general curve fitting issues with a broad range of applications. For many specific curve fitting tasks elaborate specific solutions already exist that avoid problems outlined in the previous sections. Thus the scientific literature should always be consulted in advance (which is of course a mandatory and sensible advice for all scientific endeavours to avoid a reinvention of the wheel).

Chapter 3
Clustering

```
Clear["Global`*"];
<<CIP`CalculatedData`
<<CIP`Graphics`
<<CIP`Cluster`
```

A clustering method tries to partition inputs into different groups/clusters (see chapter 1 for terminology). For an introductory example the following 1000 two-dimensional inputs are clustered (the details of construction are outlined in a minute):

```
standardDeviation=0.05;
numberOfCloudInputs=500;
centroid1={0.3,0.7};
cloudDefinition1={centroid1,numberOfCloudInputs,standardDeviation};
inputs1=CIP`CalculatedData`GetDefinedGaussianCloud[
 cloudDefinition1];
centroid2={0.7,0.3};
cloudDefinition2={centroid2,numberOfCloudInputs,standardDeviation};
inputs2=CIP`CalculatedData`GetDefinedGaussianCloud[
 cloudDefinition2];
inputs=Join[inputs1,inputs2];
labels={"x","y","Inputs to be clustered"};
plotStyle={PointSize[0.01],Blue};
points2DWithPlotStyle={inputs,plotStyle};
points2DWithPlotStyleList={points2DWithPlotStyle};
CIP`Graphics`PlotMultiple2dPoints[points2DWithPlotStyleList,labels]
```

© Springer International Publishing Switzerland 2016   157
A. Zielesny, *From Curve Fitting to Machine Learning*, Intelligent Systems
Reference Library 109, DOI 10.1007/978-3-319-32545-3_3

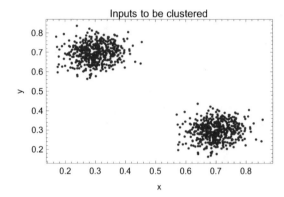

By visual inspection two distinct clusters are visible. They may be (correctly) detected by a clustering method

```
clusterInfo=CIP'Cluster'GetClusters[inputs];
```

which leads to a clustering result that can be illustrated by different coloring for each detected cluster and the display of their particular centroids (points in the middle of the two clouds):

```
indexOfCluster=1;
inputsOfCluster1=CIP'Cluster'GetInputsOfCluster[inputs,
  indexOfCluster,clusterInfo];
points2DWithPlotStyle1={inputsOfCluster1,{PointSize[0.01],Red}};
indexOfCluster=2;
inputsOfCluster2=CIP'Cluster'GetInputsOfCluster[inputs,
  indexOfCluster,clusterInfo];
points2DWithPlotStyle2={inputsOfCluster2,{PointSize[0.01],Green}};
properties={"CentroidVectors"};
centroids2D=
  CIP'Cluster'GetClusterProperty[properties,clusterInfo][[1]];
centroids2DBackground={centroids2D,{PointSize[0.035],White}};
centroids2DWithPlotStyle1={centroids2D,{PointSize[0.03],Black}};
points2DWithPlotStyleList={points2DWithPlotStyle1,
  points2DWithPlotStyle2,centroids2DBackground,
  centroids2DWithPlotStyle1};
labels={"x","y","Colored clusters and their centroids"};
CIP'Graphics'PlotMultiple2dPoints[points2DWithPlotStyleList,labels]
```

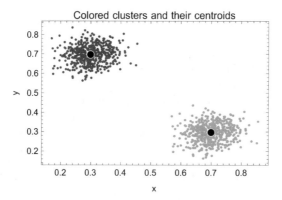

The determined centroids

```
centroids2D
```

$\{\{0.298944, 0.69648\}, \{0.698944, 0.29648\}\}$

are an optimum solution for this clustering task: The mean (euclidean) distance of all inputs to their particular corresponding centroid becomes a minimum (compare next section for details). Also note that the detected centroids are very close to the centroids used for the (above) construction of the clouds of inputs. In principle there are two extremes of clustering depending on the inputs and the clustering method used: Either the clustering process results in one single cluster that contains all inputs or the resulting number of clusters is equal to number of inputs, i.e. each cluster contains one single input. In both extreme cases clustering does not make any sense. A useful clustering result is somewhere in between: In general the number of clusters should be considerably smaller than the number of inputs but at least two. Different clustering methods differ in the way they try to achieve this goal and therefore may lead to different outcomes.

Clustering techniques are usually attributed to the unsupervised machine learning methods since the method is not told in which cluster a specific input is to be put: This decision is to be made by the clustering algorithm itself. If a method is told in a supervised manner to which group a specific input belongs the process is called (supervised) classification and the group is called a class. Classification tasks are discussed at the end of this chapter and predominantly in the machine learning chapter 4. Within the scope of this book clustering methods are separated into an own chapter since a clustering method does not really learn anything: It simply partitions.

The final number of clusters may be specified in advance or may be left open for the clustering method to decide itself. Open-categorical clustering without a fixed number of clusters seems to be the natural choice but sometimes it is sensible to assure a predefined number of clusters. If the number of clusters is not fixed a clustering methods checks various numbers clusters according to an internal decision criterion to determine the optimum number. It should be noted that there is nothing like the best or objective clustering since there is no objective way to partition inputs in general. Depending on the inputs to be clustered and the intentions of the scientists there may be intuitively correct up to totally arbitrary clustering results. This is outlined in the next sections.

Chapter 3 starts with an introduction on some basics of clustering: The partitioning of inputs into a fixed number of clusters is shown to be a global optimization task. The heuristic k-means approach is sketched and the mean-silhouette-width method for the determination of an appropriate number of clusters is explained (section 3.1). Then three intuitive situations for partitioning inputs are tackled: Unambiguous, reasonable and senseless clustering (section 3.2). The number of clusters may be a priori fixed: Corresponding examples are outlined and their consequences discussed (section 3.3). Getting a small number of representatives from a large number of inputs is an important application of clustering methods. The advantages of cluster-based representatives in comparison to randomly chosen ones are pointed out with different examples (section 3.4). Cluster occupancies of joined sets of inputs and an application to the famous iris flower inputs are described as a next step (section 3.5): They also allow the detection of white spots (empty space regions) in comparing different sets of inputs (section 3.6). Since there are numerous clustering methods in use an alternative to the CIP k-medoids default method is demonstrated: ART-2a based clustering. It leads to different results due to a different view of the world (section 3.7). Unsupervised learning may be used to construct a class predictor for new inputs. This aspect prepares the entry to supervised machine learning in chapter 4 (section 3.8). With final cookbook recipes for clustering this chapter ends (section 3.9).

3.1 Basics

As already indicated a clustering task is in fact an optimization task. This may be motivated as follows: Consider a number N of inputs that are to be partitioned into k clusters. What would be an optimum clustering result in this case?

Cluster number i with N_i inputs may be described by a so called centroid $\underline{c}_i$ that is the center of mass of its inputs $\underline{input}_1^{(i)}$ to $\underline{input}_{N_i}^{(i)}$:

$$\underline{c}_i = \frac{1}{N_i} \sum_{u=1}^{N_i} \underline{input}_u^{(i)} \text{ with } u : \text{Index of the input and } i : \text{Index of the cluster}$$

The notation $\underline{input}_u^{(i)}$ denotes the uth input of cluster i. All inputs $\underline{input}_1^{(i)}, ..., \underline{input}_{N_i}^{(i)}$ that are assigned to cluster number i possess the property that their (euclidean) distances $d_{1,i}^{(i)}, ..., d_{N_i,i}^{(i)}$ to the clusters' centroid $\underline{c}_i$

$$d_{u,i}^{(i)} = \left| \underline{input}_u^{(i)} - \underline{c}_i \right| \text{ with } u : \text{Index of the input and } i : \text{Index of the cluster}$$

is less than or equal to their distances to any other centroid $\underline{c}_j$ of other clusters, i.e.

$$d_{u,i}^{(i)} \leq d_{u,j}^{(i)} \text{ with } j \neq i \text{ and}$$

j : Index of another cluster different from i and u : Index of the input

where $d_{u,j}^{(i)}$ is the (euclidean) distance of the uth input of cluster i to another cluster's centroid $\underline{c}_j$:

$$d_{u,j}^{(i)} = \left| \underline{\text{input}}_u^{(i)} - \underline{c}_j \right| \text{ with } j \neq i \text{ and}$$

u : Index of the input and i, j : Index of the cluster

Again: What would be an optimum clustering result for partitioning N inputs into k clusters? The k centroids $\underline{c}_1$ to $\underline{c}_k$ must be chosen to globally minimize the overall mean distance of the inputs to their corresponding cluster centroids, i.e.

$$\bar{d}_i = \frac{1}{N_i} \sum_{u=1}^{N_i} d_{u,i}^{(i)} = \frac{1}{N_i} \sum_{u=1}^{N_i} \left| \underline{\text{input}}_u^{(i)} - \underline{c}_i \right| \text{ with}$$

u : Index of the input and i : Index of the cluster

$$d = \sum_{i=1}^{k} \bar{d}_i \longrightarrow \text{minimize! with}$$

$\bar{d}_i$: Mean distance of all inputs of cluster i to the centroid of cluster i

Optimum positions of the k centroids $\underline{c}_1$ to $\underline{c}_k$ correspond to the global minimum of d which thus is a function of the components of the centroids:

$$d = d(\underline{c}_1, ..., \underline{c}_k) = d\left(c_{1,1}, ..., c_{1,v}, ..., c_{k,1}, ..., c_{k,v}\right) \longrightarrow \text{minimize!}$$

with $c_{i,v}$: vth component of centroid i

In practice this global optimization procedure is rarely performed since its overall computational costs are considerable. To speed up clustering tasks mainly heuristic local optimization methods are used. But increased speed leads to decreased accuracy in general so these methods may fail to converge to the sketched global optimum. A simple and widely used heuristic approach is the so called k-means clustering (see [MacQueen 1967]). To partition inputs into k clusters this method starts with k randomly chosen points in the input's space, the initial centroids. Then it alternates between two steps ...

- **Step 1:** Assign each input to its nearest centroid. After all inputs are assigned k clusters are formed.
- **Step 2:** Calculate the center of mass of the inputs of each cluster (see above). These k center of mass points become the new centroids for the next iteration. Return to Step 1.

... until the assignment of all inputs remains unchanged: In this case the algorithm is deemed to have converged. Since k-means clustering converges quite fast in practice it is usually run multiple times with different initial centroids to scan for an optimum clustering result (which by no means is guaranteed to be equal to the real global optimum described before). Note that there are lots of alternatives to k-means clustering available, e.g. the more robust but slower partitioning around medoids method (abbreviated k-medoids) which is used as the default by the CIP GetClusters method (see [Kaufman 1990] and [GetClusters] in the references).

After the problem of partitioning inputs into k clusters is solved the issue remains how to choose k, i.e. what is the best or the optimum number of clusters for a given set of inputs? A criterion is in need that allows to assess the overall clustering quality for a chosen number of clusters. One such quality measure is the so called mean silhouette width (see [Rousseeuw 1987]). Consider the following setup: The inputs are clustered by some method (e.g. k-means or k-medoids) into k clusters. For a single input u that is assigned to cluster i two quantities are calculated: The mean (euclidean) distance $a_{u,i}$ between this input u and all other inputs that are assigned to the same cluster i

$$a_{u,i} = \frac{1}{N_i} \sum_{v=1}^{N_i} \left| \text{input}_u^{(i)} - \text{input}_v^{(i)} \right|$$

and the mean distance $b_{u,j}$ between this input u and all inputs that are assigned to another cluster j which is nearest to input u

$$b_{u,j} = \frac{1}{N_j} \sum_{v=1}^{N_j} \left| \text{input}_u^{(i)} - \text{input}_v^{(j)} \right| \text{ where } j \neq i$$

i.e. the mean distance $b_{u,j}$ that is the smallest for any other cluster j. With these two quantities $a_{u,i}$ and $b_{u,j}$ the silhouette width $s(u)$ is calculated for the single input u according to:

$$s(u) = \begin{cases} 1 - \frac{a_{u,i}}{b_{u,j}} & \text{if } a_{u,i} < b_{u,j} \\ 0 & \text{if } a_{u,i} = b_{u,j} \\ \frac{b_{u,j}}{a_{u,i}} - 1 & \text{if } a_{u,i} > b_{u,j} \end{cases}$$

The silhouette width $s(u)$ may vary from -1 to 1. A $s(u)$ value near 1 means that $a_{u,i} \ll b_{u,j}$, i.e. the mean distance of input u to the other inputs of its own cluster is far smaller than the mean distance to the inputs of its nearest cluster: This indicates a good clustering. A $s(u)$ value near -1 means the opposite: A poor clustering because input u should be assigned to its nearest cluster and not to the currently assigned cluster. A $s(u)$ value around 0 means that input u is on the border between its own and its nearest cluster. If the silhouette width is calculated for all inputs the mean silhouette width $\bar{s}_k$ for a partitioning approach with k clusters may be obtained:

$$\bar{s}_k = \frac{1}{N} \sum_{w=1}^{N} s(w) \text{ with } N : \text{Number of inputs}$$

The nearer the value of $\bar{s}_k$ is to 1 the better the overall clustering of the inputs may be appraised. And with the mean silhouette width $\bar{s}_k$ the desired decision quantity is obtained that may be used to compare clustering results for different numbers of classes k_1 and k_2: If

$$\bar{s}_{k_1} > \bar{s}_{k_2}$$

then k_1 is the more appropriate number of classes. Silhouette widths and their interpretation will be discussed throughout the whole chapter. Again note: Clustering is an important issue so there are numerous families of clustering methods with an even higher number of variants ready for use. The same holds for decision criteria to obtain optimum cluster numbers. This introduction just scratched the surface.

3.2 Intuitive clustering

```
Clear["Global`*"];
<<CIP`Cluster`
<<CIP`Graphics`
<<CIP`CalculatedData`
```

Clustering is an intuitive task if the data can be visualized, i.e. in the case that the inputs are of dimension two or three. In the following inputs of dimension two are used for simplicity (but all results may be easily generalized to an arbitrary number of dimensions: Just use the same CIP methods with higher dimensional inputs). There are three situations for intuitive clustering that may be distinguished:

- **Clustering is unambiguous or objective:** The inputs are clearly structured or grouped and may unambiguously assigned to separated clusters (like the introductory example above).
- **Clustering is overall reasonable but ambiguous in detail:** The distribution of inputs still exhibits obvious structures and clusters of inputs can be detected. But an unambiguous assignment of all inputs is no longer possible since the clusters are too closely neighbored. The clustering method thus has to generate a reasonable separation between the clusters to assign the inputs.
- **The inputs are unstructured and do not reveal any reasonable clusters.** This may be the case if the inputs are uniformly distributed.

Every clustering method must prove to generate convincing results in these three situations. In the following clustering tasks are discussed with the help of so called Gaussian clouds. A Gaussian cloud of arbitrary dimension consists of vectors that are normally distributed around a center (centroid). Here is an example of a Gaussian cloud in two dimensions: A centroid must be defined

```
centroid1={0.3,0.7};
```

together with a standard deviation of the normal distribution (which determines the size of the cloud) and the number of random cloud inputs:

```
standardDeviation=0.05;
numberOfCloudInputs=5000;
```

The inputs of the cloud are generated with a CIP CalculatedData package method

```
cloudDefinition1={centroid1,numberOfCloudInputs,standardDeviation};
inputs1=
 CIP`CalculatedData`GetDefinedGaussianCloud[cloudDefinition1];
```

and may be illustrated by the corresponding diagram:

```
labels={"x","y","Gaussian cloud in two dimensions"};
points2DWithPlotStyle={inputs1,{PointSize[0.01],Blue}};
points2DWithPlotStyleList={points2DWithPlotStyle};
CIP`Graphics`PlotMultiple2dPoints[points2DWithPlotStyleList,labels]
```

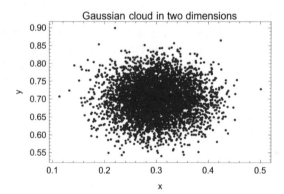

Note that the distortions of the symmetrical cloud result from the unequal golden aspect ratio of the diagram, i.e. one unit has different lengths on the x and the y axis. It is often useful to inspect the frequency distribution of each single component of the inputs: Therefore a number of intervals between the minimum and maximum component values are defined and the frequency of the component values in each interval bin is counted. This may be performed with a CIP Cluster package method:

```
indexOfComponentList={1,2};
numberOfIntervals=20;
CIP 'Cluster 'ShowComponentStatistics[inputs1,indexOfComponentList,
  ClusterOptionNumberOfIntervals -> numberOfIntervals]
```

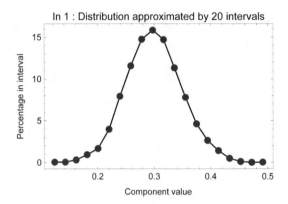

Min / Max = 1.13×10^{-1} / 5.01×10^{-1}

Mean / Median = $3. \times 10^{-1}$ / 2.99×10^{-1}

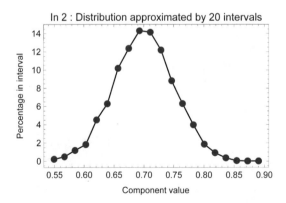

Min / Max = 5.4×10^{-1} / 8.99×10^{-1}

Mean / Median = $7. \times 10^{-1}$ / $7. \times 10^{-1}$

"In 1 " and "In 2 " denote the first and the second component of an input vector.

In the case of Gaussian clouds the frequency distribution of each component should be a normal distribution: The above approximations would converge to a perfect bell-shaped normal distribution if the number of cloud inputs and the number of intervals would be increased to infinity.

As an example for unambiguous or objective clustering two clearly separated Gaussian clouds are generated:

```
standardDeviation=0.05;
numberOfCloudInputs=500;
centroid1={0.3,0.7};
cloudDefinition1={centroid1,numberOfCloudInputs,standardDeviation};
inputs1=
  CIP'CalculatedData'GetDefinedGaussianCloud[cloudDefinition1];
centroid2={0.7,0.3};
cloudDefinition2={centroid2,numberOfCloudInputs,standardDeviation};
inputs2=
  CIP'CalculatedData'GetDefinedGaussianCloud[cloudDefinition2];
```

The two clouds are joined to form a single set of inputs

```
inputs=Join[inputs1,inputs2];
```

for the clustering process:

```
labels={"x","y","Inputs to be clustered"};
points2DWithPlotStyle={inputs,{PointSize[0.01],Blue}};
points2DWithPlotStyleList={points2DWithPlotStyle};
CIP'Graphics'PlotMultiple2dPoints[points2DWithPlotStyleList,labels]
```

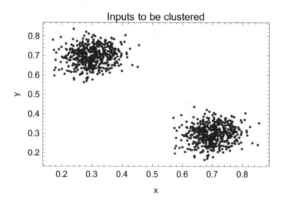

The euclidean distance of the centers of the Gaussian clouds is

```
EuclideanDistance[centroid1,centroid2]
```

0.565685

Clustering tasks are performed with the CIP Cluster package. Unsupervised open-categorical clustering without an a priori specification of the number of desired clusters with the default CIP GetClusters method (see [GetClusters] in the references for details)

```
clusterInfo=CIP`Cluster`GetClusters[inputs];
```

leads to the following (expected) result:

```
CIP`Cluster`ShowClusterResult[{"NumberOfClusters",
  "EuclideanDistanceDiagram","ClusterStatistics"},clusterInfo]
```

Number of clusters = 2

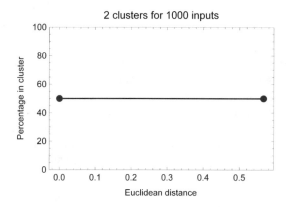

Cluster 1 : 500 members (50.%) with distance = 0.
Cluster 2 : 500 members (50.%) with distance = 0.565685

Two clusters are detected with each containing exactly 50% of the inputs: The detected euclidean distance of the center of mass centroids of both clusters perfectly agrees with the predefined centers used for cloud generation. The result may also be visualized in the space of the inputs with differently colored clusters:

```
indexOfCluster=1;
inputsOfCluster1=
  CIP`Cluster`GetInputsOfCluster[inputs,indexOfCluster,clusterInfo];
indexOfCluster=2;
inputsOfCluster2=
  CIP`Cluster`GetInputsOfCluster[inputs,indexOfCluster,clusterInfo];
labels={"x","y","Clusters in different colors"};
points2DWithPlotStyle1={inputsOfCluster1,{PointSize[0.01],Red}};
```

```
points2DWithPlotStyle2={inputsOfCluster2,{PointSize[0.01],Green}};
points2DWithPlotStyleList={points2DWithPlotStyle1,
 points2DWithPlotStyle2};
CIP`Graphics`PlotMultiple2dPoints[points2DWithPlotStyleList,labels]
```

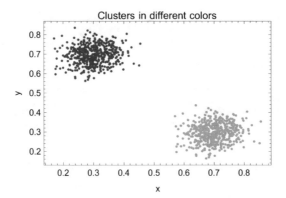

The number of detected clusters is the result of a scan with different clustering approaches where each approach used a different predefined fixed number of clusters: The different clustering results were then evaluated according to an internal decision criterion to determine the optimum number of clusters which in this case is obviously 2 (GetClusters uses the mean silhouette width as a default decision criterion - compare the previous section). This scan with different fixed numbers of clusters may be illustrated by a silhouette plot to visualize the optimization procedure:

```
minimumNumberOfClusters=2;
maximumNumberOfClusters=10;
silhouettePlotPoints=CIP`Cluster`GetSilhouettePlotPoints[
 inputs,minimumNumberOfClusters,maximumNumberOfClusters];
CIP`Cluster`ShowSilhouettePlot[silhouettePlotPoints]
```

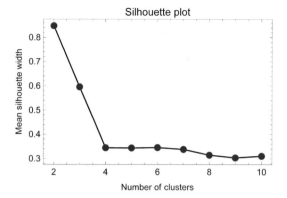

For two clusters the highest value is obtained with a mean silhouette width over 0.7 which indicates strongly structured inputs that may be well partitioned into separated clusters. We may also have a look at the different silhouette widths of each of the two optimum clusters. A plot of the sorted individual silhouette widths of each input of cluster 1

```
silhouetteStatisticsForClusters=
 CIP`Cluster`GetSilhouetteStatisticsForClusters[inputs,clusterInfo];
indexOfCluster=1;
CIP`Cluster`ShowSilhouetteWidthsForCluster[
 silhouetteStatisticsForClusters,indexOfCluster]
```

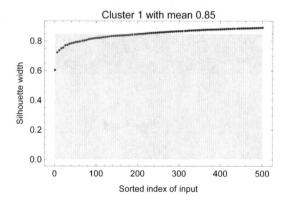

and cluster 2

```
indexOfCluster=2;
CIP`Cluster`ShowSilhouetteWidthsForCluster[
 silhouetteStatisticsForClusters,indexOfCluster]
```

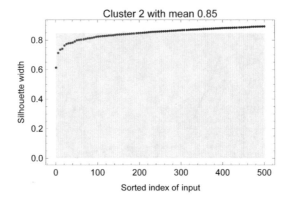

confirm the assessment of good clustering (note that the number of displayed individual silhouette widths is reduced for a better overview). The height of the rectangle corresponds to the mean silhouette width of the cluster and indicates with a mean value greater 0.7 a well separated good cluster.

The demonstrated unambiguous clustering is no longer possible if two enlarged Gaussian clouds are generated that overlap each other to a certain extent:

```
centroid1={0.3,0.7};
standardDeviation=0.175;
cloudDefinition1={centroid1,numberOfCloudInputs,standardDeviation};
inputs1=
 CIP`CalculatedData`GetDefinedGaussianCloud[cloudDefinition1];
centroid2={0.7,0.3};
cloudDefinition2={centroid2,numberOfCloudInputs,standardDeviation};
inputs2=
 CIP`CalculatedData`GetDefinedGaussianCloud[cloudDefinition2];
inputs=Join[inputs1,inputs2];
labels={"x","y","Input Vectors"};
points2DWithPlotStyle={inputs,{PointSize[0.01],Blue}};
points2DWithPlotStyleList={points2DWithPlotStyle};
CIP`Graphics`PlotMultiple2dPoints[points2DWithPlotStyleList,labels]
```

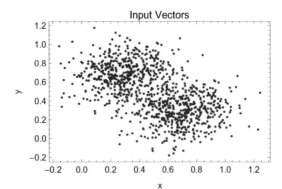

The euclidean distance between the cloud centers is unchanged

```
EuclideanDistance[centroid1,centroid2]
```

0.565685

and a structured distribution of the inputs is still visible but each input may no longer be unambiguously assigned to its specific cloud:

```
clusterInfo=CIP`Cluster`GetClusters[inputs];
CIP`Cluster`ShowClusterResult[{"NumberOfClusters",
 "EuclideanDistanceDiagram","ClusterStatistics"},clusterInfo]
```

Number of clusters = 2

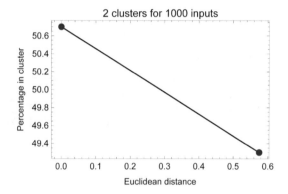

Cluster 1 : 507 members (50.7%) with distance = 0.

Cluster 2 : 493 members (49.3%) with distance = 0.574959

There are two clusters detected with the first cluster being a little bigger than the second. The euclidean distance between the centers of mass is also a little different from the predefined center distance used for the inputs generation. The separation between the two clusters becomes visible by different coloring of the different clusters:

```
indexOfCluster=1;
inputsOfCluster1=
 CIP`Cluster`GetInputsOfCluster[inputs,indexOfCluster,clusterInfo];
indexOfCluster=2;
inputsOfCluster2=
 CIP`Cluster`GetInputsOfCluster[inputs,indexOfCluster,clusterInfo];
labels={"x","y","Clusters in different colors"};
points2DWithPlotStyle1={inputsOfCluster1,{PointSize[0.01],Red}};
points2DWithPlotStyle2={inputsOfCluster2,{PointSize[0.01],Green}};
points2DWithPlotStyleList={points2DWithPlotStyle1,
```

```
      points2DWithPlotStyle2};
  CIP`Graphics`PlotMultiple2dPoints[points2DWithPlotStyleList,labels]
```

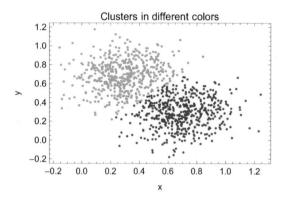

The silhouette plot confirms

```
  silhouettePlotPoints=CIP`Cluster`GetSilhouettePlotPoints[inputs,
    minimumNumberOfClusters,maximumNumberOfClusters];
  CIP`Cluster`ShowSilhouettePlot[silhouettePlotPoints]
```

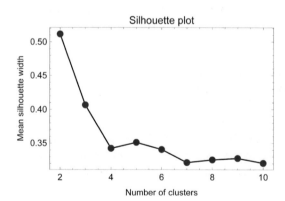

that two clusters are still the best number of clusters but note that the value of the mean silhouette width decreased in comparison to the unambiguous clustering example before: This demonstrates a deteriorated overall clustering because of the ambiguities due to the two overlapping clouds. But a mean silhouette width between 0.5 and 0.7 still indicates reasonably structured inputs where clustering is useful. If the individual silhouette widths of the two optimum clusters are inspected

```
silhouetteStatisticsForClusters=
 CIP'Cluster'GetSilhouetteStatisticsForClusters[inputs,clusterInfo];
indexOfCluster=1;
CIP'Cluster'ShowSilhouetteWidthsForCluster[
 silhouetteStatisticsForClusters,indexOfCluster]
```

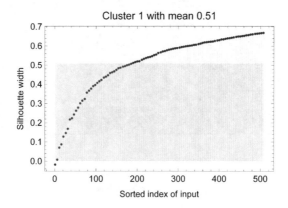

```
indexOfCluster=2;
CIP'Cluster'ShowSilhouetteWidthsForCluster[
 silhouetteStatisticsForClusters,indexOfCluster]
```

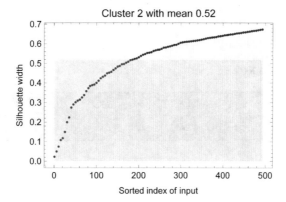

the minimum values around zero attract attention: Silhouette widths around zero
correspond to inputs that are at the borderline between the clusters and thus may not
unambiguously be assigned. This finding nicely confirms the intuition and the above

statements in a more quantitative manner. For a mean silhouette value between 0.5 and 0.7 clustering is still helpful.

If the two Gaussian clouds are further enlarged so that they nearly completely overlap the distribution of the inputs looses any structure:

```
centroid1={0.3,0.7};
standardDeviation=0.5;
cloudDefinition1={centroid1,numberOfCloudInputs,standardDeviation};
inputs1=
 CIP`CalculatedData`GetDefinedGaussianCloud[cloudDefinition1];
centroid2={0.7,0.3};
cloudDefinition2={centroid2,numberOfCloudInputs,standardDeviation};
inputs2=
 CIP`CalculatedData`GetDefinedGaussianCloud[cloudDefinition2];
inputs=Join[inputs1,inputs2];
labels={"x","y","Input Vectors"};
points2DWithPlotStyle={inputs,{PointSize[0.01],Blue}};
points2DWithPlotStyleList={points2DWithPlotStyle};
CIP`Graphics`PlotMultiple2dPoints[points2DWithPlotStyleList,labels]
```

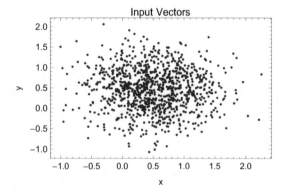

Clustering leads to

```
clusterInfo=CIP`Cluster`GetClusters[inputs];
CIP`Cluster`ShowClusterResult[{"NumberOfClusters",
 "EuclideanDistanceDiagram","ClusterStatistics"},clusterInfo]
```

Number of clusters = 3

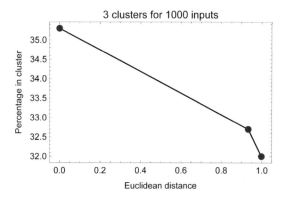

Cluster 1 : 353 members (35.3%) with distance = 0.

Cluster 2 : 327 members (32.7%) with distance = 0.932919

Cluster 3 : 320 members (32.%) with distance = 0.99795

a somewhat arbitrary result with 3 clusters of similar size

```
indexOfCluster=1;
inputsOfCluster1=CIP`Cluster`GetInputsOfCluster[inputs,
 indexOfCluster,clusterInfo];
indexOfCluster=2;
inputsOfCluster2=CIP`Cluster`GetInputsOfCluster[inputs,
 indexOfCluster,clusterInfo];
indexOfCluster=3;
inputsOfCluster3=CIP`Cluster`GetInputsOfCluster[inputs,
 indexOfCluster,clusterInfo];
labels={"x","y","Clusters in different colors"};
points2DWithPlotStyle1={inputsOfCluster1,{PointSize[0.01],Red}};
points2DWithPlotStyle2={inputsOfCluster2,{PointSize[0.01],Green}};
points2DWithPlotStyle3={inputsOfCluster3,{PointSize[0.01],Black}};
points2DWithPlotStyleList={points2DWithPlotStyle1,
 points2DWithPlotStyle2,points2DWithPlotStyle3};
CIP`Graphics`PlotMultiple2dPoints[points2DWithPlotStyleList,labels]
```

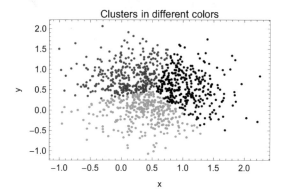

due to a silhouette plot that reveals a first maximum at 3 clusters:

```
silhouettePlotPoints=CIP`Cluster`GetSilhouettePlotPoints[inputs,
 minimumNumberOfClusters,maximumNumberOfClusters];
CIP`Cluster`ShowSilhouettePlot[silhouettePlotPoints]
```

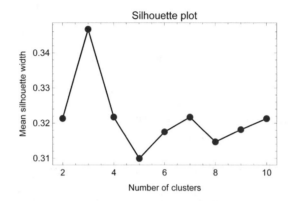

From an intuitive view this clustering result does not make sense. This view is supported by the small value of the mean silhouette width around 0.33. A mean silhouette width between 0.25 and 0.50 indicates only weakly or artificially structured inputs where clustering can not reveal any structural insights (a value below 0.25 simply means no structure). Only higher values indicate an overall reasonable clustering as was shown in the examples before. This assessment is also supported by the individual cluster inspection

```
silhouetteStatisticsForClusters=
 CIP`Cluster`GetSilhouetteStatisticsForClusters[inputs,clusterInfo];
indexOfCluster=1;
CIP`Cluster`ShowSilhouetteWidthsForCluster[
 silhouetteStatisticsForClusters,indexOfCluster]
```

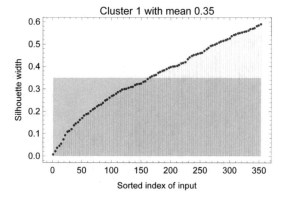

```
indexOfCluster=2;
CIP`Cluster`ShowSilhouetteWidthsForCluster[
 silhouetteStatisticsForClusters,indexOfCluster]
```

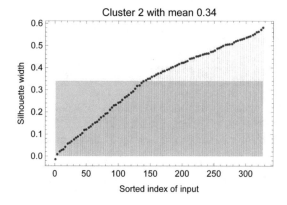

```
indexOfCluster=3;
CIP`Cluster`ShowSilhouetteWidthsForCluster[
 silhouetteStatisticsForClusters,indexOfCluster]
```

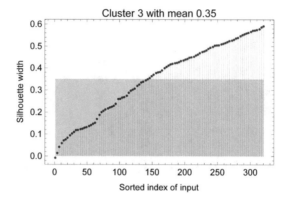

that indicates an enlarged fraction of inputs tending towards a silhouette width around zero where clustering becomes arbitrary.

3.3 Clustering with a fixed number of clusters

```
Clear["Global`*"];
<<CIP`Cluster`
<<CIP`Graphics`
<<CIP`CalculatedData`
```

A clustering process may be forced to produce an a priori fixed number of clusters. To get an impression of the consequences of this forced clustering a few examples are outlined. If the two truly separated input clouds of the last section

```
standardDeviation=0.05;
numberOfCloudInputs=500;
centroid1={0.3,0.7};
cloudDefinition1={centroid1,numberOfCloudInputs,standardDeviation};
inputs1=
 CIP`CalculatedData`GetDefinedGaussianCloud[cloudDefinition1];
centroid2={0.7,0.3};
cloudDefinition2={centroid2,numberOfCloudInputs,standardDeviation};
inputs2=
 CIP`CalculatedData`GetDefinedGaussianCloud[cloudDefinition2];
inputs=Join[inputs1,inputs2];
labels={"x","y","Inputs to be clustered"};
points2DWithPlotStyle={inputs,{PointSize[0.01],Blue}};
points2DWithPlotStyleList={points2DWithPlotStyle};
CIP`Graphics`PlotMultiple2dPoints[points2DWithPlotStyleList,labels]
```

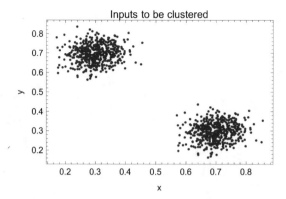

with two optimum or natural clusters are forced to be partitioned into 3 clusters

```
numberOfClusters=3;
clusterInfo=
 CIP`Cluster`GetFixedNumberOfClusters[inputs,numberOfClusters];
```

the following result is obtained:

```
CIP`Cluster`ShowClusterResult[{"NumberOfClusters",
 "EuclideanDistanceDiagram","ClusterStatistics"},clusterInfo]
```

Number of clusters = 3

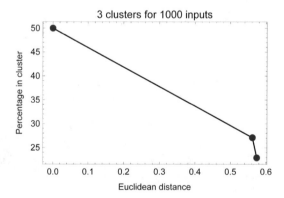

Cluster 1 : 500 members (50.%) with distance = 0.

Cluster 2 : 271 members (27.1%) with distance = 0.561643

Cluster 3 : 229 members (22.9%) with distance = 0.573776

The inputs are split into one large and two smaller neighbored clusters of similar size:

```
indexOfCluster=1;
inputsOfCluster1=
 CIP`Cluster`GetInputsOfCluster[inputs,indexOfCluster,clusterInfo];
indexOfCluster=2;
inputsOfCluster2=
 CIP`Cluster`GetInputsOfCluster[inputs,indexOfCluster,clusterInfo];
indexOfCluster=3;
inputsOfCluster3=
 CIP`Cluster`GetInputsOfCluster[inputs,indexOfCluster,clusterInfo];
labels={"x","y","Clusters in different colors"};
points2DWithPlotStyle1={inputsOfCluster1,{PointSize[0.01],Red}};
points2DWithPlotStyle2={inputsOfCluster2,{PointSize[0.01],Green}};
points2DWithPlotStyle3={inputsOfCluster3,{PointSize[0.01],Black}};
points2DWithPlotStyleList={points2DWithPlotStyle1,
 points2DWithPlotStyle2,points2DWithPlotStyle3};
CIP`Graphics`PlotMultiple2dPoints[points2DWithPlotStyleList,labels]
```

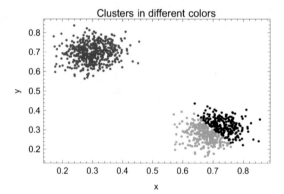

In fact the second optimum or natural cluster has simply been split in two halves. The silhouette widths inspection reveals one good cluster

```
silhouetteStatisticsForClusters=
 CIP`Cluster`GetSilhouetteStatisticsForClusters[inputs,clusterInfo];
indexOfCluster=1;
CIP`Cluster`ShowSilhouetteWidthsForCluster[
 silhouetteStatisticsForClusters,indexOfCluster]
```

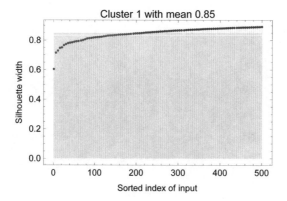

with a high mean silhouette width which is identical to the first natural cluster
and two poor clusters

```
indexOfCluster=2;
CIP`Cluster`ShowSilhouetteWidthsForCluster[
  silhouetteStatisticsForClusters,indexOfCluster]
```

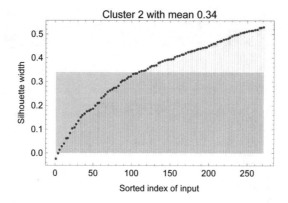

```
indexOfCluster=3;
CIP`Cluster`ShowSilhouetteWidthsForCluster[
  silhouetteStatisticsForClusters,indexOfCluster]
```

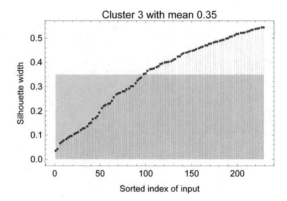

Cluster 3 with mean 0.35

with only small mean silhouette widths (compare the discussion in the previous section). If the inputs are partitioned into 4 clusters

```
numberOfClusters=4;
clusterInfo=
  CIP`Cluster`GetFixedNumberOfClusters[inputs,numberOfClusters];
```

the result may already be anticipated:

```
CIP`Cluster`ShowClusterResult[{"NumberOfClusters",
  "EuclideanDistanceDiagram","ClusterStatistics"},clusterInfo]
```

Number of clusters = 4

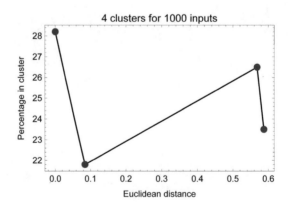

4 clusters for 1000 inputs

Cluster 1 : 282 members (28.2%) with distance = 0.

Cluster 2 : 218 members (21.8%) with distance = 0.0842652

Cluster 3 : 265 members (26.5%) with distance = 0.568283

Cluster 4 : 235 members (23.5%) with distance = 0.587472

The inputs are split into four small clusters of similar size

```
indexOfCluster=1;
inputsOfCluster1=
 CIP`Cluster`GetInputsOfCluster[inputs,indexOfCluster,clusterInfo];
indexOfCluster=2;
inputsOfCluster2=
 CIP`Cluster`GetInputsOfCluster[inputs,indexOfCluster,clusterInfo];
indexOfCluster=3;
inputsOfCluster3=
 CIP`Cluster`GetInputsOfCluster[inputs,indexOfCluster,clusterInfo];
indexOfCluster=4;
inputsOfCluster4=
 CIP`Cluster`GetInputsOfCluster[inputs,indexOfCluster,clusterInfo];
labels={"x","y","Clusters in different colors"};
points2DWithPlotStyle1={inputsOfCluster1,{PointSize[0.01],Red}};
points2DWithPlotStyle2={inputsOfCluster2,{PointSize[0.01],Green}};
points2DWithPlotStyle3={inputsOfCluster3,{PointSize[0.01],Black}};
points2DWithPlotStyle4={inputsOfCluster4,{PointSize[0.01],Blue}};
points2DWithPlotStyleList={points2DWithPlotStyle1,
 points2DWithPlotStyle2,points2DWithPlotStyle3,
 points2DWithPlotStyle4};
CIP`Graphics`PlotMultiple2dPoints[points2DWithPlotStyleList,labels]
```

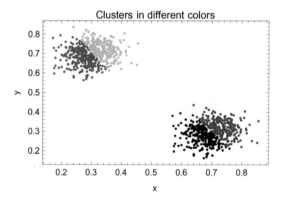

which are each the half of the two optimum natural clusters. The silhouette widths now reveal 4 poor clusters:

```
silhouetteStatisticsForClusters=
 CIP`Cluster`GetSilhouetteStatisticsForClusters[inputs,clusterInfo];
indexOfCluster=1;
CIP`Cluster`ShowSilhouetteWidthsForCluster[
 silhouetteStatisticsForClusters,indexOfCluster]
```

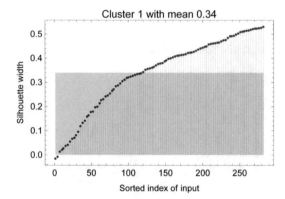

```
indexOfCluster=2;
CIP`Cluster`ShowSilhouetteWidthsForCluster[
  silhouetteStatisticsForClusters,indexOfCluster]
```

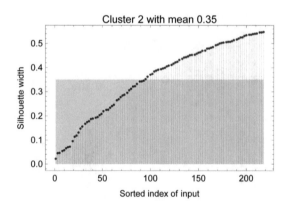

```
indexOfCluster=3;
CIP`Cluster`ShowSilhouetteWidthsForCluster[
  silhouetteStatisticsForClusters,indexOfCluster]
```

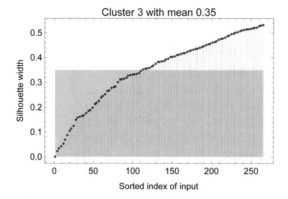

```
indexOfCluster=4;
CIP `Cluster `ShowSilhouetteWidthsForCluster[
  silhouetteStatisticsForClusters, indexOfCluster]
```

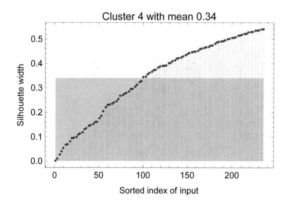

The demonstrated partitioning of inputs into an increasing number of clusters seems to be useless since the clustering quality simply decreases the more unnatural the clustering becomes. But the next section outlines an important application of this procedure.

3.4 Getting representatives

```
Clear["Global`*"];
```

```
<<CIP`Graphics`
<<CIP`Cluster`
<<CIP`CalculatedData`
```

One important application of forced clustering with a fixed number of clusters is the generation of a reduced number of representatives of a full set of inputs with a similar spatial diversity as the full set: This means that the few representatives should cover a similar input space as the full set of inputs. A purely random distribution of 5000 inputs is used for an introductory example:

```
SeedRandom[1];
inputs=Table[{RandomReal[{0.05,0.95}],RandomReal[{0.05,0.95}]},
 {5000}];
argumentRange={0.0,1.0};
functionValueRange={0.0,1.0};
labels={"x","y","Inputs"};
allInputVectorsWithPlotStyle={inputs,{PointSize[0.01],Green}};
points2DWithPlotStyleList={allInputVectorsWithPlotStyle};
CIP`Graphics`PlotMultiple2dPoints[points2DWithPlotStyleList,labels,
 GraphicsOptionArgumentRange2D -> argumentRange,
 GraphicsOptionFunctionValueRange2D -> functionValueRange]
```

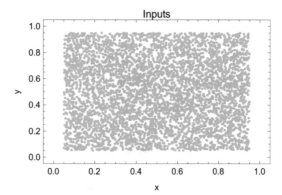

A look at the statistics for each component of the inputs with the corresponding CIP Cluster package method

```
indexOfComponentList={1,2};
numberOfIntervals=5;
argumentRange={0.0,1.0};
functionValueRange={0.0,30.0};
CIP`Cluster`ShowComponentStatistics[inputs,indexOfComponentList,
 ClusterOptionNumberOfIntervals -> numberOfIntervals,
 GraphicsOptionArgumentRange2D -> argumentRange,
 GraphicsOptionFunctionValueRange2D -> functionValueRange]
```

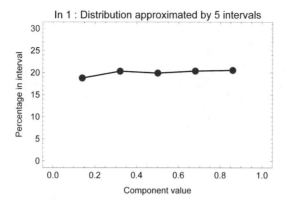

Min / Max = 5.01×10^{-2} / 9.5×10^{-1}

Mean / Median = 5.06×10^{-1} / 5.06×10^{-1}

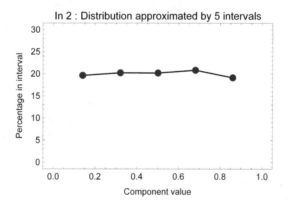

Min / Max = 5.01×10^{-2} / 9.5×10^{-1}

Mean / Median = 4.99×10^{-1} / 5.01×10^{-1}

shows an approximated uniform distribution. If a number of 20 representatives is desired (0.4 % of the total inputs) it seems to be a good choice to simply select twenty inputs by chance. This can be performed by a corresponding method of the CIP Cluster package:

```
numberOfRepresentatives=20;
randomRepresentatives=CIP`Cluster`GetRandomRepresentatives[inputs,
 numberOfRepresentatives];
labels={"x","y","Random representatives"};
argumentRange={0.0,1.0};
functionValueRange={0.0,1.0};
randomRepresentativesBackground={randomRepresentatives,
 {PointSize[0.025],White}};
```

```
randomRepresentativesWithPlotStyle={randomRepresentatives,
 {PointSize[0.02],Black}};
points2DWithPlotStyleList={allInputVectorsWithPlotStyle,
 randomRepresentativesBackground,
 randomRepresentativesWithPlotStyle};
CIP`Graphics`PlotMultiple2dPoints[points2DWithPlotStyleList,labels,
 GraphicsOptionArgumentRange2D -> argumentRange,
 GraphicsOptionFunctionValueRange2D -> functionValueRange]
```

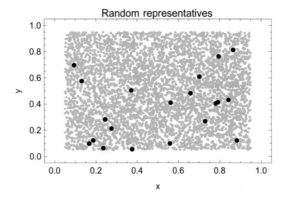

The chosen random representatives (enlarged points) are a satisfactory description of the input's space in this example. Note that randomly chosen inputs are not strictly equally spaced (this would be a single specific random result of very low probability), i.e. randomly chosen inputs always seem to cluster a little bit. An alternative to random selection is the application of a cluster-based selection. Forced clustering seems to be a sensible method: The inputs are partitioned to a number of clusters that is equal to the desired number of representatives. Then for each cluster an input is chosen that is closest to the center of mass centroid of that cluster:

```
clusterRepresentatives=CIP`Cluster`GetClusterRepresentatives[inputs,
 numberOfRepresentatives];
labels={"x","y","Cluster representatives"};
clusterRepresentativesBackground={clusterRepresentatives,
 {PointSize[0.025],White}};
clusterRepresentativesWithPlotStyle={clusterRepresentatives,
 {PointSize[0.02],Black}};
points2DWithPlotStyleList={allInputVectorsWithPlotStyle,
 clusterRepresentativesBackground,
 clusterRepresentativesWithPlotStyle};
CIP`Graphics`PlotMultiple2dPoints[points2DWithPlotStyleList,labels,
 GraphicsOptionArgumentRange2D -> argumentRange,
 GraphicsOptionFunctionValueRange2D -> functionValueRange]
```

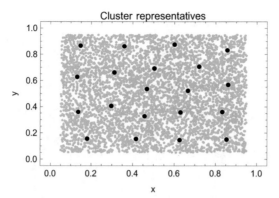

The forced clusters may in addition be visualized by different coloring:

```
numberOfClusters=numberOfRepresentatives;
clusterInfo=CIP`Cluster`GetFixedNumberOfClusters[inputs,
 numberOfClusters];
inputsOfClusterList=Table[
 GetInputsOfCluster[inputs,indexOfCluster,clusterInfo],
 {indexOfCluster,
 First[GetClusterProperty[{"NumberOfClusters"},clusterInfo]]}];
colorList={Blue,Green,Red,Yellow,Pink,Orange,Cyan,Magenta};
colorIndex=Length[colorList];
points2DWithPlotStyleList=Table[
 colorIndex++;If[colorIndex>Length[colorList],colorIndex=1];
 {inputsOfClusterList[[i]],
 {{PointSize[0.01],colorList[[colorIndex]]}}},
 {i,Length[inputsOfClusterList]}];
points2DWithPlotStyleList=Join[points2DWithPlotStyleList,
 {clusterRepresentativesBackground,
 clusterRepresentativesWithPlotStyle}];
CIP`Graphics`PlotMultiple2dPoints[points2DWithPlotStyleList,labels,
 GraphicsOptionArgumentRange2D -> argumentRange,
 GraphicsOptionFunctionValueRange2D -> functionValueRange]
```

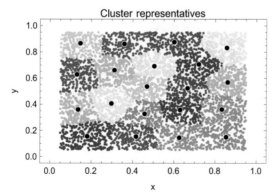

The cluster-based representatives seem to be more equally spaced as it is desired for representatives. But for this example the random and the cluster-based selection of representatives provide comparable results with the cluster-based selection just to be a bit favorable. This finding fundamentally changes if the full set of inputs has different densities in the input's space. Here is an illustrative example:

```
centroid1={0.3,0.7};
standardDeviation=0.05;
numberOfCloudInputs=470;
cloudDefinition1={centroid1,numberOfCloudInputs,standardDeviation};
inputs1=
 CIP`CalculatedData`GetDefinedGaussianCloud[cloudDefinition1];
centroid2={0.7,0.3};
cloudDefinition2={centroid2,numberOfCloudInputs,standardDeviation};
inputs2=
 CIP`CalculatedData`GetDefinedGaussianCloud[cloudDefinition2];
centroid3={0.5,0.5};
standardDeviation=0.05;
numberOfCloudInputs=10;
cloudDefinition3={centroid3,numberOfCloudInputs,standardDeviation};
inputs3=
 CIP`CalculatedData`GetDefinedGaussianCloud[cloudDefinition3];
centroid4={0.8,0.8};
cloudDefinition4={centroid4,numberOfCloudInputs,standardDeviation};
inputs4=
 CIP`CalculatedData`GetDefinedGaussianCloud[cloudDefinition4];
centroid5={0.2,0.2};
cloudDefinition5={centroid5,numberOfCloudInputs,standardDeviation};
inputs5=
 CIP`CalculatedData`GetDefinedGaussianCloud[cloudDefinition5];
inputs=Join[inputs1,inputs2,inputs3,inputs4,inputs5];
labels={"x","y","Inputs"};
allInputVectorsWithPlotStyle={inputs,{PointSize[0.01],Green}};
points2DWithPlotStyleList={allInputVectorsWithPlotStyle};
CIP`Graphics`PlotMultiple2dPoints[points2DWithPlotStyleList,labels,
 GraphicsOptionArgumentRange2D -> argumentRange,
 GraphicsOptionFunctionValueRange2D -> functionValueRange]
```

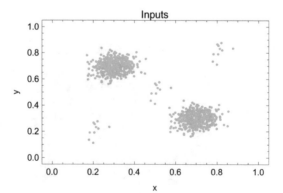

The majority of inputs (97%) is confined to 2 natural clusters. The component statistics of the inputs also show a distribution with two distinct peaks in both dimensions:

```
indexOfComponentList={1,2};
numberOfIntervals=25;
CIP'Cluster'ShowComponentStatistics[inputs,indexOfComponentList,
  ClusterOptionNumberOfIntervals -> numberOfIntervals]
```

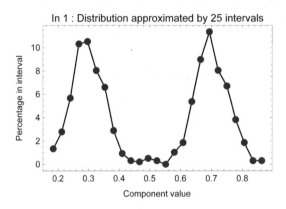

Min / Max = 1.7×10^{-1} / 8.77×10^{-1}

Mean / Median = $5. \times 10^{-1}$ / 5.06×10^{-1}

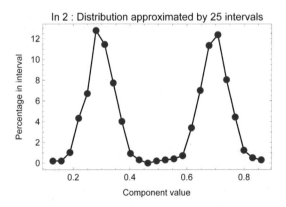

Min / Max = 1.12×10^{-1} / 8.76×10^{-1}

Mean / Median = 4.97×10^{-1} / 5.26×10^{-1}

If 20 representatives are randomly chosen

```
numberOfRepresentatives=20;
randomRepresentatives=CIP`Cluster`GetRandomRepresentatives[inputs,
 numberOfRepresentatives];
labels={"x","y","Random representatives"};
randomRepresentativesBackground={randomRepresentatives,
 {PointSize[0.025],White}};
randomRepresentativesWithPlotStyle={randomRepresentatives,
 {PointSize[0.02],Black}};
points2DWithPlotStyleList={allInputVectorsWithPlotStyle,
 randomRepresentativesBackground,
 randomRepresentativesWithPlotStyle};
CIP`Graphics`PlotMultiple2dPoints[points2DWithPlotStyleList,labels,
 GraphicsOptionArgumentRange2D -> argumentRange,
 GraphicsOptionFunctionValueRange2D -> functionValueRange]
```

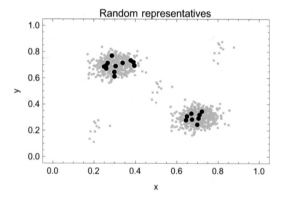

it is very likely that they are only taken from these two natural high density
regions as shown above. Thus the data space with a lower density of inputs will be
completely neglected with a high probability. In this case a cluster-based selection
becomes a distinct advantage

```
clusterRepresentatives=CIP`Cluster`GetClusterRepresentatives[inputs,
 numberOfRepresentatives];
labels={"x","y","Cluster representatives"};
clusterRepresentativesBackground={clusterRepresentatives,
 {PointSize[0.025],White}};
clusterRepresentativesWithPlotStyle={clusterRepresentatives,
 {PointSize[0.02],Black}};
points2DWithPlotStyleList={allInputVectorsWithPlotStyle,
 clusterRepresentativesBackground,
 clusterRepresentativesWithPlotStyle};
CIP`Graphics`PlotMultiple2dPoints[points2DWithPlotStyleList,labels,
 GraphicsOptionArgumentRange2D -> argumentRange,
 GraphicsOptionFunctionValueRange2D -> functionValueRange]
```

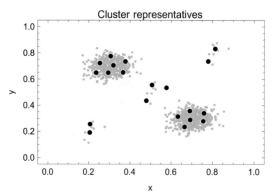

since it leads to representatives that cover the whole space of inputs:

```
numberOfClusters=numberOfRepresentatives;
clusterInfo=CIP`Cluster`GetFixedNumberOfClusters[inputs,
 numberOfClusters];
inputsOfClusterList=Table[
 GetInputsOfCluster[inputs,indexOfCluster,clusterInfo],
 {indexOfCluster,
 First[GetClusterProperty[{"NumberOfClusters"},clusterInfo]]}];
colorList={Blue,Green,Red,Yellow,Pink,Orange,Cyan,Magenta};
colorIndex=Length[colorList];
points2DWithPlotStyleList=Table[
 colorIndex++;If[colorIndex>Length[colorList],colorIndex=1];
 {inputsOfClusterList[[i]],
 {{PointSize[0.01],colorList[[colorIndex]]}}},
 {i,Length[inputsOfClusterList]}];
points2DWithPlotStyleList=Join[points2DWithPlotStyleList,
 {clusterRepresentativesBackground,
 clusterRepresentativesWithPlotStyle}];
CIP`Graphics`PlotMultiple2dPoints[points2DWithPlotStyleList,labels,
 GraphicsOptionArgumentRange2D -> argumentRange,
 GraphicsOptionFunctionValueRange2D -> functionValueRange]
```

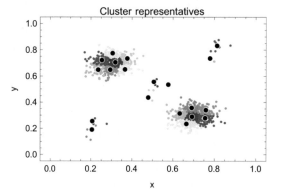

Again note: In two dimensions the representation problem is comparatively simple and could be solved by mere visual inspection. For inputs of higher dimensions this possibility is completely lost but a random or cluster-based selection is still possible with the same methods outlined above. The required number of representatives that are necessary for a good representation of a full set of inputs depends on number of issues: How diverse is the full data space? Which representation accuracy is in demand? How many representatives can be handled in further processing steps? These issues will be addressed in the cross validation oriented discussion in the machine learning chapter 4.

3.5 Cluster occupancies and the iris flower example

```
Clear["Global`*"];
<<CIP`ExperimentalData`
<<CIP`Cluster`
<<CIP`Graphics`
```

Another important application of clustering is the comparison of the spatial diversity of different sets of inputs. To compare the spatial diversity the different sets of inputs are joined and clustered as a union. Then the occupancy of a cluster with respect to each set of inputs is evaluated. As an example the famous iris flower data are used (see Appendix A). They consist of three sets of inputs for three iris flower species under investigation: Iris setosa (species 1), iris versicolor (species 2) and iris virginica (species 3). Each set of inputs consists of fifty measurements of the sepal length and width and the petal length and width so each input is of dimension four. Therefore the inputs can not simply be visually inspected. To get a feeling for the data the sepal and petal length and width components are analyzed. The iris flower data are available with the CIP ExperimentalData package:

```
inputsOfSpecies1=CIP`ExperimentalData`GetIrisFlowerInputsSpecies1[];
inputsOfSpecies2=CIP`ExperimentalData`GetIrisFlowerInputsSpecies2[];
inputsOfSpecies3=CIP`ExperimentalData`GetIrisFlowerInputsSpecies3[];
```

To get the full 150 iris flower inputs the single inputs of the three species (with 50 inputs each) are joined:

```
inputs=Join[inputsOfSpecies1,inputsOfSpecies2,inputsOfSpecies3];
```

For later use a list of the minimum and maximum index of each species in the full set of inputs is defined. Inputs 1 to 50 belong to species 1 (or class 1), inputs 51 to 100 to species 2 (or class 2) and inputs 101 to 150 to species 3 (or class 3):

```
classIndexMinMaxList={{1,50},{51,100},{101,150}};
```

The first 2 components of an input are sepal length and sepal width. The frequency distributions of these first two components are as follows: The sepal length (In 1) shows one wide peak

```
indexOfComponentList={1};
numberOfIntervals=6;
argumentRange={40.0,80.0};
functionValueRange={0.0,30.0};
CIP `Cluster`ShowComponentStatistics[inputs,indexOfComponentList,
ClusterOptionNumberOfIntervals->numberOfIntervals,
 GraphicsOptionArgumentRange2D -> argumentRange,
 GraphicsOptionFunctionValueRange2D -> functionValueRange]
```

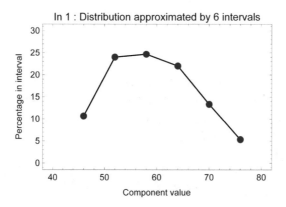

In 1 : Distribution approximated by 6 intervals

Min / Max = 4.3×10^1 / 7.9×10^1

Mean / Median = 5.84×10^1 / 5.8×10^1

and the same is true for the sepal width (In 2):

```
indexOfComponentList={2};
argumentRange={20.0,45.0};
functionValueRange={0.0,45.0};
CIP `Cluster`ShowComponentStatistics[inputs,indexOfComponentList,
 ClusterOptionNumberOfIntervals->numberOfIntervals,
 GraphicsOptionArgumentRange2D -> argumentRange,
 GraphicsOptionFunctionValueRange2D -> functionValueRange]
```

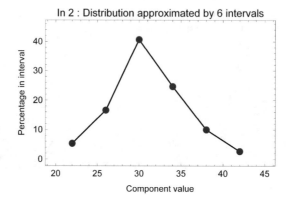

Min / Max = $2. \times 10^1$ / 4.4×10^1

Mean / Median = 3.06×10^1 / $3. \times 10^1$

Both components alone will not be able to differentiate between the three species. A plot of the sepal width (In 2) against the sepal length (In 1) reveals more insight:

```
labels={"In 1","In 2","Iris flower species 1, 2, 3"};
points2DWithPlotStyle1={inputsOfSpecies1[[All,{1,2}]],
  {PointSize[0.02],Red}};
points2DWithPlotStyle2={inputsOfSpecies2[[All,{1,2}]],
  {PointSize[0.02],Green}};
points2DWithPlotStyle3={inputsOfSpecies3[[All,{1,2}]],
  {PointSize[0.02],Black}};
points2DWithPlotStyleList={points2DWithPlotStyle1,
  points2DWithPlotStyle2,points2DWithPlotStyle3};
CIP`Graphics`PlotMultiple2dPoints[points2DWithPlotStyleList,labels]
```

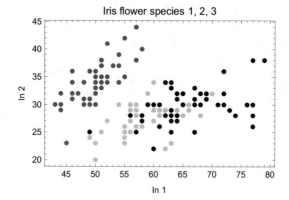

The species 1 inputs fill the upper left part and are separated from the other two species whereas the species 2 and 3 inputs occupy nearly the same space on the lower right. This finding becomes more pronounced by the analysis of the latter two components: The petal length and width. The petal length (In 3) shows two frequency peaks:

```
indexOfComponentList={3};
argumentRange={10.0,70.0};
functionValueRange={-2.0,35.0};
CIP`Cluster`ShowComponentStatistics[inputs,indexOfComponentList,
  ClusterOptionNumberOfIntervals -> numberOfIntervals,
  GraphicsOptionArgumentRange2D -> argumentRange,
  GraphicsOptionFunctionValueRange2D -> functionValueRange]
```

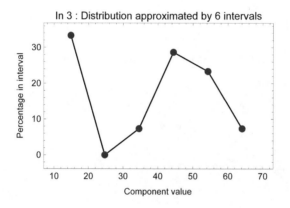

In 3 : Distribution approximated by 6 intervals

Min / Max = $1. \times 10^1$ / 6.9×10^1

Mean / Median = 3.76×10^1 / 4.35×10^1

One sharp peak at the left and a wider peak at the right. The petal width (In 4) is similarly distributed:

```
indexOfComponentList={4};
argumentRange={0.0,25.0};
functionValueRange={-2.0,35.0};
CIP`Cluster`ShowComponentStatistics[inputs,indexOfComponentList,
  ClusterOptionNumberOfIntervals -> numberOfIntervals,
  GraphicsOptionArgumentRange2D -> argumentRange,
  GraphicsOptionFunctionValueRange2D -> functionValueRange]
```

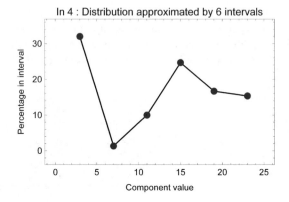

In 4 : Distribution approximated by 6 intervals

Min / Max = 1. / 2.5 × 10¹

Mean / Median = 1.2 × 10¹ / 1.3 × 10¹

A plot of the petal width (In 4) against the petal width (In 3)

```
labels={"In 3","In 4","Iris flower species 1, 2, 3"};
points2DWithPlotStyle1={inputsOfSpecies1[[All,{3,4}]],
  {PointSize[0.02],Red}};
points2DWithPlotStyle2={inputsOfSpecies2[[All,{3,4}]],
  {PointSize[0.02],Green}};
points2DWithPlotStyle3={inputsOfSpecies3[[All,{3,4}]],
  {PointSize[0.02],Black}};
points2DWithPlotStyleList={points2DWithPlotStyle1,
  points2DWithPlotStyle2,points2DWithPlotStyle3};
CIP`Graphics`PlotMultiple2dPoints[points2DWithPlotStyleList,labels]
```

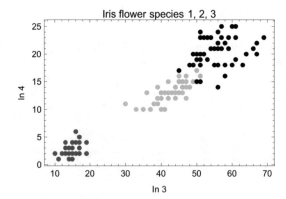

shows that the inputs of species 1 are clearly separated from the others but the inputs of species 2 and 3 do have overlap. So it can be deduced that clustering methods will not be able to differentiate perfectly between the three species. If the full inputs of all three species are clustered by the CIP default method

```
clusterInfo=CIP`Cluster`GetClusters[inputs];
CIP`Cluster`ShowClusterResult[{"NumberOfClusters",
 "EuclideanDistanceDiagram","ClusterStatistics"},clusterInfo]
```

Number of clusters = 2

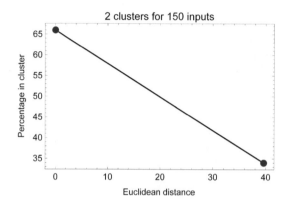

Cluster 1 : 99 members (66.%) with distance = 0.
Cluster 2 : 51 members (34.%) with distance = 39.6076

two clusters are detected. The bigger cluster 1

```
indexOfCluster=1;
CIP`Cluster`GetIndexListOfCluster[indexOfCluster,clusterInfo]
```

{51,52,53,54,55,56,57,58,59,60,61,62,63,64,65,66,67,68,69,70,71,72,73,75,76,77,78,79,80,81,82,
83,84,85,86,87,88,89,90,91,92,93,94,95,96,97,98,99,100,101,102,103,104,105,106,107,108,109,
110,111,112,113,114,115,116,117,118,119,120,121,122,123,124,125,126,127,128,129,130,131,132,
133,134,135,136,137,138,139,140,141,142,143,144,145,146,147,148,149,150}

consists of all indices of inputs that belong to species 2 and species 3 except one. The smaller cluster 2

```
indexOfCluster=2;
CIP`Cluster`GetIndexListOfCluster[indexOfCluster,clusterInfo]
```

{1,2,3,4,5,6,7,8,9,10,11,12,13,14,15,16,17,18,19,20,21,22,23,24,25,26,27,28,29,30,31,32,33,34,
35,36,37,38,39,40,41,42,43,44,45,46,47,48,49,50,74}

contains all indices of inputs that belong to species 1 and 1 input from species 2 (index 74). This finding may be graphically illustrated by so called cluster occupancies. For each cluster the percentage of inputs of the three species is obtained

```
clusterOccupancies=CIP`Cluster`GetClusterOccupancies[
  classIndexMinMaxList,clusterInfo]
```

$\{\{0.,98.,100.\},\{100.,2.,0.\}\}$

and may be visualized by a bar chart:

```
CIP`Cluster`ShowClusterOccupancies[clusterOccupancies]
```

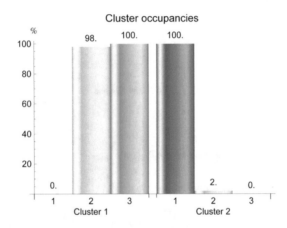

Cluster 1 contains 100% of the inputs from species 3 and 98% from species 2: One input (2%) is missing. Cluster 2 contains 100% of the inputs from species 1 and the single input from species 2 (the 2%). The silhouette plots for both clusters

```
silhouetteStatisticsForClusters=
  CIP`Cluster`GetSilhouetteStatisticsForClusters[inputs,clusterInfo];
indexOfCluster=1;
CIP`Cluster`ShowSilhouetteWidthsForCluster[
  silhouetteStatisticsForClusters,indexOfCluster]
```

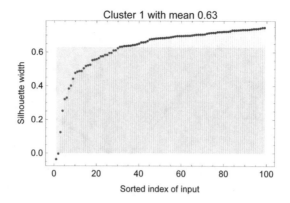

```
indexOfCluster=2;
CIP`Cluster`ShowSilhouetteWidthsForCluster[
 silhouetteStatisticsForClusters,indexOfCluster]
```

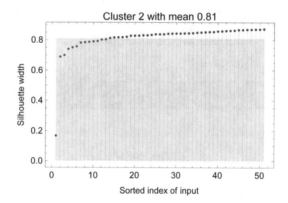

show that cluster 2 is a good cluster with only one exceptionally small silhouette value (that should correspond to the single input from species 2) in comparison to the acceptable cluster 1 (from the viewpoint of silhouettes). If the number of clusters is forced to be 3 (the natural choice for 3 species)

```
numberOfClusters=3;
clusterInfo=CIP`Cluster`GetFixedNumberOfClusters[inputs,
 numberOfClusters];
CIP`Cluster`ShowClusterResult[{"NumberOfClusters",
 "EuclideanDistanceDiagram","ClusterStatistics"},clusterInfo]
```

Number of clusters = 3

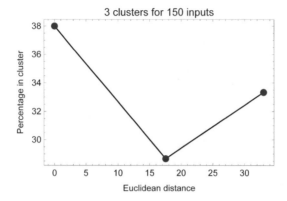

Cluster 1 : 57 members (38.%) with distance = 0.

Cluster 2 : 43 members (28.6667%) with distance = 17.5241

Cluster 3 : 50 members (33.3333%) with distance = 33.0051

the resulting cluster occupancies are:

```
clusterOccupancies=CIP`Cluster`GetClusterOccupancies[
    classIndexMinMaxList,clusterInfo]
```

$\{\{0.,90.,24.\},\{0.,10.,76.\},\{100.,0.,0.\}\}$

```
CIP`Cluster`ShowClusterOccupancies[clusterOccupancies]
```

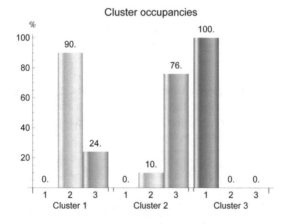

The clustering process now partitions the full inputs into three groups that mainly represent the three species: But whereas cluster 3 now consists only of inputs from

species 1 (the species that was expected to separate) the clusters 1 and 2 are dominated by species 2 and 3 respectively but still consist of inputs from the other species
due to the overlap of their inputs in space. An inspection of the cluster's silhouette
widths

```
silhouetteStatisticsForClusters=
 CIP 'Cluster 'GetSilhouetteStatisticsForClusters[inputs,clusterInfo];
indexOfCluster=1;
CIP 'Cluster 'ShowSilhouetteWidthsForCluster[
 silhouetteStatisticsForClusters, indexOfCluster]
```

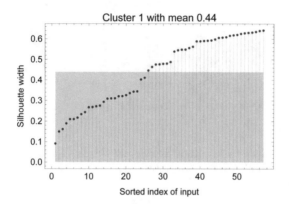

```
indexOfCluster=2;
CIP 'Cluster 'ShowSilhouetteWidthsForCluster[
 silhouetteStatisticsForClusters, indexOfCluster]
```

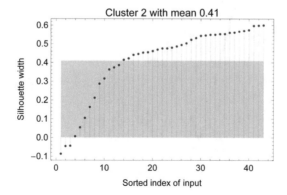

```
indexOfCluster=3;
CIP`Cluster`ShowSilhouetteWidthsForCluster[
  silhouetteStatisticsForClusters,indexOfCluster]
```

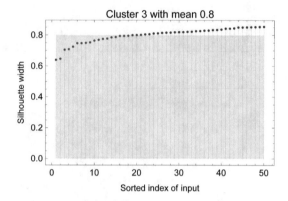

confirms that cluster 3 is a good cluster and clusters 1 and 2 are only poor. The situation remains similar if the number of fixed clusters is further increased:

```
numberOfClusters=6;
clusterInfo=CIP`Cluster`GetFixedNumberOfClusters[inputs,
  numberOfClusters];
CIP`Cluster`ShowClusterResult[{"NumberOfClusters",
  "EuclideanDistanceDiagram","ClusterStatistics"},clusterInfo]
```

Number of clusters = 6

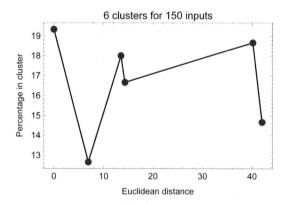

Cluster 1 : 29 members (19.3333%) with distance = 0.

Cluster 2 : 19 members (12.6667%) with distance = 6.94575

Cluster 3 : 27 members (18.%) with distance = 13.4926

Cluster 4 : 25 members (16.6667%) with distance = 14.305

Cluster 5 : 28 members (18.6667%) with distance = 40.1159

Cluster 6 : 22 members (14.6667%) with distance = 41.9742

Two clusters (5 and 6) are spatially more separated from the others with larger distances

```
clusterOccupancies=CIP `Cluster `GetClusterOccupancies[
  classIndexMinMaxList,clusterInfo]
```

$\{\{0.,14.,44.\},\{0.,38.,0.\},\{0.,0.,54.\},\{0.,48.,2.\},\{56.,0.,0.\},\{44.,0.,0.\}\}$

```
CIP `Cluster `ShowClusterOccupancies[clusterOccupancies]
```

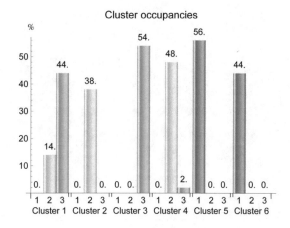

and contain all inputs of species 1. Two other cluster (2 and 3) do only contain inputs of one species: Cluster 2 only inputs from species 2 and cluster 3 only those of species 3. But there are still two overlap clusters (1 and 4) that contain inputs from species 2 as well as species 3. In summary it becomes clear that simple clustering is not able to distinguish between the iris flower species 2 and 3 due to their overlap.

3.6 White-spot analysis

```
Clear["Global`*"];
<<CIP`CalculatedData`
<<CIP`Graphics`
<<CIP`Cluster`
```

Cluster occupancies may be used to systematically reveal white spots in a specific set of inputs in comparison to an alternative set, i.e. spatial areas where a specific set of inputs does not contain a relevant number of data but the alternative set of inputs does. The following example shows two set of inputs (A and B) of different size and different spatial distribution:

```
centroid1={0.3,0.7};
standardDeviation=0.05;
numberOfCloudInputs=200;
cloudDefinition1={centroid1,numberOfCloudInputs,standardDeviation};
inputs1=
 CIP`CalculatedData`GetDefinedGaussianCloud[cloudDefinition1];
centroid2={0.7,0.3};
cloudDefinition2={centroid2,numberOfCloudInputs,standardDeviation};
inputs2=
 CIP`CalculatedData`GetDefinedGaussianCloud[cloudDefinition2];
centroid3={0.3,0.3};
cloudDefinition3={centroid3,numberOfCloudInputs,standardDeviation};
inputs3=
 CIP`CalculatedData`GetDefinedGaussianCloud[cloudDefinition3];
inputsA=Join[inputs1,inputs2,inputs3];
centroid1={0.35,0.65};
standardDeviation=0.05;
numberOfCloudInputs=50;
cloudDefinition1={centroid1,numberOfCloudInputs,standardDeviation};
inputs1=
 CIP`CalculatedData`GetDefinedGaussianCloud[cloudDefinition1];
centroid2={0.75,0.25};
cloudDefinition2={centroid2,numberOfCloudInputs,standardDeviation};
inputs2=
 CIP`CalculatedData`GetDefinedGaussianCloud[cloudDefinition2];
centroid3={0.7,0.7};
cloudDefinition3={centroid3,numberOfCloudInputs,standardDeviation};
inputs3=
 CIP`CalculatedData`GetDefinedGaussianCloud[cloudDefinition3];
inputsB=Join[inputs1,inputs2,inputs3];
argumentRange={0.0,1.0};
functionValueRange={0.0,1.0};
labels={"x","y","Inputs A and B"};
inputsAwithPlotStyle={inputsA,{PointSize[0.01],Green}};
inputsBwithPlotStyle={inputsB,{PointSize[0.01],Red}};
points2DWithPlotStyleList={inputsAwithPlotStyle,
 inputsBwithPlotStyle};
CIP`Graphics`PlotMultiple2dPoints[points2DWithPlotStyleList,labels,
 GraphicsOptionArgumentRange2D -> argumentRange,
 GraphicsOptionFunctionValueRange2D -> functionValueRange]
```

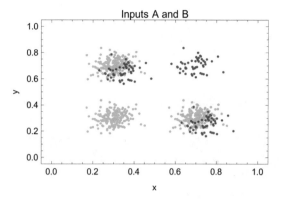

If the input sets A and B are joined

```
inputs=Join[inputsA,inputsB];
classIndexMinMaxList={{1,Length[inputsA]},{Length[inputsA]+1,
  Length[inputsA]+Length[inputsB]}}
```

$\{\{1,600\},\{601,750\}\}$

the first 600 inputs belong to inputs A and the next 150 inputs to inputs B. If the joined inputs are partitioned into four clusters (the natural choice from visual inspection)

```
numberOfClusters=4;
clusterInfo=CIP`Cluster`GetFixedNumberOfClusters[inputs,
  numberOfClusters];
CIP`Cluster`ShowClusterResult[{"NumberOfClusters",
  "EuclideanDistanceDiagram","ClusterStatistics"},clusterInfo]
```

Number of clusters = 4

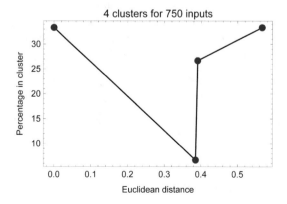

Cluster 1 : 250 members (33.3333%) with distance = 0.

Cluster 2 : 50 members (6.66667%) with distance = 0.385247

Cluster 3 : 200 members (26.6667%) with distance = 0.390721

Cluster 4 : 250 members (33.3333%) with distance = 0.565685

the expected result is obtained: Two of the four clusters are equally occupied (cluster 1 and 4)

```
clusterOccupancies=CIP 'Cluster 'GetClusterOccupancies[
  classIndexMinMaxList,clusterInfo]
```

$\{\{33.3,33.3\},\{0.,33.3\},\{33.3,0.\},\{33.3,33.3\}\}$

```
CIP 'Cluster 'ShowClusterOccupancies[clusterOccupancies]
```

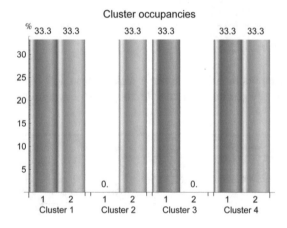

whereas clusters 2 and 3 do only contain inputs of one set. An inputs set is said to contain a white spot if it occupies a detected cluster less than a threshold value in comparison to another set of inputs. With a white spot threshold of 80% a white spot is detected for inputs A (index 1) in cluster 2

```
threshold=80.0;
indexOfInputs=1;
CIP 'Cluster 'GetWhiteSpots[clusterOccupancies,indexOfInputs,
  threshold]
```

{2}

and for inputs B (index 2) in cluster 3:

```
indexOfInputs=2;
CIP`Cluster`GetWhiteSpots[clusterOccupancies,indexOfInputs,
 threshold]
```

{3}

White spots or gaps detected in this way may advise further research strategies or indicate subtle problems.

3.7 Alternative clustering with ART-2a

```
Clear["Global`*"];
<<CIP`Graphics`
<<CIP`Cluster`
<<CIP`ExperimentalData`
<<CIP`CalculatedData`
```

As already mentioned there are numerous different clustering techniques available based on entirely different principles. As an alternative to the k-medoids clustering this section describes ART-2a clustering that is derived from neural-network-based Adaptive Resonance Theory (ART, see [Carpenter 1991] and [Wienke 1994] for details). ART-2a belongs to the open-categorical clustering techniques and is guided by a so called vigilance parameter: Depending on an a priori defined vigilance a corresponding number of clusters is created. The vigilance parameter can be varied between 0 (rough clustering with little vigilance and a small number of resulting clusters) and 1 (fine clustering with high vigilance and many resulting clusters). The CIP default value is 0.1 which means low vigilant/relatively rough clustering. If the number of clusters is a priori fixed for this method the corresponding vigilance parameter that produces this a priori defined number of clusters is determined by an iterative procedure. To get a quick insight about how ART-2a works equally distributed two-dimensional inputs

```
SeedRandom[1];
inputs=Table[{RandomReal[{0.05,0.95}],RandomReal[{0.05,0.95}]},
 {5000}];
argumentRange={0.0,1.0};
functionValueRange={0.0,1.0};
labels={"x","y","Inputs"};
allInputVectorsWithPlotStyle={inputs,{PointSize[0.01],Green}};
points2DWithPlotStyleList={allInputVectorsWithPlotStyle};
CIP`Graphics`PlotMultiple2dPoints[points2DWithPlotStyleList,labels,
 GraphicsOptionArgumentRange2D -> argumentRange,
 GraphicsOptionFunctionValueRange2D -> functionValueRange]
```

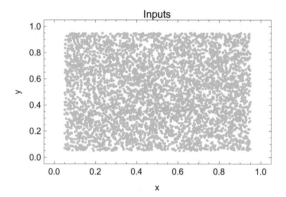

are forced to be grouped into three clusters:

```
numberOfClusters=3;
clusterMethod="ART2a";
clusterInfo=CIP`Cluster`GetFixedNumberOfClusters[inputs,
  numberOfClusters,ClusterOptionMethod -> clusterMethod];
CIP`Cluster`ShowClusterResult[{"NumberOfClusters",
  "ART2aDistanceDiagram","ART2aClusterStatistics"},clusterInfo]
```

Number of clusters = 3

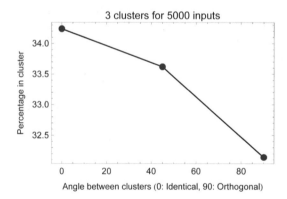

Cluster 1 : 1712 members (34.24%) with angle = 0.

Cluster 2 : 1681 members (33.62%) with angle = 44.9804

Cluster 3 : 1607 members (32.14%) with angle = 90.

ART-2a produces 3 clusters of similar size. The difference between the clusters is expressed in an angle value where a value of 0 means identity and a value of 90 means orthogonal clusters with maximum separation. This at first sight strange terminology becomes clear if the 3 clusters are visualized:

```
inputsOfClusterList=Table[
 CIP'Cluster'GetInputsOfCluster[inputs,indexOfCluster,clusterInfo],
 {indexOfCluster,numberOfClusters}];
colorList={Blue,Green,Red};
colorIndex=Length[colorList];
points2DWithPlotStyleList=Table[
 colorIndex++;If[colorIndex>Length[colorList],colorIndex=1];
 {inputsOfClusterList[[i]],{PointSize[0.01],
 colorList[[colorIndex]]}},{i,Length[inputsOfClusterList]}];
labels={"x","y","Clusters in different colors"};
CIP'Graphics'PlotMultiple2dPoints[points2DWithPlotStyleList,labels,
 GraphicsOptionArgumentRange2D -> argumentRange,
 GraphicsOptionFunctionValueRange2D -> functionValueRange]
```

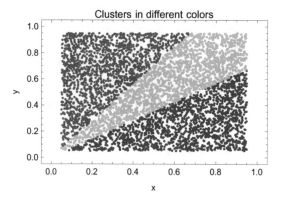

The three cluster angles of approximately 0, 45 and 90 degree are now intuitive: ART-2a has only a radial view of the world. All inputs are projected internally to a unit hypersphere around the origin so only radial differences between them are taken into account (moreover the inputs are initially transformed so that the maximum angle between clusters is 90 degree and not 180). Therefore ART-2a will be rather restricted for low dimensional clustering like the 2D example above. But when it comes to high dimensional clustering tasks in hugh spaces this restriction will become less important and ART-2a can demonstrate its strength: Speed! ART-2a is a dramatically faster clustering method compared to k-medoids or k-means. And speed becomes a critical parameter if very large data volumes with millions or billions of inputs are to be clustered. If the iris flower inputs

```
inputsOfSpecies1=CIP'ExperimentalData'GetIrisFlowerInputsSpecies1[];
inputsOfSpecies2=CIP'ExperimentalData'GetIrisFlowerInputsSpecies2[];
inputsOfSpecies3=CIP'ExperimentalData'GetIrisFlowerInputsSpecies3[];
inputs=Join[inputsOfSpecies1,inputsOfSpecies2,inputsOfSpecies3];
classIndexMinMaxList={{1,50},{51,100},{101,150}};
```

are clustered with ART-2a and the CIP default vigilance parameter of 0.1

```
clusterInfo=CIP `Cluster `GetClusters[inputs,
 ClusterOptionMethod -> clusterMethod];
CIP `Cluster `ShowClusterResult[{"NumberOfClusters",
 "ART2aDistanceDiagram","ART2aClusterStatistics"},clusterInfo]
```

Number of clusters = 3

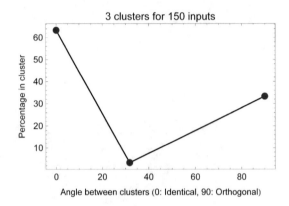

Cluster 1 : 95 members (63.3333%) with angle = 0.

Cluster 2 : 5 members (3.33333%) with angle = 31.7532

Cluster 3 : 50 members (33.3333%) with angle = 90.

3 clusters are obtained. With a scan of the sensitivity of the resulting number of clusters to the chosen vigilance parameter (which may vary between 0 and 1)

```
minimumVigilanceParameter=0.01;
maximumVigilanceParameter=0.99;
numberOfScanPoints=30;
art2aScanInfo=CIP `Cluster `GetVigilanceParameterScan[inputs,
 minimumVigilanceParameter,maximumVigilanceParameter,
 numberOfScanPoints];
CIP `Cluster `ShowVigilanceParameterScan[art2aScanInfo]
```

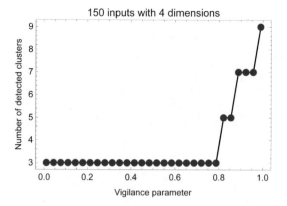

a clear plateau for 3 clusters can be detected, i.e. 3 clusters is the natural choice of the method for a wide range of vigilance values. A check of the cluster occupancies of the 3 clusters

```
clusterOccupancies=CIP`Cluster`GetClusterOccupancies[
  classIndexMinMaxList,clusterInfo]
```

$\{\{0.,90.,100.\},\{0.,10.,0.\},\{100.,0.,0.\}\}$

```
CIP`Cluster`ShowClusterOccupancies[clusterOccupancies]
```

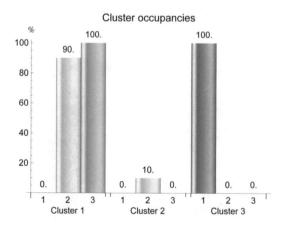

shows that species 1 is clearly separated into one cluster (3). One small cluster (2) contains only some members of species 2 but the biggest cluster (1) contains

nearly all members of species 2 and 3. So ART-2a is not able to distinguish between
these two species. A look at the silhouette widths of the clusters

```
silhouetteStatisticsForClusters=
 CIP`Cluster`GetSilhouetteStatisticsForClusters[inputs,clusterInfo];
indexOfCluster=1;
CIP`Cluster`ShowSilhouetteWidthsForCluster[
 silhouetteStatisticsForClusters,indexOfCluster]
```

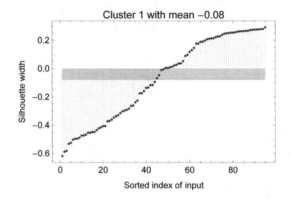

```
indexOfCluster=2;
CIP`Cluster`ShowSilhouetteWidthsForCluster[
 silhouetteStatisticsForClusters,indexOfCluster]
```

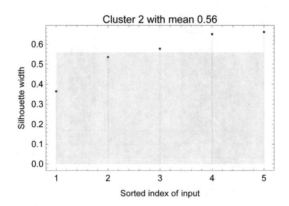

```
indexOfCluster=3;
CIP 'Cluster 'ShowSilhouetteWidthsForCluster[
  silhouetteStatisticsForClusters, indexOfCluster]
```

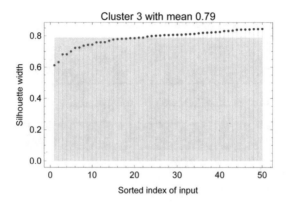

show one good cluster (3) for the separable species 1, one medium cluster (2) with only a few fairly separable inputs of species 2 and a very poor cluster (1) with the mixed species 2 and 3 inputs. If the number of clusters is enforced to be six

```
numberOfClusters=6;
clusterInfo=CIP 'Cluster 'GetFixedNumberOfClusters[inputs,
  numberOfClusters,ClusterOptionMethod -> clusterMethod];
CIP 'Cluster 'ShowClusterResult[{"NumberOfClusters",
  "ART2aDistanceDiagram","ART2aClusterStatistics"},clusterInfo]
```

Number of clusters = 6

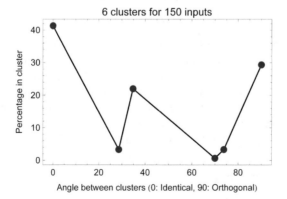

Cluster 1 : 62 members (41.3333%) with angle = 0.

Cluster 2 : 5 members (3.33333%) with angle = 28.647

Cluster 3 : 33 members (22.%) with angle = 34.761

Cluster 4 : 1 members (0.666667%) with angle = 69.9521

Cluster 5 : 5 members (3.33333%) with angle = 73.796

Cluster 6 : 44 members (29.3333%) with angle = 90.

the situation does not improve significantly:

```
clusterOccupancies=CIP`Cluster`GetClusterOccupancies[
  classIndexMinMaxList,clusterInfo]
```

$\{\{0.,58.,66.\},\{0.,10.,0.\},\{0.,32.,34.\},\{2.,0.,0.\},\{10.,0.,0.\},\{88.,0.,0.\}\}$

```
CIP`Cluster`ShowClusterOccupancies[clusterOccupancies]
```

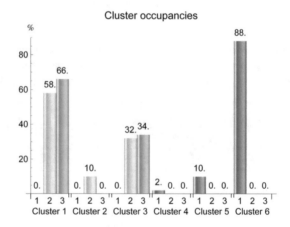

Inputs of species 1 now purely occupy three neighbored clusters (4 to 6) but a better separation of species 2 and 3 can not be achieved: They are now both distributed among two clusters (1 and 3). Compared to the CIP k-medoids default clustering of the iris flower inputs before the ART-2a based clustering is inferior with respect to the separation of species 2 and 3: A result that is expected from its radial view of the world for a problem with only 4 dimensions. A final example shows that things can change if clustering is performed in an input's space with many dimensions. Four Gaussian clouds

```
numberOfGaussianClouds=4;
```

with a small standard deviation of 0.05 and 200 inputs each

```
standardDeviation=0.05;
numberOfCloudInputs=200;
cloudVectorNumberList=Table[
 numberOfCloudInputs,{numberOfGaussianClouds}];
```

are generated in a 50 dimensional input's space at random positions:

```
numberOfDimensions=50;
inputs=CIP`CalculatedData`GetRandomGaussianCloudsInputs[
 cloudVectorNumberList,numberOfDimensions,standardDeviation];
```

The first 200 inputs belong to cloud 1 (or class 1), the next 200 inputs belong to cloud 2 (or class 2) etc. so the list with the min/max indices of the inputs for every cloud (class) is as follows:

```
classIndexMinMaxList={{1,200},{201,400},{401,600},{601,800}};
```

The whole setup can no longer be visually inspected because of the 50 dimensions but four small clouds of inputs in a hugh space should be easy to partition. If we cluster the inputs with the CIP default method with an open number of clusters

```
clusterInfo=CIP`Cluster`GetClusters[inputs];
CIP`Cluster`ShowClusterResult[{"NumberOfClusters"},clusterInfo]
```

Number of clusters = 4

we get four clusters where each cluster contains the inputs of one specific cloud as may be shown by the cluster occupancies:

```
clusterOccupancies=CIP`Cluster`GetClusterOccupancies[
 classIndexMinMaxList,clusterInfo];
CIP`Cluster`ShowClusterOccupancies[clusterOccupancies]
```

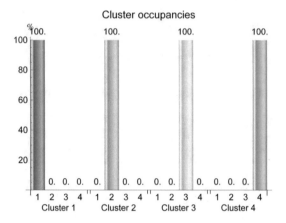

The overall minimum, mean and maximum silhouette values

```
silhouetteStatistics=CIP`Cluster`GetSilhouetteStatistics[inputs,
  clusterInfo]
```

{0.783168, 0.823311, 0.848463}

reveal a perfect separation of the clusters which may in addition be confirmed by the individual silhouette widths of the inputs of cluster 1:

```
silhouetteStatisticsForClusters=
  CIP`Cluster`GetSilhouetteStatisticsForClusters[inputs,clusterInfo];
  indexOfCluster=1;
  CIP`Cluster`ShowSilhouetteWidthsForCluster[
  silhouetteStatisticsForClusters,indexOfCluster]
```

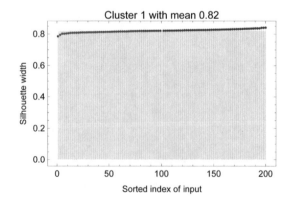

Now consider the ART-2a method: Its application with the default vigilance parameter of 0.1

```
clusterMethod="ART2a";
clusterInfo=CIP`Cluster`GetClusters[inputs,
 ClusterOptionMethod -> clusterMethod];
CIP`Cluster`ShowClusterResult[{"NumberOfClusters"},clusterInfo]
```

Number of clusters = 4

also leads to four perfect clusters which are identical to those found before (but of course in another order)

```
clusterOccupancies=CIP`Cluster`GetClusterOccupancies[
 classIndexMinMaxList,clusterInfo];
CIP`Cluster`ShowClusterOccupancies[clusterOccupancies]
```

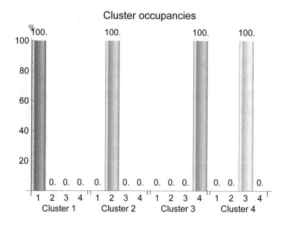

with the same silhouette statistics:

```
silhouetteStatistics=CIP`Cluster`GetSilhouetteStatistics[inputs,
 clusterInfo]
```

{0.783168, 0.823311, 0.848463}

The sensitivity of the detected number of classes to a change in the vigilance parameter is extremely low, i.e. there is a wide plateau region for four clusters

```
minimumVigilanceParameter=0.01;
maximumVigilanceParameter=0.90;
```

```
numberOfScanPoints=20;
art2aScanInfo=CIP`Cluster`GetVigilanceParameterScan[inputs,
 minimumVigilanceParameter,maximumVigilanceParameter,
 numberOfScanPoints];
CIP`Cluster`ShowVigilanceParameterScan[art2aScanInfo]
```

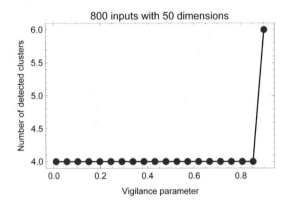

which indicates that four clusters are the natural choice. In this case any infe-riority of the ART-2a method in comparison to the k-medoids method due to its restricted radial perception of the world does no longer play a crucial role - but its higher speed in fact does.

3.8 Clustering and class predictions

```
Clear["Global`*"];
<<CIP`CalculatedData`
<<CIP`ExperimentalData`
<<CIP`Utility`
<<CIP`Graphics`
<<CIP`Cluster`
<<CIP`DataTransformation`
```

Clustering can be used to construct a class predictor, i.e. a tool that returns a class number for an arbitrary input. Predictive tools are the holy grail of machine learning and will be discussed in detail for supervised learning in chapter 4. But also an unsupervised learning approach like clustering can be predictive to a certain extent. To get a clear understanding of how a class predictor may be achieved consider the following data:

```
centroidVector1={0.2,0.2};
```

```
numberOfCloudVectors=150;
standardDeviation=0.3;
cloudDefinition1={centroidVector1,numberOfCloudVectors,
  standardDeviation};
inputs1=CIP 'CalculatedData 'GetDefinedGaussianCloud[
  cloudDefinition1];
centroidVector2={0.8,0.8};
cloudDefinition2={centroidVector2,numberOfCloudVectors,
  standardDeviation};
inputs2=CIP 'CalculatedData 'GetDefinedGaussianCloud[
  cloudDefinition2];
points2DWithPlotStyle1={inputs1,{PointSize[0.02],Red}};
points2DWithPlotStyle2={inputs2,{PointSize[0.02],Green}};
points2DWithPlotStyleList={points2DWithPlotStyle1,
  points2DWithPlotStyle2};
labels={"x","y","Inputs and their corresponding color classes"};
inputsGraphics=CIP 'Graphics 'PlotMultiple2dPoints[
  points2DWithPlotStyleList,labels]
```

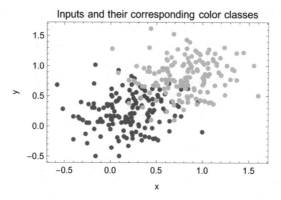

There are two classes of inputs where the inputs' cloud of class 1 overlaps with the inputs' cloud of class 2. A class predictor tries to correctly predict the class that corresponds to a specific input, i.e.corresponds to the coordinates of a specific point:

```
inputs=Join[inputs1,inputs2];
labels={"x","y","Inputs to be classified: Class 1 or 2?"};
plotStyle={PointSize[0.02],Blue};
points2DWithPlotStyle={inputs,plotStyle};
points2DWithPlotStyleList={points2DWithPlotStyle};
CIP 'Graphics 'PlotMultiple2dPoints[points2DWithPlotStyleList,labels]
```

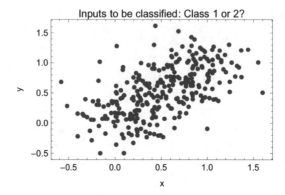

For the current task a predictor with a success rate of 100% is obviously impossible due to the clouds' overlap. The best possible result seems to be a success rate of about 90% correct predictions since the class predictions for the inputs in the overlapping region will be ambiguous which inevitably leads to classification errors. A classification data set for both classes is generated (see chapter 1 for details)

```
cloudDefinitions={cloudDefinition1,cloudDefinition2};
classificationDataSet=
 CIP `CalculatedData `GetGaussianCloudsDataSet[
 cloudDefinitions];
```

and sorted ascending according its two classes

```
sortResult=CIP `DataTransformation`SortClassificationDataSet[
 classificationDataSet];
sortedClassificationDataSet=sortResult[[1]];
```

with a corresponding min/max index list for the inputs assignment to their particular class:

```
classIndexMinMaxList=sortResult[[2]]
```

$\{\{1, 150\}, \{151, 300\}\}$

The length of the min/max list simply is the number of classes:

```
numberOfClusters=Length[classIndexMinMaxList]
```

2

As a next step the pure inputs of the classification data set are obtained

```
inputs=CIP`Utility`GetInputsOfDataSet[sortedClassificationDataSet];
```

and clustered according to the desired number of classes:

```
clusterInfo=CIP`Cluster`GetFixedNumberOfClusters[inputs,
 numberOfClusters];
```

The resulting two clusters may be visualized with their centroids

```
indexOfCluster=1;
inputsOfCluster1=CIP`Cluster`GetInputsOfCluster[inputs,
 indexOfCluster,clusterInfo];
points2DWithPlotStyle1={inputsOfCluster1,{PointSize[0.02],Red}};
indexOfCluster=2;
inputsOfCluster2=CIP`Cluster`GetInputsOfCluster[inputs,
 indexOfCluster,clusterInfo];
points2DWithPlotStyle2={inputsOfCluster2,{PointSize[0.02],Green}};
properties={"CentroidVectors"};
centroids2D=
 CIP`Cluster`GetClusterProperty[properties,clusterInfo][[1]];
centroids2DBackground={centroids2D,{PointSize[0.035],White}};
centroids2DWithPlotStyle1={centroids2D,{PointSize[0.03],Black}};
points2DWithPlotStyleList={points2DWithPlotStyle1,
 points2DWithPlotStyle2,centroids2DBackground,
 centroids2DWithPlotStyle1};
labels={"x","y","Clusters in different colors and their centroids"};
CIP`Graphics`PlotMultiple2dPoints[points2DWithPlotStyleList,labels]
```

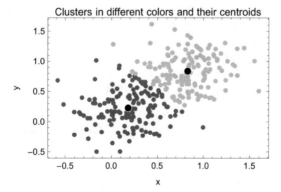

and the predictivity inspected:

```
clusterOccupancies=CIP`Cluster`GetClusterOccupancies[
 classIndexMinMaxList,clusterInfo];
CIP`Cluster`ShowClusterOccupancies[clusterOccupancies]
```

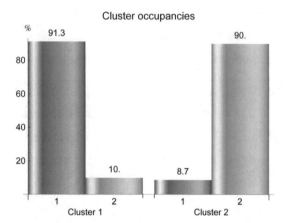

As a result it becomes obvious that cluster 1 contains around 90% of the members of class 1 and cluster 2 around 90% of the members of class 2. It is this finding that may be utilized to construct a class predictor. If each cluster is simply characterized by its centroid

```
CIP`Cluster`GetClusterProperty[{"CentroidVectors"},clusterInfo]
```

$\{\{\{0.182319, 0.22501\}, \{0.829217, 0.835822\}\}\}$

and the centroid of cluster 1 is most predictive for class 1 then it can be assigned to class 1. The centroid of cluster 2 is most predictive for class 2 so it is assigned to this class. Now a class assignment for any arbitrary input may be obtained by simply calculating its nearest centroid with a minimum euclidean distance. The attached class assignment of this nearest centroid is then the predictive output. All the above steps are collected in a single general method FitCluster

```
clusterInfo=CIP`Cluster`FitCluster[classificationDataSet];
```

that generates a clusterInfo which can be used for class predictions, e.g.

```
input={0.1,0.5};
CIP`Cluster`CalculateClusterClassNumber[input,clusterInfo]
```

1

```
input={0.9,0.5};
CIP`Cluster`CalculateClusterClassNumber[input,clusterInfo]
```

2

Input $(0.1, 0.5)$ belongs to class 1 and input $(0.9, 0.5)$ to class 2. The overall class prediction success rate for the whole data set

```
CIP`Cluster`ShowClusterSingleClassification[
  {"CorrectClassification"},classificationDataSet,clusterInfo]
```

90.7% correct classifications

is around 90% as expected. The predictivity for each single class of the data set

```
CIP`Cluster`ShowClusterSingleClassification[
  {"CorrectClassificationPerClass"},classificationDataSet,clusterInfo]
```

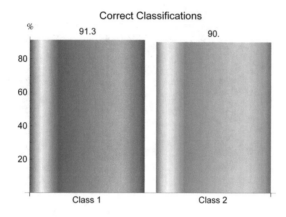

is equally good since the classification task is symmetrical. For this specific example the class predictor obtained from unsupervised learning is the best we can get from the data: Also the in general more powerful supervised machine learning methods will not perform any better in this case as will be shown in chapter 4. If the iris flower classification problem is revisited

```
classificationDataSet=
  CIP`ExperimentalData`GetIrisFlowerClassificationDataSet[];
```

a clustering-based class predictor may be constructed in the same manner:

```
clusterInfo=CIP`Cluster`FitCluster[classificationDataSet];
```

The overall classification success rate

```
CIP `Cluster`ShowClusterSingleClassification[
  {"CorrectClassification"},classificationDataSet,clusterInfo]
```

89.3% correct classifications

is found to be about 90% but it is quite different for the three species (classes):

```
CIP `Cluster`ShowClusterSingleClassification[
  {"CorrectClassificationPerClass"},classificationDataSet,
  clusterInfo]
```

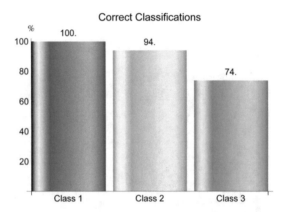

Whereas species (or class) 1 is 100% correctly predicted the results for the other two species (classes) are clearly worse. This again is expected from the findings with the iris flower data above that revealed overlap of the inputs for species 2 and 3. If the ART-2a clustering method is used instead of the default k-medoids method

```
clusterMethod="ART2a";
clusterInfo=CIP `Cluster`FitCluster[classificationDataSet,
  ClusterOptionMethod -> clusterMethod];
```

the overall classification success rate

```
CIP `Cluster`ShowClusterSingleClassification[
  {"CorrectClassification"},classificationDataSet,clusterInfo]
```

87.3% correct classifications

is a little inferior as expected and the class-based results

```
CIP`Cluster`ShowClusterSingleClassification[
  {"CorrectClassificationPerClass"},classificationDataSet,
  clusterInfo]
```

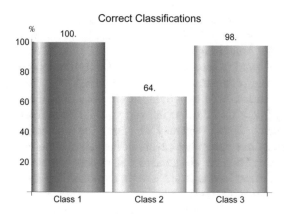

exhibit a particularly poor prediction for species (class) 2. Note that the correct classification rates are not necessarily equal to the corresponding cluster occupancies: A class is predicted on the basis of the simple (euclidean) distance of an input to its nearest cluster centroid whereas a clustering method is not necessarily centroid based (e.g. the default k-medoids method is not and ART-2a isn't either). The sketched construction of a class predictor based on unsupervised learning is very robust, i.e. it works almost always for a classification data set without any problems. But its predictive quality depends on the success of clustering with regard to the desired class assignments: If this assignment is unambiguous the predictor is perfect - otherwise the predictivity is limited up to extremely poor. A supervised learning method is in general able to extract more from the data - with the risk of extracting too much: Then the supervised learning method overfits the data which inevitably leads to a loss of predictability. In contrast the robust unsupervised learning is not prone to overfitting at all. It therefore can be regarded as a good start for a more detailed data analysis as far as a classification task is in question. As a final remark it should be noticed that the outlined construction process of a clustering-based class predictor may be modified in numerous ways: A family of variants could be created easily where each family member could be superior to another for a specific classification task. The data analysis community loves this comparison of variants so do not take this game too serious: It's usually accompanied by a tremendous effort for only incremental improvements which are most often not significant to a practitioner in the lab. Therefore it is the basic ideas that count.

3.9 Cookbook recipes for clustering

Clustering is a common first step in data analysis so the discussed topics may be condensed into a few cookbook recipes:

- **The start:** Take the inputs and perform an open-categorical clustering with your default method of choice. The clustering result suggests an optimum number of clusters. If you have any feeling or even knowledge about the true number of clusters you get some insight about success of failure of the clustering method chosen. Use alternative clustering methods: Do the results coincide? If not, why not? For example a comparison of k-medoids with ART-2a may reveal interesting aspects of the structural features of the inputs in question due to the fundamental differences of both methods. At the end you should at least have a structural feeling about your inputs - and may it be that there is nothing like a structure.
- **Cluster inspection:** Assess the quality of the detected clusters e.g. with silhouette widths plots. There may be good as well as poor clusters with all graduations in between. A closer look at each cluster is also sensible if the number of clusters is varied to different fixed values.
- **Representatives:** Clustering may be utilized to get a reduced set of representatives for the whole inputs. Cluster-based representatives have the advantage of being adequately distributed over the whole inputs' space so that they cover approximately the same spatial diversity as the complete inputs. But be aware that any method of data reduction looses information: This loss may be crucial for later failure.
- **Comparison of inputs:** Inputs of different sources may be compared by clustering techniques to reveal similarities as well as white spots. Both may have an important heuristic relevance and motivate further research.
- **Classification tasks:** If your final goal is a class predictor on the basis of a classification data set try pure unsupervised learning to construct one. A clustering-based class predictor is usually a good start for the more elaborate supervised machine learning methods. And note: If the clustering-based class predictor performs well there is no need for the more elaborate methods: You are perfectly done!

A final statement of warning may always be kept in mind: A clustering method may yield an intuitive, expected and reasonable result in some cases but not in others. That means that a clustering process may not be helpful at all or even worse: It may be misleading! Since clustering always works it always generates a result - no matter how appropriate this is. A principle problem is the fact that visual inspection is usually not possible for clustering inputs of dimensions higher than 2 or 3. The application of a specific clustering method should therefore be validated as thorough as possible.

Chapter 4
Machine Learning

```
Clear["Global`*"];
<<CIP`CalculatedData`
<<CIP`Graphics`
<<CIP`SVM`
```

Machine learning methods are applied when K input/output (I/O) pairs

$$\left(\underline{x}_1, \underline{y}_1\right), ..., \left(\underline{x}_K, \underline{y}_K\right)$$

of inputs

$$\underline{x}_k = (x_{k1}, x_{k2}, ..., x_{kM}) \; ; \; k = 1, ..., K$$

and corresponding outputs

$$\underline{y}_k = (y_{k1}, y_{k2}, ..., y_{kN}) \; ; \; k = 1, ..., K$$

are available but the model functions f_i that map the input vectors onto the output vectors

$$y_i = f_i(x_1, ..., x_M) \; ; \; i = 1, ..., N$$

or in compact vector notation

$$\underline{y} = \underline{f}(\underline{x})$$

are completely unknown: Machine learning tries to approximate these unknown model functions f_i on the basis of the provided I/O data. This situation is comparable

© Springer International Publishing Switzerland 2016
A. Zielesny, *From Curve Fitting to Machine Learning*, Intelligent Systems
Reference Library 109, DOI 10.1007/978-3-319-32545-3_4

to 2D data smoothing discussed in chapter 2 but now takes place in many more dimensions. To illustrate an example a function with two arguments $f(x, y)$ is used to generate 100 normally distributed erroneous data around it (see [Cherkassy 1996]):

```
pureOriginalFunction=Function[{x,y},
 1.9*(1.35+Exp[x]*Sin[13.0*(x-0.6)^2]*Exp[-y]* Sin[7.0*y])];
xRange={0.0,1.0};
yRange={0.0,1.0};
numberOfDataPerDimension=10;
standardDeviationRange={0.1,0.1};
dataSet3D=CIP`CalculatedData`Get3dFunctionBasedDataSet[
 pureOriginalFunction,xRange,yRange,numberOfDataPerDimension,
 standardDeviationRange];
labels={"x","y","z"};
CIP`Graphics`Plot3dDataSetWithFunction[dataSet3D,
 pureOriginalFunction,labels]
```

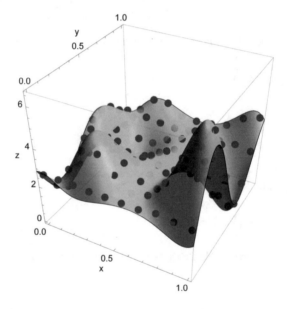

The task of the machine learning process is to create an approximate model function from the pure I/O data

```
CIP`Graphics`Plot3dDataSet[dataSet3D,labels]
```

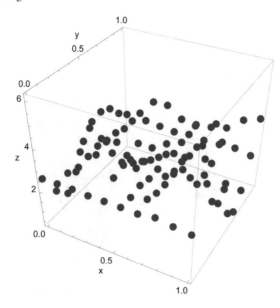

without any knowledge about the original function used for the data generation (which is the fundamental difference to pure curve fitting of chapter 2 where the model function was at least structurally known). As a possible method for machine learning a so called support vector machine (with a specific so called kernel function) is chosen to perform the fitting task :

```
kernelFunction={"Wavelet",0.3};
svmInfo=CIP`SVM`FitSvm[dataSet3D,kernelFunction];
```

Since no error messages were thrown the successful machine learning result is condensed in a svmInfo data structure which can be used for further analysis. The approximated model function may be checked by visual inspection

```
pureSvm3dFunction=
 Function[{x,y},CIP`SVM`CalculateSvm3dValue[x,y,svmInfo]];
CIP`Graphics`Plot3dDataSetWithFunction[dataSet3D,
 pureSvm3dFunction,labels]
```

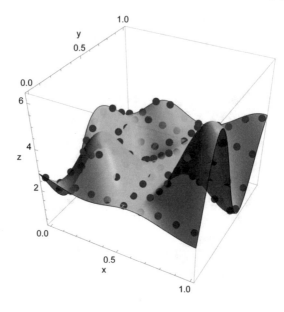

and related quality-of-fit plots

```
CIP`SVM`ShowSvmSingleRegression[{"ModelVsDataPlot",
 "AbsoluteSortedResidualsPlot","RMSE"},dataSet3D,svmInfo]
```

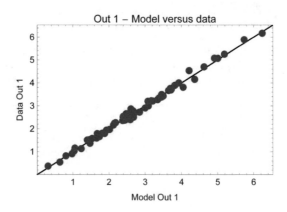

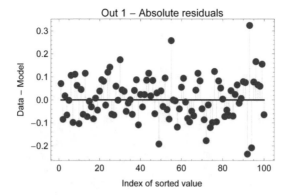

Root mean squared error (RMSE) = 9.279×10^{-2}

to be of excellent quality: The residuals (i.e. the deviations between the calculated values of approximated model function and the initially generated erroneous output values of the data set) and the RMSE are in perfect agreement with the standard deviation (error) of 0.1 used above for the data generation. In addition there are no systematic deviations in the residuals plot. Original and approximated model function may also be overlayed to reveal only minor deviations:

```
originalFunctionGraphics3D=CIP `Graphics `Plot3dFunction[
  pureOriginalFunction, xRange, yRange, labels];
approximatedFunctionGraphics3D=CIP `Graphics `Plot3dFunction[
  pureSvm3dFunction, xRange, yRange, labels];
Show[originalFunctionGraphics3D, approximatedFunctionGraphics3D]
```

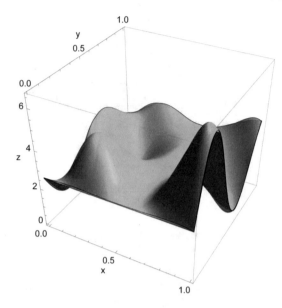

This confirms the impressive approximation success of the used machine learning method. But any enthusiasm should be complemented with a direct statement of warning: If a different support vector machine is used (with an inappropriate kernel function - and there is in general no way to know a good kernel function in advance)

```
kernelFunction={"Wavelet",2.0};
svmInfo=CIP'SVM'FitSvm[dataSet3D,kernelFunction];
```

the result is a mere disaster

```
pureSvm3dFunction=Function[
 {x,y},CalculateSvm3dValue[x,y,svmInfo]];
CIP'Graphics'Plot3dDataSetWithFunction[dataSet3D,pureSvm3dFunction,
 labels]
```

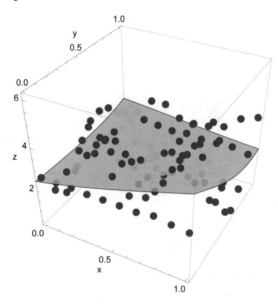

with a completely unsatisfying quality:

```
CIP'SVM'ShowSvmSingleRegression[{"ModelVsDataPlot",
 "AbsoluteSortedResidualsPlot","RMSE"},dataSet3D,svmInfo]
```

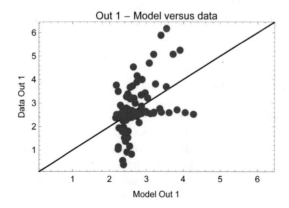

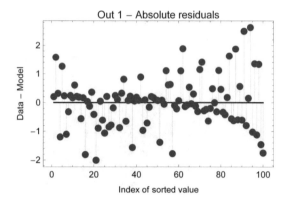

Root mean squared error (RMSE) = 8.796×10^{-1}

Welcome to machine learning! Always keep in mind that machine learning is (and unfortunately remains) a lot of laborious trial and error. It opens a space with fantastic opportunities but may often lead to dramatic failure. Since pairs of inputs $\underline{x}_k = (x_{k1}, x_{k2}, ..., x_{kM})$ and corresponding outputs $\underline{y}_k = (y_{k1}, y_{k2}, ..., y_{kN})$ are mandatory for machine learning its training (or optimization) process is called supervised learning where the learning is controlled (or supervised) by the known outputs - in contrast to the unsupervised clustering of inputs without any control discussed in chapter 3. Machine learning may address a regression or a classification task and usually involves multidimensional input and output vectors (i.e. N and M are usually substantially larger than one). As already mentioned it may be regarded as a generalization of 2D data smoothing (with N and M equal to one, see chapter 2) to multiple dimensions. Note that errors of the y data are usually not taken into account since machine learning lacks a sound statistical basis due to the missing knowledge of the model functions' structures. In this chapter different machine learning methods are sketched: Multiple linear and polynomial regression (MLR, MPR), three-layer feed-forward neural networks (three-layer perceptrons) and support vector machines (SVM). MLR and MPR are usually not accounted as machine learning techniques but perfectly fit into this chapter as a (fast) start. Perceptrons and SVMs are prominent machine learning methods and are widely used (not only) in science and engineering. For a successful application of machine learning unfortunately a lot of subtleties have to be taken into consideration and a lot can go wrong. But when these issues can be successfully tackled and the provided data are suitable machine learning can reveal its magic to the practitioner. It may provide substantial support in finding intricate relationships and hidden optima.

Chapter 4 starts with a brief sketch of the basic principles of machine learning (section 4.1). The different machine learning methods used in this chapter, i.e. MLR, MPR, three-layer perceptrons and SVMs, are summarized afterwards (section 4.2). The assessment of the goodness of a machine learning result is an essential step so some quantities and plots are described that are helpful for analyzing the outcomes of regression and classification tasks (section 4.3 and 4.4). The necessity of non-linear machine learning methods is established with a real world modelling example at the borderline of non-linearity: A fit to adhesive kinetics data (section 4.5). Also the phenomenon of overfitting is encountered and illustrated. Non-linear decision surfaces for classification tasks are demonstrated afterwards (section 4.6). Supervised classification does not necessarily aim to be a 100% correct. For ambiguous data a reduced success rate may be the superior choice. This general insight is outlined and directs the discussion to the problem of validating predictions (section 4.7). The partitioning of a data set into a training and a test set to address the validation issue evokes basic questions regarding the size and the selection of these sets. This is explored with a closer look at different selection heuristics and their success or failure (section 4.8). There are always different methods available for a specific machine learning task: Comparative aspects are sketched and discussed (section 4.9). The relevance of each component of an I/O pair's input can be analyzed by a successive leave-one-out or component-inclusion strategy. In this way the number of input components may be reduced which not only simplifies learning but allows the construction of minimal models (section 4.10). Pattern recognition is an important application of supervised machine learning. A simple example concerning the detection of face types sketches possible issues and subtleties (section 4.11). Machine learning as an optimization process is guided by several technical parameters. Their crucial influence for successful learning is exemplified (section 4.12). Final cookbook recipes for machine learning and an appendix with two scientific applications that collect several pieces described before close this chapter (sections 4.13 and 4.14).

4.1 Basics

```
Clear["Global`*"];
<<CIP`Perceptron`
<<CIP`Graphics`
<<CIP`CalculatedData`
<<CIP`DataTransformation`
```

In order to get a principle understanding about how machine learning methods work the inverse procedure to machine learning is followed and illustrated in two dimensions for simplicity (it may readily be generalized to an arbitrary number of dimensions of course). The starting point is an elementary function in form of a so called bump: A bump has a value greater than zero in its bump region and a value close to zero elsewhere. Here is an example of a two-dimensional bump around $x = 6$:

```
interval={5.,7.};
pureSigmoid1=Function[x,SigmoidFunction[x-interval[[1]]]];
pureSigmoid2=Function[x,SigmoidFunction[x-interval[[2]]]];
```

```
pureBump=Function[x,BumpFunction[x,interval]];
pureFunctions={pureSigmoid1,pureSigmoid2,pureBump};
argumentRange={-5.0,20.0};
plotRange={0.0,1.0};
plotStyle={{Thickness[0.001],Orange},{Thickness[0.001],Orange},
 {Thickness[0.005],Green}};
labels={"x","y","Sigmoid functions and resulting bump"};
CIP`Graphics`Plot2dFunctions[pureFunctions,argumentRange,plotRange,
 plotStyle,labels]
```

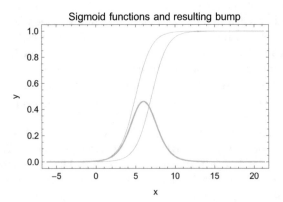

The bump is the difference between two sigmoid threshold functions on the left and right which themselves perform a transition from zero to one at a distinct x value. Layered feed-forward neural networks (perceptrons) are composed of logical neurons that work with sigmoid threshold functions: If logical neurons are combined in a network they are able to produce bumps. By added overlay of several bumps (or other elementary functions in general) an arbitrarily complex non-linear result function can be created:

```
interval1={0.,8.};
pureBump1=Function[x,BumpFunction[x,interval1]];
interval2={11.,13.};
pureBump2=Function[x,BumpFunction[x,interval2]];
interval3={10.,22.};
pureBump3=Function[x,BumpFunction[x,interval3]];
interval4={25.,29.};
pureBump4=Function[x,BumpFunction[x,interval4]];
pureBumpSum=Function[x,BumpFunction[x,interval1]+
 BumpFunction[x,interval2]+BumpFunction[x,interval3]+
 BumpFunction[x,interval4]];
pureFunctions={pureBump1,pureBump2,pureBump3,pureBump4,pureBumpSum};
argumentRange={-10,40};
plotRange={0.0,1.5};
plotStyle={{Thickness[0.001],Green},{Thickness[0.001],Green},
 {Thickness[0.001],Green},{Thickness[0.001],Green},
 {Thickness[0.005],Red}};
labels={"x","y","Result function of added bumps"};
CIP`Graphics`Plot2dFunctions[pureFunctions,argumentRange,plotRange,
 plotStyle,labels]
```

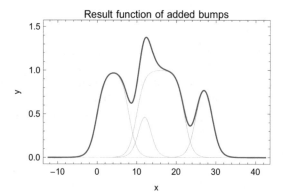

As a last step we simulate 100 error-biased data points around the result function. Since this demonstration is in 2 dimensions xy-error data with the CIP Calculated-Data package are generated with an error (standard deviation) of 0.025:

```
numberOfData=100;
standardDeviationRange={0.025,0.025};
simulatedDataArgumentRange={0.0,30.0};
xyErrorData=CIP`CalculatedData`GetXyErrorData[pureBumpSum,
  simulatedDataArgumentRange,numberOfData,standardDeviationRange];
labels={"x","y","Simulated data around result function"};
pointSize=0.02;
CIP`Graphics`PlotXyErrorDataAboveFunctions[xyErrorData,
  pureFunctions,argumentRange,plotRange,plotStyle,labels,
  GraphicsOptionPointSize -> pointSize]
```

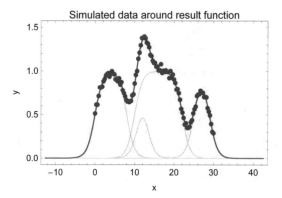

All machine learning methods try to perform the opposite work flow: They start with the data (pairs of input/ouput vectors $(\underline{x}, \underline{y})$ in general) and construct a combination of elementary functions like bumps to produce approximate model functions

f_i that describe the input/output mapping $\underline{y} = \underline{f}(\underline{x})$ of the data. The different machine learning methods only differ in the elementary functions they use and in the way they construct adequate model functions f_i with them in the data region. From a mathematical point of view the construction process is an optimization procedure. For a rough schematic picture an approximate model function f_i may be written as a weighted sum of the known elementary functions $g_{iv}(\underline{x})$ (but note that a specific machine learning method may not use a simple weighted sum at all, see below):

$$f_i(\underline{x}) = \textstyle\sum_v c_{iv} g_{iv}(\underline{x})$$

The optimization procedure then tries to find the optimum coefficients (weights) c_{iv}^{opt} for an optimum combination to obtain an optimum approximated model function f_i^{opt} which describes the input/output mapping $\underline{y} = \underline{f}(\underline{x})$ as good as possible:

$$f_i^{\mathrm{start}}(\underline{x}) = \textstyle\sum_v c_{iv}^{\mathrm{start}} g_{iv}(\underline{x}) \xrightarrow{\text{Optimization Procedure}} f_i^{\mathrm{opt}}(\underline{x}) = \textstyle\sum_v c_{iv}^{\mathrm{opt}} g_{iv}(\underline{x})$$

To perform the optimization the concrete iterative procedure of a machine learning method has to search for the global optimum of a specific hyper surface that is determined by the method's principal setup. This kind of unconstrained or constrained non-linear optimization in many dimensions belongs to the most demanding mathematical tasks known today and the development of new optimization methods is an active field of research. To close the circle of the example above the simulated erroneous data are used to approximate the result function with a machine learning method. Since xy-error data can not be used as an input of CIP machine learning methods they are transformed to a data set structure by a conversion method from the CIP DataTransformation package:

```
dataSet=
  CIP 'DataTransformation 'TransformXyErrorDataToDataSet[xyErrorData];
```

In this form the error-biased data are fitted by a three-layer perceptron (all details will be discussed in a minute)

```
numberOfHiddenNeurons=8;
perceptronInfo=
  CIP 'Perceptron 'FitPerceptron[dataSet,numberOfHiddenNeurons];
```

to create an approximated model function in the data region:

```
purePerceptronFunction=
  Function[x,CalculatePerceptron2dValue[x,perceptronInfo]];
AppendTo[pureFunctions,purePerceptronFunction];
AppendTo[plotStyle,{Thickness[0.005],Black}];
labels={"x","y","Approximated model function"};
CIP 'Graphics 'PlotXyErrorDataAboveFunctions[xyErrorData,
  pureFunctions,argumentRange,plotRange,plotStyle,labels,
```

```
GraphicsOptionPointSize -> pointSize]
```

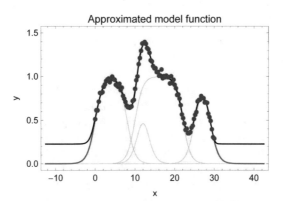

It is apparent that the approximated model function describes the data very well: There is no visible difference to the original result function inside the data region (but of course outside). A plot of the absolute sorted residuals support the assessment of a perfect fit:

```
CIP`Perceptron`ShowPerceptronSingleRegression[
   {"AbsoluteSortedResidualsPlot","RMSE"},dataSet,perceptronInfo]
```

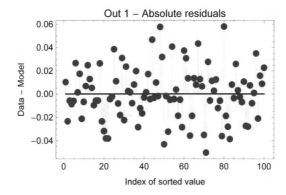

Root mean squared error (RMSE) = 2.256×10^{-2}

Also the RMSE is in very good agreement with the standard deviation (error) of 0.025 used above to generate the normally distributed data. Note again that no information about the original result function was passed to the machine learning method: It just used the data and a structural hyperparameter for the network topology (the number of hidden neurons). From the sketched nature of the model function construction process performed by machine learning some fundamental insights may be deduced:

- **Universal function approximation:** The combination of adequate elementary functions (like the bumps above) may describe an arbitrarily complex non-linear result function, i.e. machine learning techniques can in principle model everything no matter how non-linear or complex the model function is ought to be: In this sense they are computationally universal, i.e. machine learning methods perform universal function approximation. Although mathematically there are specific restrictions for different machine learning methods for all practical purposes this fundamental statement remains valid. It explains the power of these methods and their wide range of applicability.
- **Structural Failure:** Number or nature of the elementary functions may be inadequate for a specific modelling problem so the modelling effort will fail inevitably. This problem is in close connection to ...
- **The problem of overfitting:** Think about the simplest solution for a machine learning method to perfectly describe the input/output mapping of data: Just build a sharp bump for every datum. Despite the fact that the resulting model function would be perfect for the (training) data it would be in general completely useless for predictions. But satisfactory predictions are the final goal of machine learning methods: A machine learning method that builds a model function that describes the data well but that has no generalization or prediction abilities is said to overfit the (training) data.
- **Technical Failure:** Non-linear optimization techniques as iterative numerical algorithms may fail to find the global or even a local optimum for numerous reasons (compare chapter 1).
- **No extrapolation abilities:** Since elementary functions like bumps are zero or have arbitrary values outside the data region where they are constructed it is obvious that model functions created by machine learning methods can not be used for extrapolation purposes. So only interpolation may be performed with success.

Failure and overfitting are severe problems of all machine learning methods: Some strategies to tackle these issues are discussed throughout this chapter.

4.2 Machine learning methods

Machine learning methods require the a priori definition of two types of parameters for successful operation:

- **Structural hyperparameters:** These parameters determine fundamental structural features of the method like the kernel function for support vector machines

or the number of hidden neurons for three-layer feed-forward neural networks. Within CIP these parameters must be explicitly passed via a method's signature.
- **Technical optimization parameters:** They guide the technical details of the optimization process like the maximum-number-of-iterations parameter which sets an absolute upper bound to the number of optimization steps. CIP has default values for all optimization parameters but they may be modified via options. Unfortunately the modification of optimization parameters is often necessary since the default values can not be optimum choices for all cases.

The setting of parameters is guided by experience and rules of thumb since a theoretically based choice is not possible in general. So machine learning is inevitably a lot of educated trial and error.

4.2.1 Multiple linear and polynomial regression (MLR, MPR)

```
Clear["Global`*"];
<< CIP`ExperimentalData`
<< CIP`Graphics`
<< CIP`MPR`
```

As already mentioned multiple linear regression (MLR) is usually not accounted as a machine learning technique since this method is not able to construct non-linear model functions in principal. Thus its applicability is extremely limited but a valuable point of start to dig into a non-linear regression or classification task. MLR fits into the general scheme of input/output mapping

$$\underline{y} = \underline{f}(\underline{x})$$

but the model functions f_i it is able to construct are restricted to be hyperplanes. So each MLR model function f_i can be written as

$$y_i = f_i(x_1, x_2, ..., x_M) = \sum_{h=1}^{M} a_{ih} x_h + a_{iM+1} \; ; \; i = 1, ..., N$$

the general form a hyperplane in $M + 1$ dimensions. MLR may be regarded as the multidimensional analog to fitting a straight line in two dimensions. The model parameters a_{ih} are determined by least squares minimization with the linear model functions f_i (compare the curve fitting chapter 2 and see [Edwards 1976], [Edwards 1979] and [Chatterjee 2000] for details)

$$\sum_{k=1}^{K} \left(y_i^{(k)} - f_i\left(x_1^{(k)}, x_2^{(k)}, ..., x_M^{(k)}\right) \right)^2 \longrightarrow \text{minimize!}$$

$$\sum_{k=1}^{K} \left(y_i^{(k)} - \left(\sum_{h=1}^{M} a_{ih} x_h^{(k)} + a_{iM+1} \right) \right)^2 \longrightarrow \text{minimize!}$$

where K denotes the number of I/O pairs of the data set, $y_i^{(k)}$ is the output component i of the I/O pair k and $x_h^{(k)}$ is the input component h of the I/O pair k. I/O pair k is $\left(\underline{x}^{(k)}, \underline{y}^{(k)} \right)$ with input vector $\underline{x}^{(k)} = \left(x_1^{(k)}, x_2^{(k)}, ..., x_M^{(k)} \right)$ and output vector $\underline{y}^{(k)} = \left(y_1^{(k)}, y_2^{(k)}, ..., y_M^{(k)} \right)$. Note that the inflation of indices is an unlovely necessity to uniquely characterize every quantity in use: The only ray of hope is that they are used consistently throughout this chapter - but the situation is even getting harder in the next subsection.

MLR does not contain any structural hyperparameters since the structure of its model function is fixed to be a hyperplane. The linear restriction on the other hand implies that MLR is not prone to overfitting (but compare the appendix of this chapter) - a severe problem of the non-linear methods already mentioned above and discussed thoroughly below. In addition MLR may be used for extrapolation purposes which may not be tackled with the non-linear methods in principle.

Multiple polynomial regression (MPR) allows a higher polynomial degree - its structural hyperparameter - where MPR with a polynomial degree of one is nothing but MLR. Higher polynomial degrees may be used to extend the linear MLR approach to the non-linear realm. A polynomial degree of two allows the model functions f_i to adopt parabolic shapes (with a single optimum)

$$y_i = f_i(x_1, x_2, ..., x_M) = \sum_{h=1}^{M} \sum_{l=1}^{M} a_{ihl} x_h x_l + \sum_{h=1}^{M} a_{ih} x_h + a_{iM+1} \; ; \; i = 1, ..., N$$

and further increased degrees lead to corresponding polynomial hyper surfaces (with multiple optima). As for MLR all model parameters are determined by linear least squares minimization

$$\sum_{k=1}^{K} \left(y_i^{(k)} - \left(\sum_{h=1}^{M} \sum_{l=1}^{M} a_{ihl} x_h^{(k)} x_l^{(k)} + \sum_{h=1}^{M} a_{ih} x_h^{(k)} + a_{iM+1} \right) \right)^2 \longrightarrow \text{minimize!}$$

Note that MPR becomes increasingly prone to overfitting for higher polynomial degrees since the number of model parameters grows (polynomially). As an example the adhesive kinetics data set (see Appendix A)

```
dataSet = CIP`ExperimentalData`GetAdhesiveKineticsDataSet[];
CIP`Graphics`ShowDataSetInfo[{"IoPairs", "InputComponents",
   "OutputComponents"}, dataSet]
```

Number of IO pairs = 73

Number of input components = 3

Number of output components = 1

with input vectors $\underline{x}^{(k)} = \left(x_1^{(k)}, x_2^{(k)}, x_3^{(k)} \right)$ which consist of three input components requires four model parameters

```
polynomialDegree = 1;
CIP`MPR`GetMprNumberOfParameters[dataSet, polynomialDegree]
```

4

for a MLR approach (with a polynomial degree of one) but already ten model parameters

```
polynomialDegree = 2;
CIP`MPR`GetMprNumberOfParameters[dataSet, polynomialDegree]
```

10

for a parabolic MPR approach (with a polynomial degree of two) and remarkable 286 model parameters

```
polynomialDegree = 10;
CIP`MPR`GetMprNumberOfParameters[dataSet, polynomialDegree]
```

286

for a MPR approach with a polynomial degree of ten.

Due to the linearity in their model parameters a MLR or MPR approach is very fast on today's computers (often performed within seconds thus being orders of magnitude faster than the more powerful non-linear methods like perceptrons or SVMs). On the other hand MLR and MPR are not computationally universal: Their range of applicability is extremely (for MLR) up to rather limited (for MPR) in general. So as a rule of thumb a MLR or MPR based regression or classification approach should always be a first (fast) step before applying the more powerful machine learning methods. If MLR or MPR is successful then there is simply no need for the slower, more subtle and more error-prone methods. If not one may at least get a feeling of the degree of non-linearity involved in the regression or classification task in question. Note that for every component of an output vector one single MLR/MPR minimization is performed, e.g. for outputs of dimension five (i.e. output vectors with five components) there are five MLR/MPR minimizations to be performed. With CIP all MLR/MPR tasks are performed with the FitMlr/FitMpr command (see [FitMlr] and [FitMpr] in the references for implementation details).

4.2.2 Three-layer feed-forward neural networks

```
Clear["Global`*"];
<<CIP`Perceptron`
<<CIP`Graphics`
<<CIP`CalculatedData`
<<CIP`DataTransformation`
```

Layered feed-forward neural networks (perceptrons) consist of multiple layers of logical neurons (see [Hertz 1991], [Freeman 1993], [Rojas 1996] and [Murphy 2012]), e.g. a three-layer perceptron consists of three neuron layers denoted input, hidden and output layer. The neurons interact feed-forward only, i.e. within a three-layer perceptron each neuron of the input layer is exclusively connected to (all) the neurons of the hidden layer and each neuron in the hidden layer exclusively to (all) the neurons of the output layer. Thus the flow of information is restricted to happen from the input to the hidden and from the hidden to the output layer. The most simple feed-forward architecture would consist of only two layers (input and output) with a direct connection of input neurons and output neurons (without any hidden neurons in between) but this elementary perceptron is not computationally universal, i.e. it does not accomplish universal function approximation (in practice it already fails to learn a simple XOR logic). A three-layer perceptron with one layer of hidden neurons overcomes this fundamental deficiency and may be regarded as the most shallow computationally universal network. Deep feed-forward neural networks consist of additional hidden layers which may allow for improved learning abilities (deep learning) but give rise to additional technical optimization problems like vanishing gradients. Thus only three-layer perceptrons are discussed throughout this chapter.

A logical neuron of the hidden or the output layer is mathematically simply characterized: It sums up its weighted inputs $w_l u_l$ (i.e. the outputs from all of its preceding layer neurons), then subtracts a threshold Θ and passes the result to a so-called activation function g to calculate its own output a:

$$a = g\left(\sum_l w_l u_l - \Theta\right)$$

The (generally non-linear) activation function may be the sigmoid function

$$g(x) = \frac{1}{1+\exp\{-x\}}$$

already sketched in the previous section so the whole logical neuron acts as a non-linear threshold element: Its argument

$$\sum_l w_l u_l - \Theta$$

determines the position of the threshold. If an input $\underline{x}^{(k)} = \left(x_1^{(k)}, x_2^{(k)}, ..., x_M^{(k)}\right)$ of the kth I/O pair of the data set is fed into a three-layer perceptron each of the M neurons of the input layer take their corresponding input component from $x_1^{(k)}$ to $x_M^{(k)}$ as their output (it may be noticed that again the index overkill starts: These details may be skipped and only paid attention to by those who want to follow them. This confusing notation is one of the reasons why mathematicians invented the more abstract notation of linear algebra - which on the other hand is too abstract for most scientists). The outputs of the following L hidden layer neurons can then be computed from the M outputs of the input layer neurons

$$a_j^{(k)(\text{hidden})} = g\left(\sum_{h=1}^{M} w_{hj}^{(\text{input}\rightarrow\text{hidden})} x_h^{(k)} - \Theta_j^{(\text{hidden})}\right) \; ; j = 1, ..., L$$

where $a_j^{(k)(\text{hidden})}$ is the output of the jth neuron of the hidden layer for the input $\underline{x}^{(k)}$ of the kth I/O pair of the data set. The matrix of weights $w_{hj}^{(\text{input}\rightarrow\text{hidden})}$ and the thresholds $\Theta_j^{(\text{hidden})}$ set the specific connections between the input and the hidden layer neurons. In a final step the outputs of the N output neurons are calculated with the L hidden layer neurons outputs

$$\underline{a}^{(k)(\text{hidden})} = \left(a_1^{(k)(\text{hidden})}, a_2^{(k)(\text{hidden})}, ..., a_L^{(k)(\text{hidden})}\right)$$

in the same manner

$$b_i^{(k)(\text{output})} = g\left(\sum_{j=1}^{L} w_{ji}^{(\text{hidden}\rightarrow\text{output})} a_j^{(k)(\text{hidden})} - \Theta_i^{(\text{output})}\right) \; ; i = 1, ..., N$$

where $b_l^{(k)(\text{output})}$ is the output of the lth neuron of the output layer for the input $\underline{x}^{(k)}$ of the kth I/O pair of the data set. By inserting the expression for the $a_j^{(k)(\text{hidden})}$ from above the cumulative computational formula for a three-layer-perceptron results to

$$b_i^{(k)(\text{output})} = g\left(\sum_{j=1}^{L} w_{ji}^{(\text{hidden}\rightarrow\text{output})} g\left(\sum_{h=1}^{M} w_{hj}^{(\text{input}\rightarrow\text{hidden})} x_h^{(k)} - \Theta_j^{(\text{hidden})}\right) - \Theta_i^{(\text{output})}\right)$$

$$i = 1, ..., N$$

where every network output $b_i^{(k)(\text{output})}$ should approximate the corresponding I/O pairs's output component $y_i^{(k)}$ of the data set. The apparent sum of weighted sigmoid threshold functions within the cumulative computational formula for a three-layer-perceptron

$$\sum_{j=1}^{L} w_{ji}^{(\text{hidden}\rightarrow\text{output})} g(...)$$

is the mathematical basis for the internal construction of the bumps mentioned in the previous section where the network parameters (the weights w and thresholds Θ) determine their forms and positions. If we remember the basic task of a machine learning method to approximate the unknown model functions f_i (see above)

$$y_i = f_i(x_1,...,x_M) \; ; \; i = 1,...,N$$

the concrete approach of a three-layer-perceptron can now be identified: The unknown model functions are expressed as

$$f_i(x_1,...,x_M) = b_i^{(k)(\text{output})}$$

$$f_i(x_1,...,x_M) = g\left(\sum_{j=1}^{L} w_{ji}^{(\text{hidden}\rightarrow\text{output})} g\left(\sum_{h=1}^{M} w_{hj}^{(\text{input}\rightarrow\text{hidden})} x_h - \Theta_j^{(\text{hidden})}\right) - \Theta_i^{(\text{output})}\right)$$

$$i = 1,...,N$$

where it may be shown by rigorous mathematical proof that this approach is capable of approximating any arbitrarily complex and difficult function to any desired degree of accuracy (with some negligible restrictions for most practical purposes): This is what the above mentioned computational universality essentially means. Note that the previous section used the more plausible bump illustration to demonstrate this same finding. The entire computed output of the network $\underline{b}^{(k)(\text{output})} = \left(b_1^{(\text{output})}, b_2^{(\text{output})},...,b_M^{(\text{output})}\right)$ for an input $\underline{x}^{(k)} = \left(x_1^{(k)}, x_2^{(k)},...,x_M^{(k)}\right)$ can finally be compared to the corresponding output $\underline{y}^{(k)} = \left(y_1^{(k)}, y_2^{(k)},...,y_M^{(k)}\right)$ of the data set. It is common to define a type of cost function C like the mean squared error MSE

$$C_{\text{MSE}} = \frac{1}{KN} \sum_{k=1}^{K} \sum_{i=1}^{N} \left(y_i^{(k)} - b_i^{(k)(\text{output})}\right)^2$$

to quantify the difference between the outputs $\underline{b}^{(k)(\text{output})}$ of the network and the desired outputs $\underline{y}^{(k)}$ for the whole data set. Note that the cost function C_{MSE} is a function of the network parameters, i.e. a function of all the weights w and thresholds Θ of the network:

$$C_{\text{MSE}} = C_{\text{MSE}}(w_{11}^{(\text{input}\rightarrow\text{hidden})},...,w_{ML}^{(\text{input}\rightarrow\text{hidden})},$$

$$\Theta_1^{(\text{hidden})},...,\Theta_L^{(\text{hidden})},$$

$$w_{11}^{(\text{hidden}\rightarrow\text{output})},...,w_{LN}^{(\text{hidden}\rightarrow\text{output})},$$

$$\Theta_1^{(\text{output})}, ..., \Theta_N^{(\text{output})})$$

The smaller the value of the cost function C_{MSE} the better the perceptron approximates the desired outputs $y^{(k)}$ of the data set. So the network parameters should be adjusted to minimize C_{MSE}

$$C_{\text{MSE}} = C_{\text{MSE}}(\text{network parameters}) \longrightarrow \text{minimize!}$$

or: Supervised learning with a perceptron is nothing but an unconstrained global minimization of the hyper surface C_{MSE} (see Appendix A and [FitPerceptron] in the references for details about the algorithms used by the CIP method FitPerceptron). The number of internal network parameters of a three-layer perceptron is

$(M \times L)$ weights $w_{ij}^{(\text{input} \to \text{hidden})}$ plus

L thresholds $\Theta_j^{(\text{hidden})}$ plus

$(L \times N)$ weights $w_{jl}^{(\text{hidden} \to \text{output})}$ plus

N thresholds $\Theta_N^{(\text{output})}$

so a network with 3 input neurons, 10 hidden neurons and 2 output neurons contains

$$(3 \times 10) + 10 + (10 \times 2) + 2 = 62$$

internal network parameters. Hence the minimization of C_{MSE} is an unconstrained global minimization problem in 62 dimensions in this case. Since the number of neurons of the input and output layer is equal to the number of components of the inputs and outputs of the data set's I/O pairs the central structural hyperparameter of a three-layer perceptron is the number of neurons of the hidden layer. The larger this number the more bumps the perceptron is able to create but also the more internal parameters are to be optimized. Thus the learning task (the unconstrained global minimization of C_{MSE}) will become more difficult. In general a more difficult model function to approximate requires an increasing number of hidden neurons. But an increase of the number of hidden neurons also boosts the network's tendency to overfitting. Therefore in practice this central structural perceptron hyperparameter should be kept as small as possible but large enough to fulfill the learning task. To demonstrate the crucial role of the number of hidden neurons the perfect perceptron fit of the previous section is performed with a reduced number of hidden neurons. After restoration of the settings

```
interval1={0.,8.};
pureBump1=Function[x,BumpFunction[x,interval1]];
interval2={11.,13.};
pureBump2=Function[x,BumpFunction[x,interval2]];
interval3={10.,22.};
pureBump3=Function[x,BumpFunction[x,interval3]];
interval4={25.,29.};
pureBump4=Function[x,BumpFunction[x,interval4]];
pureBumpSum=Function[x,BumpFunction[x,interval1]+
 BumpFunction[x,interval2]+BumpFunction[x,interval3]+
 BumpFunction[x,interval4]];
pureFunctions={pureBump1,pureBump2,pureBump3,pureBump4,pureBumpSum};
argumentRange={-10,40};
plotRange={0.0,1.5};
numberOfData=100;
standardDeviationRange={0.025,0.025};
simulatedDataArgumentRange={0.0,30.0};
xyErrorData=CIP`CalculatedData`GetXyErrorData[pureBumpSum,
 simulatedDataArgumentRange,numberOfData,standardDeviationRange];
dataSet=CIP`DataTransformation`TransformXyErrorDataToDataSet[
 xyErrorData];
```

the training (optimization) with a reduced number of three hidden neurons

```
numberOfHiddenNeurons=3;
perceptronInfo=
 CIP`Perceptron`FitPerceptron[dataSet,numberOfHiddenNeurons];
```

is successful but an adequate fit of the data fails which can be detected by visual inspection of the approximated model function:

```
purePerceptronFunction=
 Function[x,CalculatePerceptron2dValue[x,perceptronInfo]];
AppendTo[pureFunctions,purePerceptronFunction];
plotStyle={{Thickness[0.001],Green},{Thickness[0.001],Green},
 {Thickness[0.001],Green},{Thickness[0.001],Green},
 {Thickness[0.005],Red},{Thickness[0.005],Black}};
labels={"x","y","Approximated model function"};
pointSize=0.02;
CIP`Graphics`PlotXyErrorDataAboveFunctions[xyErrorData,
 pureFunctions,argumentRange,plotRange,plotStyle,labels,
 GraphicsOptionPointSize -> pointSize]
```

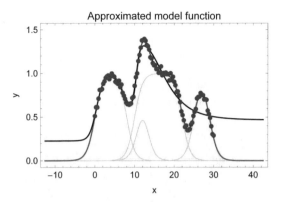

This failure can not be traced to technical optimization problems in this case: The resulting model function is the optimum of a perceptron with the defined topology, i.e. the defined number of hidden neurons. It is the reduced structural complexity of the perceptron that does not allow the construction of an adequate approximate model function to successfully describe the data. The network is simply not able to produce enough bumps. But if three hidden neurons are not enough and eight hidden neurons are sufficient what is the minimum value of this structural hyper-parameter for a satisfactory fit? Unfortunately this has to be evaluated by educated trial and error since there is in general no way to calculate the necessary number of hidden neurons from theoretical considerations in advance (this is why this problem remains an active field of research). Finally there are two ways a perceptron training may be performed: One perceptron is trained for every output component of an output vector, i.e. each single perceptron has only one output neuron that corresponds to one single component of an I/O pair's output, or one perceptron is trained with a complete output layer where each output neuron corresponds to a component of an I/O pair's output: Then the number of output neurons is equal to the number of components of an I/O pair's output. The first choice (and the default choice in CIP) is in general more powerful but also computationally more demanding (but may be especially accelerated by parallelized calculation, see Appendix A). As already shown within CIP the FitPerceptron command performs all perceptron related machine learning operations (see [FitPerceptron] in the references for implementation details).

4.2.3 Support vector machines (SVM)

```
Clear["Global`*"];
<<CIP`Graphics`
<<CIP`SVM`
```

A (regression) support vector machine (SVM) places an a priori defined elementary function at the position of every input of the data set to be learned. In SVM terminology an admissible elementary function is called a kernel function since it must satisfy a specific mathematical condition (Mercer's condition) to be usable as a kernel. The already touched Wavelet kernel (with a width parameter of 0.1) may be illustrated in three dimensions (note the mexican-hat shape)

```
a=0.1;
pureWaveletKernel3D=
  Function[{x,y},CIP`SVM`KernelWavelet[{x},{y},a]];
xRange={-0.2,0.2};
yRange={-0.2,0.2};
labels={"x","y","z"};
viewPoint3D={-1.3,-2.5,1.5};
CIP`Graphics`Plot3dFunction[pureWaveletKernel3D,xRange,yRange,
  labels,GraphicsOptionViewPoint3D -> viewPoint3D]
```

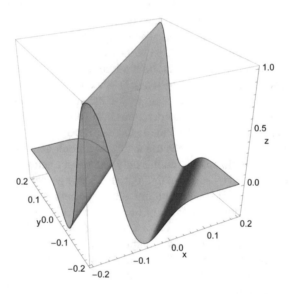

which reduces to

```
x=0.0;
pureWaveletKernel2D=Function[y,CIP`SVM`KernelWavelet[{x},{y},a]];
argumentRange={-0.7,0.7};
functionValueRange={-0.4,1.1};
labels={"x","z","Wavelet kernel function"};
CIP`Graphics`Plot2dFunction[pureWaveletKernel2D,argumentRange,
  functionValueRange,labels]
```

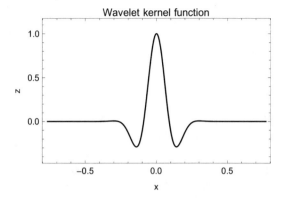

Wavelet kernel function

if the Wavelet kernel is fixed to a position (in this case to $x = 0$): Its bump character is obvious. The wavelet parameter a controls the width of the bump, i.e. a higher value of the wavelet width parameter a leads to a widened bump:

```
a=0.3;
waveletKernel=Cos[1.75*(x-y)/a]* Exp[-(x-y)^2/(2*a^2)];
x=0.0;
pureWaveletKernel2D=Function[x1,waveletKernel/.y->x1];
CIP`Graphics`Plot2dFunction[pureWaveletKernel2D,argumentRange,
  functionValueRange,labels]
```

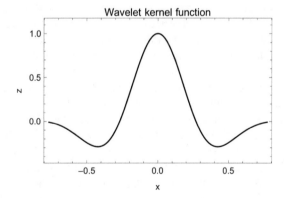

Wavelet kernel function

The necessary a priori definition of the kernel function with a fixed width parameter a thus influences the possible smoothness of the resulting model function, e.g. a higher value for a inevitably leads to a smoother model function. A (regression) SVM approximates the unknown model functions f_i with weighted sums of the kernel functions $g_i^{(\text{kernel})}\left(\underline{x}^{(k)},\underline{x}\right)$ at all K input positions $\underline{x}^{(k)}$ of the data set plus a constant value b (the regression bias):

$$y_i = f_i(x_1, x_2, ..., x_M) = \sum_{k=1}^{K} \alpha_{ik} g_i^{(kernel)} \left(x_1^{(k)}, x_2^{(k)}, ..., x_M^{(k)}, x_1, x_2, ..., x_M \right) + b_i$$

$$i = 1, ..., N$$

or more compact:

$$\underline{y} = \underline{f}(\underline{x}) = \underline{\underline{\alpha}} \cdot \underline{g}^{(kernel)} \left(\underline{x}^{(k)}, \underline{x} \right) + \underline{b}$$

Since the kernel function must be a priori chosen (it is a precondition to the method) the task of a (regression) SVM is to determine optimum weights α_{ik}^{opt} and optimum regression biases b_i^{opt} to approximate the unknown model functions $f_i(\underline{x})$. Compared to the perceptron approach above the SVM's strategy seems to be intuitively simple and straightforward - but there is a solid theoretical basis (statistical learning theory) with a sophisticated mathematical machinery (quadratic programming) behind the scenes (see [Vapnik 1995], [Vapnik 1998], [Schölkopf 1998], [Gunn 1998], [Schölkopf 1999], [Cristianini 2000], [Schölkopf 2002], [Bishop 2006] and [Murphy 2012] for details). The magic of a SVM comes from the fact that it allows the definition of a constrained objective function which can be globally maximized without the risk of being trapped in a local maximum in order to successfully determine (structurally) optimum values of the parameters α_{ik}^{opt} and b_i^{opt}. This is a fundamental difference to the (empirical) cost function minimization process for perceptrons where local minima traps bob up consistently. With having a theoretically well-defined and working global optimization strategy up one's sleeve the only central unknown structural hyperparameter of a SVM is the type of kernel function to be used. The proper choice of the kernel function decides about success or failure of this machine learning method (as was shown with the introductory example at the beginning of this chapter) - and again the type of kernel function can not simply be deduced from theoretical considerations in general: Educated trial and error is the only path to success. Also note that the number of weights α_k to optimize is equal to the number of input vectors in the data set: This means that the SVM's optimization task becomes more complex (and therefore more difficult and time consuming in general) with an increasing data set size. This is different to a perceptron where the number of internal coefficients to optimize is determined solely by the network topology. On the other hand the number of weights α_k of a SVM does not depend on the dimension of the inputs (i.e. the number of components of an input vector) but the network topology of a perceptron does since this dimension determines the number of input neurons. Therefore the number of internal parameters to optimize of a perceptron increases with an increase of the input's dimension which leads to a more difficult optimization task in general. Like MLR and MPR for every component of an I/O pair's output one single SVM optimization is to be performed, e.g. for outputs of dimension 5 (an output with 5 components) there are 5 SVM optimizations to be performed (where each optimization may be processed in parallel, see Appendix A). As already demonstrated SVM related computations with CIP use the FitSvm command (see [FitSvm] in the references for details).

A short intermezzo about machine learning history: Neural networks and support vector machines ...

In the late 1980s and early 1990s there was a real neural network hype. They invaded the different scientific communities and attracted a lot of attention of a broad scientific and non-scientific audience. One reason for this exploding popularity was (besides their innovative features) that they heavily used a biological terminology for their description instead of a purely mathematical one: The logical neuron was almost always motivated on the basis of its biological predecessor and not by its mathematical features - although a logical and a biological neuron do not have too much in common. The impression emerged that computers finally started to model the human brain and human intelligence - an event that nobody wanted to miss. In this sense the development of SVMs in the late 1990s on the "obscure grounds" of statistical learning theory may be regarded as a revenge of the mathematicians: Quadratic programming and kernel tricks is nothing to write home about for non-specialists. But it was the neural network hype that paved the road for machine learning applications to a broader community so the emerging SVMs could rapidly spread through the established channels. The question of computational intelligence will be addressed in more detail in chapter 5.

What is the best machine learning method? MLR and MPR almost always work fast (within seconds on today's computers) and technically without any problems but are limited for a successful practical application due to their linear/polynomial nature. The more powerful non-linear methods are computationally universal but they are comparatively slow and often subject to severe technical and structural problems. SVMs are more recent and become increasingly popular. They possess attractive features like the path to successful global optimization and are often argued to be superior to the somewhat older perceptron-type neural networks: They seem to be stronger in classification whereas perceptrons gain ground in regression tasks, SVMs are attributed a reduced tendency to overfitting, the SVM's structural risk is preferable to the perceptron's empirical risk - but honestly there is no final definite answer to the best-method question: It simply depends ... and thus machine learning is a vivid and active field of research and there are daily claims of new superior algorithmic variants (e.g. see [Platt 1999], [Joachims 1999], [Keerthi 2002], [Fan 2005] or [Glasmachers 2006] for SVM related global optimization improvements). They all should be regarded as parts of a growing tool box that stimulates further progress. For the practitioner the already sketched basic problems of machine learning play the predominant role: What type of kernel function (SVM) or network topology (perceptron) is to be used? How can the immense computational efforts that are necessary to tackle large data sets be reduced (besides simply waiting for Moore's law, i.e. faster computers, to do the job)? How can overfitting be avoided and how reliable are machine learning results in practice? How can they be validated? In the following sections several aspects of these issues will be illustrated and explored.

4.3 Evaluating the goodness of regression

```
Clear["Global`*"];
<<CIP`CalculatedData`
<<CIP`Graphics`
<<CIP`MLR`
```

In analogy to curve fitting a simple linear example is used to introduce machine learning and the evaluation of the goodness of a regression result. Therefore error-biased data are simulated around a plane in three dimensions

$$z = f(x,y) = 1 + 2x + 3y$$

with an x and y argument range of [0, 1] and an absolute standard deviation of 0.2 (which leads to corresponding relative errors from approximately 3% around point (1, 1) to approximately 20% around point (0, 0)) with the CIP CalculatedData package:

```
pureOriginalFunction=Function[{x,y},1.0+2.0*x+3.0*y];
xRange={0.0,1.0};
yRange={0.0,1.0};
numberOfDataPerDimension=20;
standardDeviationRange={0.2,0.2};
dataSet3D=CIP`CalculatedData`Get3dFunctionBasedDataSet[
  pureOriginalFunction,xRange,yRange,numberOfDataPerDimension,
  standardDeviationRange];
labels={"x","y","z"};
pointSize=0.02;
CIP`Graphics`Plot3dDataSet[dataSet3D,labels,
  GraphicsOptionPointSize -> pointSize]
```

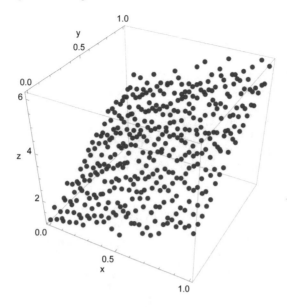

As a machine learning technique Multiple Linear Regression (MLR) is used which is perfectly capable of fitting linear data in multiple dimensions:

```
mlrInfo=CIP `MLR`FitMlr[dataSet3D];
```

The root mean squared error of a machine learning method is defined as

$$\text{RMSE} = \sqrt{\frac{1}{KN} \sum_{k=1}^{K} \sum_{i=1}^{N} \left(y_i^{(k)} - f_i\left(\underline{x}^{(k)}\right) \right)^2}$$

and its value

```
CIP `MLR`ShowMlrSingleRegression[{"RMSE"},dataSet3D,
  mlrInfo];
```

Root mean squared error (RMSE) = 2.043×10^{-1}

and the (absolute) residual statistics

```
CIP `MLR`ShowMlrSingleRegression[
  {"AbsoluteResidualsStatistics","RelativeResidualsStatistics"},
  dataSet3D,mlrInfo];
```

Definition of 'Residual (absolute)': Data - Model

Out 1 : Residual (absolute): Mean/Median/Maximum Value = 1.61×10^{-1} / 1.28×10^{-1} / 7.71×10^{-1}

Definition of 'Residual (percent)': 100*(Data - Model)/Data

Out 1 : Residual (percent): Mean/Median/Maximum Value = 5.15 / 4.07 / 3.28 × 10^1

correspond perfectly to the standard deviation of 0.2 used for the data simulation. The model-versus-data plot and the sorted-model-versus-data plot

```
CIP`MLR`ShowMlrSingleRegression[
 {"ModelVsDataPlot","SortedModelVsDataPlot",
  "CorrelationCoefficient"},dataSet3D,mlrInfo,
  GraphicsOptionPointSize -> pointSize];
```

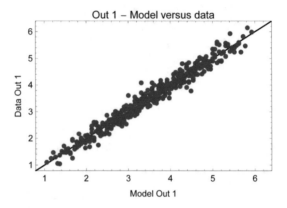

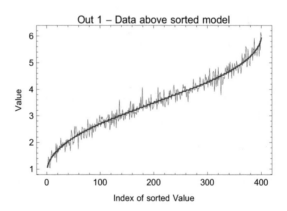

Out 1 : Correlation coefficient = 0.982142

show the expected behavior: The simulated data in the first diagram scatter around the diagonal (on the diagonal machine and model output are identical), the data line of the second diagram crawls statistically around the model line defined by the sorted machine learning outputs. A correlation coefficient may be used to condense the agreement between data and model into a single quantity (with a value near one meaning a desired high correlation - but compare the discussion in chapter 2 and below). The absolute and relative residuals plots

```
CIP`MLR`ShowMlrSingleRegression[
  {"AbsoluteSortedResidualsPlot","RelativeSortedResidualsPlot"},
  dataSet3D,mlrInfo,GraphicsOptionPointSize -> pointSize];
```

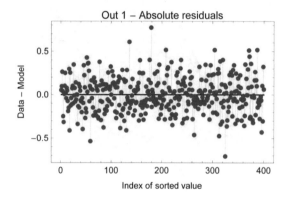

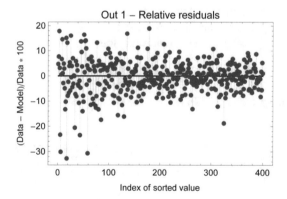

show statistically distributed residuals with no systematic deviation patterns within the expected magnitudes: Excellent! And last but not least the visual 3D inspection

```
pureMlr3dFunction=
  Function[{x,y},CalculateMlr3dValue[x,y,mlrInfo]];
CIP`Graphics`Plot3dDataSetWithFunction[dataSet3D,pureMlr3dFunction,
  labels,GraphicsOptionPointSize -> pointSize]
```

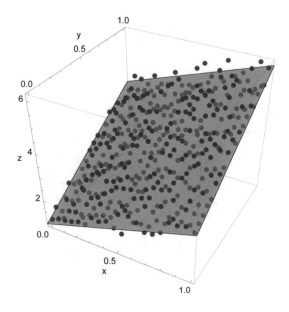

is convincing. Note that the latter is usually not available for machine learning problems in many dimensions: This is a severe disadvantage since human beings are very powerful in judging visual representations at a glance.

4.4 Evaluating the goodness of classification

```
Clear["Global`*"];
<<CIP`CalculatedData`
<<CIP`Utility`
<<CIP`Graphics`
<<CIP`MLR`
<<CIP`DataTransformation`
```

The goodness of a machine learning method's classification result may be stated straight forward: The correctly and incorrectly classified I/O pairs are simply counted and displayed. In three dimensions it is furthermore possible to visualize the decision surfaces. Here a perfect classification example with decision surfaces that consists of 3D planes is demonstrated: Two clearly separated clouds of two-dimensional inputs

```
centroid1={0.2,0.2};
numberOfCloudVectors1=50;
standardDeviation1=0.1;
singleCloudDefinition1={centroid1,numberOfCloudVectors1,
 standardDeviation1};
centroid2={0.8,0.8};
numberOfCloudVectors2=50;
standardDeviation2=0.1;
singleCloudDefinition2={centroid2,numberOfCloudVectors2,
 standardDeviation2};
cloudDefinitions={singleCloudDefinition1,singleCloudDefinition2};
classificationDataSet=
 CIP`CalculatedData`GetGaussianCloudsDataSet[
 cloudDefinitions];
labels={"x","y","Inputs"};
CIP`Graphics`Plot2dPoints[
 CIP`Utility`GetInputsOfDataSet[classificationDataSet],labels]
```

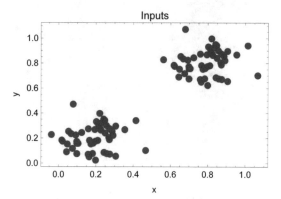

are classified (note that this classification task could be perfectly performed by a mere unsupervised clustering-based class predictor as shown in chapter 3 so it is no real challenge for supervised machine learning). The inputs are combined with an adequately coded output for classification, e.g. the first input/output (I/O) pair

```
ioPair1=classificationDataSet[[1]]
```

$\{\{0.248568, 0.240914\}, \{1., 0.\}\}$

is attributed to class 1: It has a value of 1.0 at position 1 for class 1 in the output vector and a value of 0.0 at position 2 for class 2. The last I/O pair of the generated data set

```
ioPair1=classificationDataSet[[Length[classificationDataSet]]]
```

$\{\{0.891354, 0.821637\}, \{0., 1.\}\}$

is attributed to class 2 respectively (with value of 0.0 at position 1 for class 1 and a value of 1.0 at position 2 for class 2). The linear MLR fit generates a decision surface

```
mlrInfo=CIP`MLR`FitMlr[classificationDataSet];
CIP`MLR`ShowMlrSingleClassification[
 {"CorrectClassification","CorrectClassificationPerClass"},
 classificationDataSet,mlrInfo]
```

100.% correct classifications

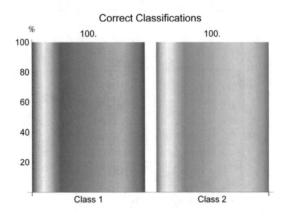

that allows 100% correct classification of all input vectors for both classes. Since the output of every I/O pair has dimension two there are two planes, one for every output, that are combined for decision (a distinct input is attributed to the class that corresponds to the surface with the higher output value at this point). Both planes can be visualized independently: The plane for output 1 with the corresponding subset of data

```
classificationDataSet3DList=
 CIP`DataTransformation`TransformDataSetToMultipleDataSet[
 classificationDataSet];
indexOfInput1=1;
indexOfInput2=2;
indexOfOutput=1;
input={0.0,0.0};
pureMlr3dFunction=Function[{x,y},
 CIP`MLR`CalculateMlr3dValue[x,y,indexOfInput1,
 indexOfInput2,indexOfOutput,input,mlrInfo]];
labels={"In 1","In 2","Out 1"};
CIP`Graphics`Plot3dDataSetWithFunction[
 classificationDataSet3DList[[1]],pureMlr3dFunction,labels]
```

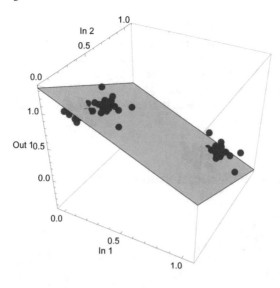

and the other plane for output 2

```
indexOfInput1=1;indexOfInput2=2;indexOfOutput=2;input={0.0,0.0};
pureMlr3dFunction=Function[{x,y},
 CIP`MLR`CalculateMlr3dValue[x,y,indexOfInput1,
 indexOfInput2,indexOfOutput,input,mlrInfo]];
labels={"In 1","In 2","Out 2"};
CIP`Graphics`Plot3dDataSetWithFunction[
 classificationDataSet3DList[[2]],pureMlr3dFunction,labels]
```

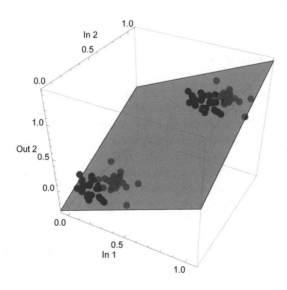

illustrate the perfect classification behavior of the machine learning method for the data in question. Note that the fit process generated just one set of an infinite number of possible decision surfaces for the data clouds. It should also be noted that linear decision surfaces will not be powerful in general since they need data that can be clearly separated by mere planes. The non-linear machine learning methods like perceptrons and SVMs will be able to construct arbitrarily curved decision surfaces as will be shown below.

4.5 Regression: Entering non-linearity

```
Clear["Global`*"];
<<CIP`ExperimentalData`
<<CIP`Graphics`
<<CIP`MPR`
<<CIP`SVM`
<<CIP`Perceptron`
```

In this section a model function for an experimental adhesive kinetics data set is approximated with different approaches and common pitfalls. The experimental data describe the dependence of a kinetics parameter on the composition of an adhesive polymer mixture and are outlined in detail in Appendix A. The data set is provided by the CIP ExperimentalData package:

```
dataSet=CIP`ExperimentalData`GetAdhesiveKineticsDataSet[];
CIP`Graphics`ShowDataSetInfo[{"IoPairs","InputComponents",
  "OutputComponents"},dataSet]
```

Number of IO pairs = 73

Number of input components = 3

Number of output components = 1

The 73 I/O pairs are four-dimensional (i.e. contain inputs with three components and outputs with one component) so the complete data set can not be displayed with 2D or 3D graphics. But due to the experimental measurement setup it is possible to obtain subsets of data that are suitable for visual inspection in 3D. The experimental errors of the data are reported to be in the order of 10% to 20% with some outliers which is an essential information for the assessment of the goodness of regression in the following. The machine learning task is initially tackled by MLR (multiple linear regression which is identical to multiple polynomial regression with a polynomial degree of 1)

```
polynomialDegree=1;
```

which contains 4 model parameters

```
CIP `MPR `GetMprNumberOfParameters[dataSet,polynomialDegree]
```

4

and leads to a regression result

```
mprInfo=CIP `MPR `FitMpr[dataSet,polynomialDegree];
```

with obvious systematic deviations between data and model (positive deviations for small and large output values and negative deviations in between) as is illustrated by the model-versus-data and relative residuals plot:

```
CIP `MPR `ShowMprSingleRegression[{"ModelVsDataPlot",
  "RelativeSortedResidualsPlot"},dataSet,mprInfo];
```

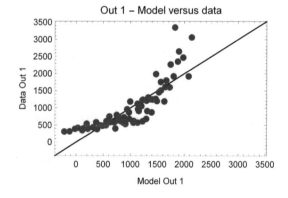

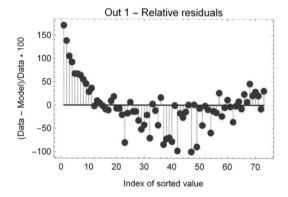

The relative residuals statistics

```
CIP`MPR`ShowMprSingleRegression[{"RelativeResidualsStatistics",
  "CorrelationCoefficient"},dataSet,mprInfo];
```

Out 1 : Relative residuals (100*(Data - Model)/Data):

Mean/Median/Maximum Value in % = 3.44×10^1 / 2.02×10^1 / 1.71×10^2

Out 1 : Correlation coefficient = 0.840707

show that the magnitude of the deviations (over 30%) is obviously above the reported experimental errors of 10 to 20% and the correlation coefficient is poor in addition. So it can be deduced that the adhesive kinetics data can not be satisfactorily modelled by a simple hyperplane. This may also be illustrated by the 3D display of a subset of the data that corresponds to a specific polymer mass ratio of the mixture:

```
polymerMassRatio="80";
dataSet3D =
 CIP`ExperimentalData`GetAdhesiveKinetics3dDataSet[
  polymerMassRatio];
indexOfInput1=2;
indexOfInput2=3;
indexOfOutput=1;
input = {80.0,0.0,0.0};
pureMpr3dFunction=
 Function[{x,y},
  CIP`MPR`CalculateMpr3dValue[x,y,indexOfInput1,indexOfInput2,
   indexOfOutput,input,mprInfo]];
labels={"In 2","In 3","Out 1"};
CIP`Graphics`Plot3dDataSetWithFunction[
 dataSet3D,pureMpr3dFunction,labels]
```

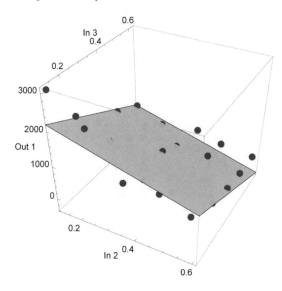

The linear plane is only a poor approximate description of the data. On the other hand the adhesive kinetics data are not dramatically non-linear- the true relation seems to be a slightly curved surface. Thus the linear MLR approach may be slightly extended into the non-linear region by an a priori/a posteriori data transformation with a logarithmic/exponential function: The output components are transformed by a logarithmic function before the MLR fit, the function values of the MLR generated model functions are then afterwards inversely transformed by an exponential function (see the CIP code for details):

```
dataTransformationMode="Log";
mprInfo =
 CIP`MPR`FitMpr[dataSet,polynomialDegree,
  MprOptionDataTransformationMode -> dataTransformationMode];
CIP`MPR`ShowMprSingleRegression[{"ModelVsDataPlot",
 "RelativeSortedResidualsPlot","RelativeResidualsStatistics",
 "CorrelationCoefficient"},dataSet,mprInfo];
```

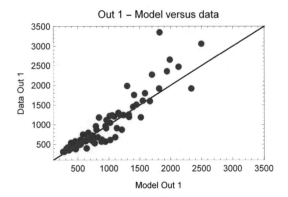

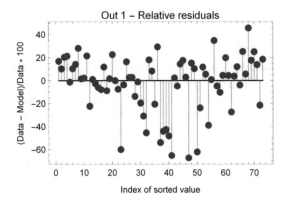

Out 1 : Relative residuals (100*(Data - Model)/Data):

Mean/Median/Maximum Value in % = 1.84×10^1 / 1.41×10^1 / 6.74×10^1

Out 1 : Correlation coefficient = 0.912715

The systematic deviations between data and model are now confined to the larger output value region and the residuals statistics are more acceptable with a value around 18% on average. Also the correlation coefficient increased. The 3D plot of the subset of data with the approximated model function

```
pureMpr3dFunction=
 Function[{x,y},
  CIP`MPR`CalculateMpr3dValue[x,y,indexOfInput1,indexOfInput2,
   indexOfOutput,input,mprInfo]];
CIP`Graphics`Plot3dDataSetWithFunction[
 dataSet3D,pureMpr3dFunction,labels]
```

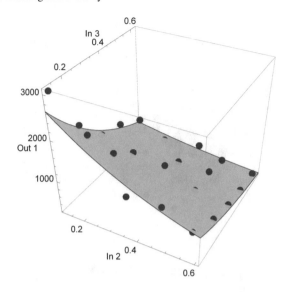

demonstrates the improvement. The adhesive kinetics data seem to be at the borderline for a successful application of a non-linear enhanced MLR approach. To further improve the modelling result a switch to non-linear machine learning methods is indicated with a polynomial expansion as a sensible first step. If a general quadratic polynomial (i.e. a polynomial degree of 2) is chosen

```
polynomialDegree=2;
CIP`MPR`GetMprNumberOfParameters[dataSet,polynomialDegree]
```

10

the model now contains 10 parameters to be estimated. Compared to pure MLR the polynomial expansion leads to a clear improvement:

```
mprInfo=CIP`MPR`FitMpr[dataSet,polynomialDegree];
CIP`MPR`ShowMprSingleRegression[{"ModelVsDataPlot",
  "RelativeSortedResidualsPlot","RelativeResidualsStatistics",
  "CorrelationCoefficient"},dataSet,mprInfo];
```

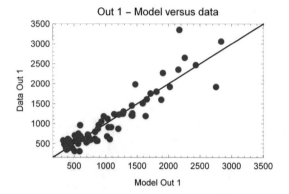

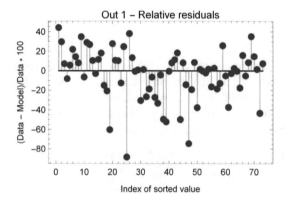

Out 1 : Relative residuals (100*(Data - Model)/Data):

Mean/Median/Maximum Value in % = 1.9×10^1 / 1.44×10^1 / 8.85×10^1

Out 1 : Correlation coefficient = 0.92732

The systematic deviations are less pronounced and the residuals are at the upper experimental limit. The improvement is also directly visible by the 3D plot of the subset of data:

```
pureMpr3dFunction=
 Function[{x,y},
  CIP`MPR`CalculateMpr3dValue[x,y,indexOfInput1,indexOfInput2,
   indexOfOutput,input,mprInfo]];
CIP`Graphics`Plot3dDataSetWithFunction[
 dataSet3D,pureMpr3dFunction,labels]
```

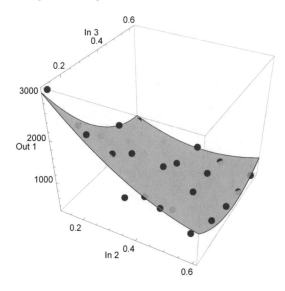

An additional logarithmic/exponential transformation

```
mprInfo=
 CIP`MPR`FitMpr[dataSet,polynomialDegree,
  MprOptionDataTransformationMode -> dataTransformationMode];
CIP`MPR`ShowMprSingleRegression[{"ModelVsDataPlot",
 "RelativeSortedResidualsPlot","RelativeResidualsStatistics",
 "CorrelationCoefficient"},dataSet,mprInfo];
```

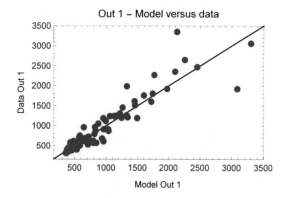

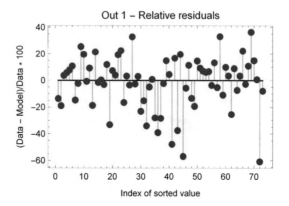

Out 1 : Relative residuals (100*(Data - Model)/Data):

Mean/Median/Maximum Value in % = 1.49×10^1 / 1.09×10^1 / 6.08×10^1

Out 1 : Correlation coefficient = 0.921525

leads to an even improved and overall satisfying result:

```
pureMpr3dFunction=
 Function[{x,y},
  CIP`MPR`CalculateMpr3dValue[x,y,indexOfInput1,indexOfInput2,
   indexOfOutput,input,mprInfo]];
quadraticLogMprGraphics3D=
 CIP`Graphics`Plot3dDataSetWithFunction[
  dataSet3D,pureMpr3dFunction,labels]
```

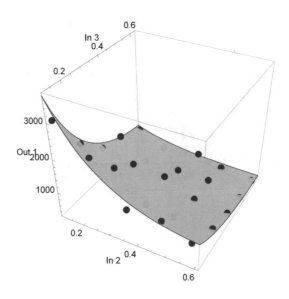

The achieved quadratic-log MPR model is probably near the best we can get for the adhesive kinetics regression problem - and it is convincing and helpful to the adhesive scientists, e.g. for predictive interpolation calculations. Note that an increase of the polynomial degree (polynomially) increases the number of model related parameters to be estimated by the fitting procedure and thus allows the model function to be more non-linearly curved. This may lead to an unwanted overfitting of the data as may be illustrated with a polynomial degree of 6 for the current regression task:

```
polynomialDegree=6;
CIP `MPR `GetMprNumberOfParameters[dataSet,polynomialDegree]
```

84

The number of 84 model parameters now exceeds the number of 73 I/O pairs in the data set and the corresponding MPR fit

```
mprInfo=CIP `MPR `FitMpr[dataSet,polynomialDegree];
CIP `MPR `ShowMprSingleRegression[{"ModelVsDataPlot",
 "RelativeSortedResidualsPlot","RelativeResidualsStatistics",
 "CorrelationCoefficient"},dataSet,mprInfo];
```

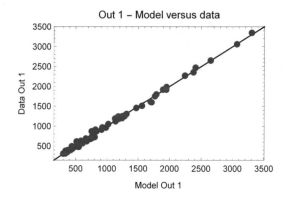

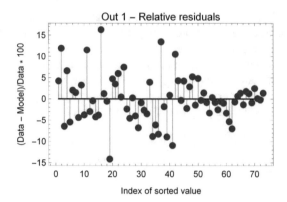

Out 1 : Relative residuals (100*(Data - Model)/Data):

Mean/Median/Maximum Value in % = 3.97 / 3.24 / 1.63×10^1

Out 1 : Correlation coefficient = 0.998134

leads to a good description of the data (i.e. a kind of look-up table) but to an unwanted overfitted model without any predictability, i.e. it would be useless or even misleading for predictive interpolation calculations:

```
pureMpr3dFunction=
 Function[{x,y},
  CIP`MPR`CalculateMpr3dValue[x,y,indexOfInput1, indexOfInput2,
   indexOfOutput,input,mprInfo]];
CIP`Graphics`Plot3dDataSetWithFunction[
 dataSet3D,pureMpr3dFunction,labels]
```

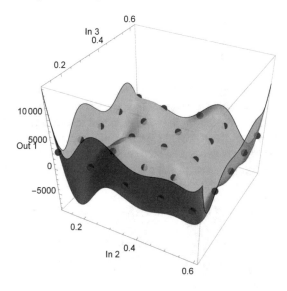

Since a convincing and predictive model could be achieved by MPR there would be no need to apply the more powerful non-linear methods. But for comparison results of SVMs and perceptrons may be finally investigated. Since a SVM needs a kernel function as its structural hyperparameter (besides several optimization parameters which will be left unchanged at their default values) the earlier successful wavelet kernel with a width parameter of 0.1 (see the beginning of this chapter) is used. After the fit is performed

```
kernel={"Wavelet",0.1};
svmInfo = CIP`SVM`FitSvm[dataSet,kernel];
CIP`SVM`ShowSvmSingleRegression[{"ModelVsDataPlot",
  "RelativeResidualsStatistics","CorrelationCoefficient"},dataSet,
  svmInfo];
```

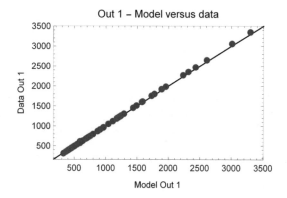

Out 1 : Relative residuals (100*(Data - Model)/Data):

Mean/Median/Maximum Value in % = 1.67 / 1.43 / 4.74

Out 1 : Correlation coefficient = 0.999977

the result exhibits nearly perfectly modelled data - the diagonal looks like a rope of pearls with a perfect correlation (indicated by a correlation coefficient of practically 1) and the relative deviations between data and model around 2% are an order of magnitude lower than the reported experimental errors. These findings again indicate a clear overfitting of the data which can be well illustrated by the 3D display of the subset of data together with the overfitted model function (where each data point seems to have its own bump)

```
pureSvm3dFunction=
  Function[{x,y},
    CIP'SVM'CalculateSvm3dValue[x,y,indexOfInput1,indexOfInput2,
      indexOfOutput,input,svmInfo]];
CIP'Graphics'Plot3dDataSetWithFunction[
  dataSet3D,pureSvm3dFunction,labels]
```

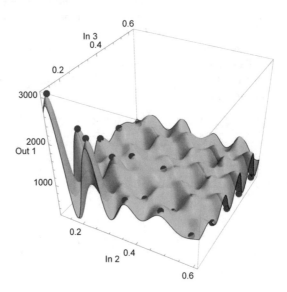

As already pointed out before this overfitted model function is completely useless for any predictive interpolation tasks. If a perceptron is used for non-linear modelling the number of hidden neurons must be a priori defined as its structural hyperparameter (again the several default technical optimization parameters are not touched). If 15 hidden neurons are arbitrarily chosen

```
numberOfHiddenNeurons=15;
perceptronInfo=
  CIP `Perceptron `FitPerceptron[dataSet,numberOfHiddenNeurons];
CIP `Perceptron `ShowPerceptronSingleRegression[{"ModelVsDataPlot",
  "RelativeResidualsStatistics","CorrelationCoefficient"},dataSet,
  perceptronInfo];
```

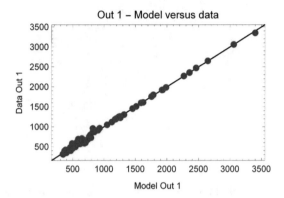

Out 1 : Relative residuals (100*(Data - Model)/Data):

Mean/Median/Maximum Value in % = 4.15 / 2.11 / 1.76×10^1

Out 1 : Correlation coefficient = 0.998328

the result is similar to the SVM approach: A clear overfitting is detected and visible in the 3D display of the subset of data.

```
purePerceptron3dFunction=
  Function[{x,y},
    CIP `Perceptron `CalculatePerceptron3dValue[x,y,indexOfInput1,
      indexOfInput2,indexOfOutput,input,perceptronInfo]];
CIP `Graphics `Plot3dDataSetWithFunction[
  dataSet3D,purePerceptron3dFunction,labels]
```

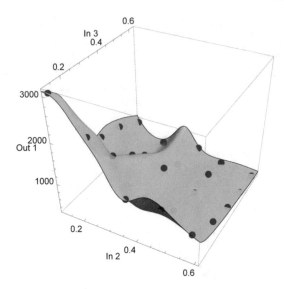

These failed attempts show that the choice of the structural hyperparameters is essential: They were simply not adequate for the machine learning task in question. And this is exactly where the trial and error begins: The kernel function or the number of hidden neurons respectively has to be adjusted to improve the approximated model functions and to allow reasonable predictive interpolations. This can be done by single smart guesses (least computationally demanding if successful), systematic variation of the structural hyperparameters (more computationally demanding) or an exhaustive search with e.g. an evolutionary strategy (extremely demanding). Since the adhesive kinetics data are near linear it can be deduced from experience that for the SVM based approach the width parameter a of the wavelet kernel should be increased. A tenfold increase from 0.1 to 1.0

```
kernelFunction={"Wavelet",1.0};
svmInfo = CIP`SVM`FitSvm[dataSet,kernelFunction];
CIP`SVM`ShowSvmSingleRegression[{"ModelVsDataPlot",
  "RelativeResidualsStatistics","CorrelationCoefficient",
  "RelativeSortedResidualsPlot"},dataSet,svmInfo];
```

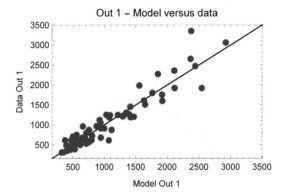

Out 1 : Relative residuals (100*(Data - Model)/Data):

Mean/Median/Maximum Value in % = 1.46×10^1 / 1.05×10^1 / 7.58×10^1

Out 1 : Correlation coefficient = 0.949581

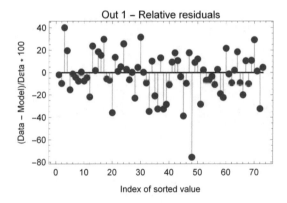

leads to an overall satisfactory modelling result with well-sized non-systematic deviations. To avoid overfitting for the perceptron approach the number of hidden neurons has to be decreased - again from experience a very small number of 2 should be sufficient for a near linear modelling problem:

```
numberOfHiddenNeurons=2;
perceptronInfo=
 CIP `Perceptron`FitPerceptron[dataSet,numberOfHiddenNeurons];
CIP `Perceptron`ShowPerceptronSingleRegression[{"ModelVsDataPlot",
 "RelativeResidualsStatistics","CorrelationCoefficient",
 "RelativeSortedResidualsPlot"},dataSet,perceptronInfo];
```

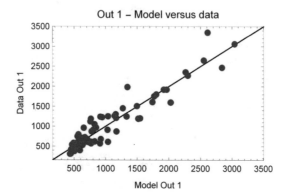

Out 1 : Relative residuals (100*(Data - Model)/Data):

Mean/Median/Maximum Value in % = 1.76×10^1 / 1.35×10^1 / $7. \times 10^1$

Out 1 : Correlation coefficient = 0.949762

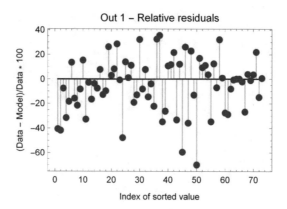

RMSE, residuals and the correlation coefficient are comparable to the SVM result before so again an acceptable result is achieved. This may finally be illustrated by the 3D overlay of the subset of data and the approximated smooth and balancing model functions of the quadratic-log MPR, SVM and perceptron approach

```
pureSvm3dFunction=
 Function[{x,y},
  CIP`SVM`CalculateSvm3dValue[x,y,indexOfInput1,indexOfInput2,
   indexOfOutput,input,svmInfo]];
svmGraphics3D=
 CIP`Graphics`Plot3dDataSetWithFunction[
  dataSet3D,pureSvm3dFunction,labels];
purePerceptron3dFunction=
 Function[{x,y},
```

```
  CIP 'Perceptron'CalculatePerceptron3dValue[x,y,indexOfInput1,
    indexOfInput2,indexOfOutput,input,perceptronInfo]];
perceptronGraphics3D=
 CIP 'Graphics'Plot3dDataSetWithFunction[dataSet3D,
  purePerceptron3dFunction,labels];
Show[{quadraticLogMprGraphics3D,svmGraphics3D,perceptronGraphics3D}]
```

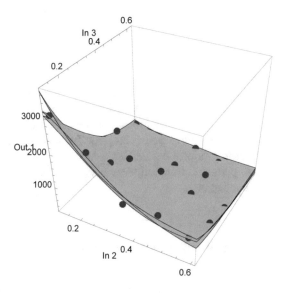

which all adequately fit the data - the one to choose for a predictive interpolation task is thus a mere matter of taste. In practice the quadratic-log MPR approach would of course be preferred since this fitting procedure clearly outperforms the two more powerful non-linear methods with regard to performance ...

- MPR:

```
Print["Time in s: ",
 AbsoluteTiming[
  CIP 'MPR'FitMpr[dataSet,polynomialDegree,
   MprOptionDataTransformationMode -> dataTransformationMode]][[1]]]
```

Time in s: 0.0661322

- SVM:

```
Print["Time in s: ",
 AbsoluteTiming[CIP 'SVM'FitSvm[dataSet,kernelFunction]][[1]]]
```

Time in s: 47.4564

- Three-layer perceptron:

```
Print ["Time in s: ",
 AbsoluteTiming[
  CIP `Perceptron`FitPerceptron[
   dataSet, numberOfHiddenNeurons]]][[1]]]
```

Time in s: 60.0174

... due to its linear nature (which is a general rule of thumb: If MLR or MPR lead to a satisfactory result you are done and saved a lot of time and trouble). Finally note that a correlation coefficient can indicate a better model (its values for the adequate quadratic-log MPR, SVM and perceptron models are higher than those of the linear and linear-log models before) but a higher value (especially one that is very close to 1, i.e. perfect correlation) may also mean undesired overfitting as already encountered in chapter 2.

4.6 Classification: Non-linear decision surfaces

```
Clear["Global`*"];
<<CIP `ExperimentalData`
<<CIP `DataTransformation`
<<CIP `Graphics`
<<CIP `MLR`
<<CIP `SVM`
```

A classification task in general requires the construction of non-linear curved decision surfaces beyond the simplicity of linear planes. A nice example which may be visually inspected is the classification of intertwined spirals (see Appendix A). A corresponding classification data set with a defined number of I/O pairs for each spiral can be obtained from the CIP ExperimentalData package:

```
numberOfSingleSpiralIoPairs=30;
classificationDataSet60=
 CIP `ExperimentalData`GetSpiralsClassificationDataSet[
 numberOfSingleSpiralIoPairs];
```

The inputs of the classification data set are 2D points which may be visualized with their class assignment denoted by their colors:

```
classIndex=1;
inputsOfSpiral1=
 CIP 'DataTransformation 'GetInputsForSpecifiedClass[
 classificationDataSet60,classIndex];
classIndex=2;
inputsOfSpiral2=
 CIP 'DataTransformation 'GetInputsForSpecifiedClass[
 classificationDataSet60,classIndex];
points2DWithPlotStyle1={inputsOfSpiral1,{PointSize[0.02],Red}};
points2DWithPlotStyle2={inputsOfSpiral2,{PointSize[0.02],Green}};
points2DWithPlotStyleList={points2DWithPlotStyle1,
 points2DWithPlotStyle2};
labels={"x","y","Intertwined spirals"};
CIP 'Graphics 'PlotMultiple2dPoints[points2DWithPlotStyleList,labels]
```

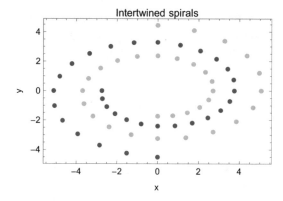

It is obvious that a highly non-linear curved decision surfaces in three dimensions is necessary to separate the two spirals for a successful classification. Thus linear MLR as a machine learning method

```
mlrInfo=CIP 'MLR 'FitMlr[classificationDataSet60];
CIP 'MLR 'ShowMlrSingleClassification[
 {"CorrectClassification"},classificationDataSet60,mlrInfo]
```

62.9% correct classifications

is clearly ruled out with only about 63% correct classifications. A non-linear method is advised and a SVM approach with an adequate kernel function

```
kernel={"Wavelet",0.1};
svmInfo=CIP 'SVM 'FitSvm[classificationDataSet60,kernel];
CIP 'SVM 'ShowSvmSingleClassification[
 {"CorrectClassification"},classificationDataSet60,svmInfo];
```

100.% correct classifications

leads to a perfect 100% correct classifications. The interpolating predictivity of the SVM classifier may be tested with an enlarged classification data set which consists of additional I/O pairs within each spiral

```
numberOfSingleSpiralIoPairs=100;
classificationDataSet200=
 CIP`ExperimentalData`GetSpiralsClassificationDataSet[
 numberOfSingleSpiralIoPairs];
classIndex=1;
inputsOfSpiral1=
 CIP`DataTransformation`GetInputsForSpecifiedClass[
 classificationDataSet200,classIndex];
classIndex=2;
inputsOfSpiral2=
 CIP`DataTransformation`GetInputsForSpecifiedClass[
 classificationDataSet200,classIndex];
points2DWithPlotStyle1={inputsOfSpiral1,{PointSize[0.02],Red}};
points2DWithPlotStyle2={inputsOfSpiral2,{PointSize[0.02],Green}};
points2DWithPlotStyleList={points2DWithPlotStyle1,
 points2DWithPlotStyle2};
CIP`Graphics`PlotMultiple2dPoints[points2DWithPlotStyleList,labels]
```

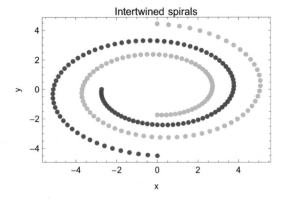

to result again in a perfect 100% classification:

```
CIP`SVM`ShowSvmSingleClassification[
 {"CorrectClassification"},classificationDataSet200,svmInfo];
```

100.% correct classifications

Therefore it can be deduced that the decision surfaces of the SVM classifier are predictive and not overfitted. The latter would also lead to an initial perfect 100% classification result but to a poor predictivity for new unknown data afterwards (this problem is discussed in detail below). This finding may be finally verified with a visual inspection of the decision surface for each class (combined with the data of the enlarged classification data set). The highly non-linear decision surface for class 1

```
classificationDataSet3DList=
 CIP`DataTransformation`TransformDataSetToMultipleDataSet[
 classificationDataSet200];
indexOfInput1=1;indexOfInput2=2;indexOfOutput=1;input={0.0,0.0};
plotStyle3D=Directive[Green,Specularity[White,40],Opacity[0.6]];
pureSvm3dFunction=Function[{x,y},
 CIP`SVM`CalculateSvm3dValue[x,y,indexOfInput1,
 indexOfInput2,indexOfOutput,input,svmInfo]];
labels={"In 1","In 2","Out 1"};
CIP`Graphics`Plot3dDataSetWithFunction[
 classificationDataSet3DList[[1]],pureSvm3dFunction,labels,
 GraphicsOptionPlotStyle3D -> plotStyle3D]
```

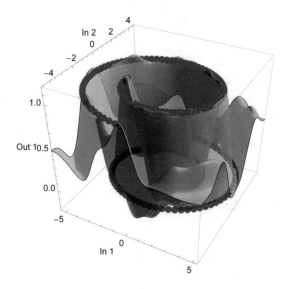

as well as the decision surface for class 2

```
indexOfInput1=1;indexOfInput2=2;indexOfOutput=2;input={0.0,0.0};
pureSvm3dFunction=Function[{x,y},
 CIP`SVM`CalculateSvm3dValue[x,y,indexOfInput1,
 indexOfInput2,indexOfOutput,input,svmInfo]];
labels={"In 1","In 2","Out 2"};
CIP`Graphics`Plot3dDataSetWithFunction[
 classificationDataSet3DList[[2]],pureSvm3dFunction,labels,
 GraphicsOptionPlotStyle3D -> plotStyle3D]
```

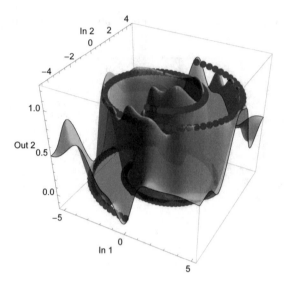

allow a perfect assignment of each input to its corresponding class. Note that perfect decision planes would perfectly model the spirals in 3D (with values to be exactly one at the spiral positions and zero elsewhere). The machine learning result is just an approximant with deviations to these ideal curves but close enough to allow for proper decisions.

4.7 Ambiguous classification

```
Clear["Global`*"];
<<CIP`CalculatedData`
<<CIP`Utility`
<<CIP`Graphics`
<<CIP`MLR`
<<CIP`SVM`
<<CIP`Cluster`
<<CIP`DataTransformation`
```

Consider the following two overlapping Gaussian clouds where each point is attributed to its class indicated by color:

```
centroidVector1={0.2,0.2};
numberOfCloudVectors=50;
standardDeviation=0.3;
cloudDefinition1={centroidVector1,numberOfCloudVectors,
  standardDeviation};
inputs1=CIP`CalculatedData`GetDefinedGaussianCloud[
  cloudDefinition1];
centroidVector2={0.8,0.8};
```

```
cloudDefinition2={centroidVector2,numberOfCloudVectors,
 standardDeviation};
inputs2=CIP 'CalculatedData 'GetDefinedGaussianCloud[
 cloudDefinition2];
points2DWithPlotStyle1={inputs1,{PointSize[0.02],Black}};
points2DWithPlotStyle2={inputs2,{PointSize[0.02],Blue}};
points2DWithPlotStyleList={points2DWithPlotStyle1,
 points2DWithPlotStyle2};
labels={"x","y","Inputs and their corresponding color classes"};
inputsGraphics=CIP 'Graphics 'PlotMultiple2dPoints[
 points2DWithPlotStyleList,labels]
```

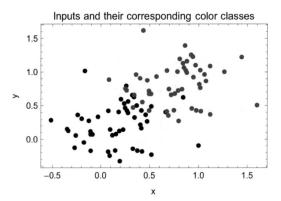

If the points are to be classified according to their class memberships a decision line must be constructed to separate the class areas. It is obvious that in the current case a perfect (100%) correct classification is not desirable since the clouds penetrate each other. So an ordinary human solution for the required separation would look like this

```
labels={"x","y","Human separation line"};
inputsGraphics=CIP 'Graphics 'PlotMultiple2dPoints[
 points2DWithPlotStyleList,labels];
lineGraphics=Graphics[{Thick,
 Yellow,Line[{{1.25,-0.3},{-0.55,1.4}}]}];
Show[inputsGraphics,lineGraphics]
```

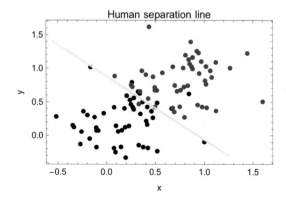

with the rule: Below the line = class 1, above the line = class 2. The decision line classifies about 90% of the points in a correct manner (just count) and this is roughly the best we can reasonably get. If a classification method performs significantly better this would be suspicious (i.e. indicate overfitting), if it performs significantly poorer the method would not be appropriate or failed due to technical reasons. To perform a classification task a classification data set is constructed from the cloud definitions

```
cloudDefinitions={cloudDefinition1,cloudDefinition2};
classificationDataSet=
 CIP`CalculatedData`GetGaussianCloudsDataSet[
 cloudDefinitions];
```

and split to a set of classification data sets with each containing one output component of the original data set

```
classificationDataSet3DList=
 CIP`DataTransformation`TransformDataSetToMultipleDataSet[
 classificationDataSet];
```

to allow graphical illustrations in the following. Again note that every I/O pair of the classification data set is attributed to its class by a 0/1 coding at the corresponding position of its output, e.g. the first I/O pair that belongs to class 1 contains a 1 at position 1 of its output and a 0 at position 2

```
firstIoPair=classificationDataSet[[1]]
```

$\{\{0.345704, 0.322742\}, \{1., 0.\}\}$

whereas the last I/O pair that belongs to class 2 contains a 0 at position 1 and a 1 at position 2:

```
lastIoPair=classificationDataSet[[Length[classificationDataSet]]]
```

$\{\{1.07406, 0.86491\}, \{0., 1.\}\}$

As a good start an unsupervised clustering-based class predictor can be constructed

```
clusterInfo=CIP`Cluster`FitCluster[classificationDataSet];
```

that achieves an overall success rate of about 90% correct predictions

```
CIP`Cluster`ShowClusterSingleClassification[
  {"CorrectClassification"},classificationDataSet,clusterInfo]
```

92.% correct classifications

and a satisfying prediction result for both classes (compare chapter 3):

```
CIP`Cluster`ShowClusterSingleClassification[
  {"CorrectClassificationPerClass"},classificationDataSet,
  clusterInfo]
```

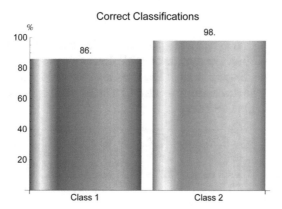

If we compare the cluster separation of the unsupervised learning result with the human straight line

```
inputs=CIP`Utility`GetInputsOfDataSet[classificationDataSet];
numberOfClusters=2;
clusterInfo=CIP`Cluster`GetFixedNumberOfClusters[inputs,
  numberOfClusters];
```

```
indexOfCluster=1;
inputsOfCluster1=CIP`Cluster`GetInputsOfCluster[inputs,
 indexOfCluster,clusterInfo];
points2DWithPlotStyle1={inputsOfCluster1,{PointSize[0.02],Green}};
indexOfCluster=2;
inputsOfCluster2=CIP`Cluster`GetInputsOfCluster[inputs,
 indexOfCluster,clusterInfo];
points2DWithPlotStyle2={inputsOfCluster2,{PointSize[0.02],Red}};
points2DWithPlotStyleList={points2DWithPlotStyle1,
 points2DWithPlotStyle2};
labels={"x","y","Clusters and the human separation line"};
clusterGraphics=CIP`Graphics`PlotMultiple2dPoints[
 points2DWithPlotStyleList,labels];
Show[clusterGraphics,lineGraphics]
```

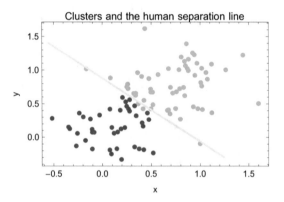

we see a very good agreement - and in practice the classification task would
be successfully fulfilled. But for the current context we proceed into the realm of
supervised learning. Since a straight line does the separation job properly the classification task may be successfully tackled by a linear MLR approach. A MLR classification results in a success rate

```
mlrInfo=CIP`MLR`FitMlr[classificationDataSet];
CIP`MLR`ShowMlrSingleClassification[
 {"CorrectClassification"},classificationDataSet,mlrInfo]
```

92.% correct classifications

of about 90% percent correct classifications as expected. The decision surface for
class 1 confirms the proper result:

```
pureMlr3dFunction=Function[{x,y},
 CalculateMlr3dValue[x,y,mlrInfo]];
labels={"In 1","In 2","Out 1"};
CIP`Graphics`Plot3dDataSetWithFunction[
 classificationDataSet3DList[[1]],pureMlr3dFunction,labels]
```

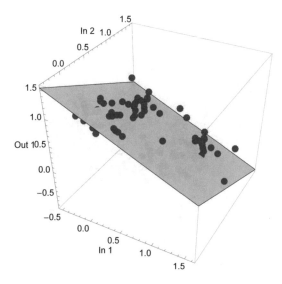

Again note that a linear technique has no structural flexibility: It simply succeeds or fails since there is nothing to be tuned. If a non-linear method like a SVM is used for the very same classification task things may get more difficult because non-linear methods are far more flexible, i.e. they allow the construction of complex and highly non-linear curved decision surfaces as already shown above. Their behavior is fundamentally guided by their structural hyperparameters - and they deserve adequate structural (as well as technical) parameters' settings to work properly. If the following inappropriate wavelet kernel function is arbitrarily chosen for the current classification task (the wavelet width parameter a is set to a very small value so the generated model function may be extremely curved)

```
kernelFunction={"Wavelet",0.05};
svmInfo=CIP`SVM`FitSvm[classificationDataSet,kernelFunction];
CIP`SVM`ShowSvmSingleClassification[
  {"CorrectClassification"},classificationDataSet,svmInfo]
```

100.% correct classifications

we get a suspicious 100% correct classifications with a SVM's decision surface for class 1

```
pureSvm3dFunction=Function[{x,y},
  CIP`SVM`CalculateSvm3dValue[x,y,svmInfo]];
CIP`Graphics`Plot3dDataSetWithFunction[
  classificationDataSet3DList[[1]],pureSvm3dFunction,labels]
```

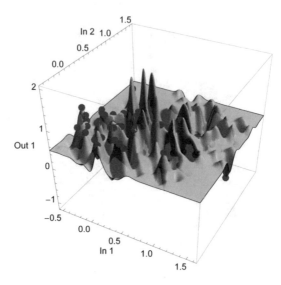

that is the mathematical analog of a pure look-up table: Every input is nearly individually classified by its own bump which means pure overfitting (again: This was possible since a very small wavelet width parameter was chosen which allowed these small bumps). This overfitted decision surface is of course suboptimal for any class predictions of new inputs. Thus the powerful non-linear method utterly failed to perform a simple linear classification task because of its inappropriate structural settings. In this situation a kind of manual tuning of the kernel function could be applied to arrive at a classification result of human quality but this is not feasible in general where visual inspection is not available so the optimum 90% classification result would not be known in advance. In general the machine learning procedure itself should be able to come to this decision. As a solution strategy it seems reasonable to facilitate the uselessness of the overfitted decision function to proceed: Therefore the original data set is split into a training set and a test set where the first is used for machine learning and the second to evaluate its predictivity after training. Then it becomes possible to monitor the performance of learning in dependence of the structural settings (here: the kernel function) used. The test set validates the training set in each step and overfitting will become apparent by a significant difference between the classification results for the training and the test set. As a start the original data set is randomly split into a training and test set of equal size

```
trainingFraction=0.50;
trainingAndTestSet=CIP`Cluster`GetRandomTrainingAndTestSet[
  classificationDataSet,trainingFraction];
trainingSet=trainingAndTestSet[[1]];
testSet=trainingAndTestSet[[2]];
trainingSet3DList=
  CIP`DataTransformation`TransformDataSetToMultipleDataSet[
  trainingSet];
```

what seems to be an unbiased and fair partitioning which may be visually controlled (with the training inputs in green/light gray and the test inputs in red/dark gray):

```
inputsOfTrainingSet=CIP`Utility`GetInputsOfDataSet[trainingSet];
inputsOfTestSet=CIP`Utility`GetInputsOfDataSet[testSet];
trainingPoints2DWithPlotStyle={inputsOfTrainingSet,
 {PointSize[0.02],Green}};
testPoints2DWithPlotStyle={inputsOfTestSet,{PointSize[0.02],Red}};
points2DWithPlotStyleList={testPoints2DWithPlotStyle,
 trainingPoints2DWithPlotStyle};
labels={"x","y","Training inputs (green), Test inputs (red)"};
inputsGraphics=CIP`Graphics`PlotMultiple2dPoints[
 points2DWithPlotStyleList,labels]
```

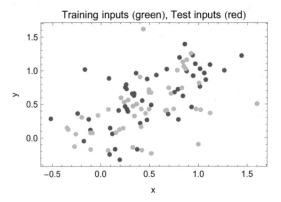

Training and test data are equally distributed over the input's space. If again a MLR classification is performed with the training set

```
mlrInfo=CIP`MLR`FitMlr[trainingSet];
CIP`MLR`ShowMlrClassificationResult[
 {"CorrectClassification"},trainingAndTestSet,mlrInfo]
```

Training Set:

94.% correct classifications

Test Set:

90.% correct classifications

we get the expected classification success rates for training and test set (training/test = 94/90) with a test set result being a bit inferior (90% < 94%) to the training sets outcome. So the MLR result is predictive. If now the SVM classification is explored with different kernel functions in a systematic manner (where the first attempt will again be the catastrophic look-up table case from above) the classification success for training and test set may be compared for each setting (with the training result in green/light gray and the test result in red/dark gray):

```
kernelFunctionList=Table[{"Wavelet",kernelParameter},
 {kernelParameter,0.05,2.0,0.05}];
svmInfoList=CIP`SVM`FitSvmSeries[trainingSet,kernelFunctionList];
svmSeriesClassificationResult=
 CIP`SVM`GetSvmSeriesClassificationResult[trainingAndTestSet,
 svmInfoList];
CIP`SVM`ShowSvmSeriesClassificationResult[
 svmSeriesClassificationResult]
```

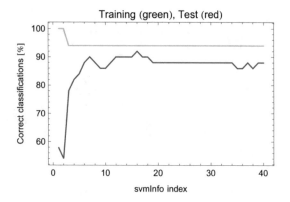

Best test set classification with svmInfo index = {16}

For settings with a small index the classification obviously takes place in the realm of overfitting with distinct differences between training and test set performance (100% training success but a clear test failure). For the following settings the SVM arrives at the expected classification quality of the human solution. The best SVM solution with the highest test set predictivity

```
svmInfo=svmInfoList[[16]];
CIP`SVM`GetKernelFunction[svmInfo]
```

{Wavelet,0.8}

yields a classification result

```
CIP`SVM`ShowSvmClassificationResult[{"CorrectClassification"},
 trainingAndTestSet,svmInfo]
```

Training Set:

94.% correct classifications

Test Set:

92.% correct classifications

with a little superior test set predictivity (training/test = 94/92) compared to the MLR result (training/test = 94/90) above (2% or one I/O pair more is correctly classified). But this difference is irrelevant: Both methods perform equally well on the data set. As far as optimum solutions are concerned the classification task is successfully tackled. But with regard to the outlined strategy of partitioning the data set into a training and a test set the findings of chapter 3 could be taken into account: There it was demonstrated that cluster representatives (abbreviated CR in the following) are a promising description of an input's space spatial diversity. In some cases CRs are similar to their random brothers but in general they are often superior. If the data set is again split half by half into a training and a test set on the basis of CRs of the I/O pair's inputs

```
trainingFraction=0.5;
trainingAndTestSet=CIP`Cluster`GetClusterBasedTrainingAndTestSet[
 classificationDataSet,trainingFraction];
trainingSet=trainingAndTestSet[[1]];
testSet=trainingAndTestSet[[2]];
trainingSet3DList=
 CIP`DataTransformation`TransformDataSetToMultipleDataSet[
 trainingSet];
```

the result seems to be not much different from the random partitioning above:

```
inputsOfTrainingSet=CIP`Utility`GetInputsOfDataSet[trainingSet];
inputsOfTestSet=CIP`Utility`GetInputsOfDataSet[testSet];
trainingPoints2DWithPlotStyle={inputsOfTrainingSet,
 {PointSize[0.02],Green}};
testPoints2DWithPlotStyle={inputsOfTestSet,{PointSize[0.02],Red}};
points2DWithPlotStyleList={testPoints2DWithPlotStyle,
 trainingPoints2DWithPlotStyle};
labels={"x","y","Training inputs (green), Test inputs (red)"};
inputsGraphics=CIP`Graphics`PlotMultiple2dPoints[
 points2DWithPlotStyleList,labels]
```

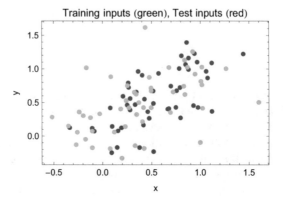

But if the same kernel function settings are scanned as before

```
kernelFunctionList=Table[{"Wavelet",kernelParameter},
 {kernelParameter,0.05,2.0,0.05}];
svmInfoList=CIP`SVM`FitSvmSeries[trainingSet,kernelFunctionList];
svmSeriesClassificationResult=
 CIP`SVM`GetSvmSeriesClassificationResult[trainingAndTestSet,
 svmInfoList];
CIP`SVM`ShowSvmSeriesClassificationResult[
 svmSeriesClassificationResult]
```

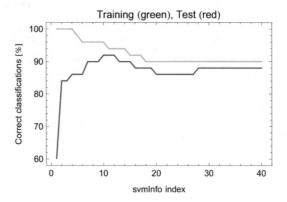

Best test set classification with svmInfo index = {10, 11, 12}

a kind of improvement can be recognized: The training and test set curves are closer together over a wide range of kernel function settings which indicates a more similar distribution of the training and test set's inputs. The best SVM settings like

```
svmInfo=svmInfoList[[11]];
CIP`SVM`GetKernelFunction[svmInfo]
```

{Wavelet, 0.55}

perform equally well

```
CIP`SVM`ShowSvmClassificationResult[{"CorrectClassification"},
 trainingAndTestSet,svmInfo]
```

Training Set:

94.% correct classifications

Test Set:

92.% correct classifications

in comparison to their best predecessor before (which also arrived at training/test = 94/92). A plot of a best SVM's decision surface for class 1

```
pureSvm3dFunction=Function[{x,y},
 CIP`SVM`CalculateSvm3dValue[x,y,svmInfo]];
labels={"In 1","In 2","Out 1"};
CIP`Graphics`Plot3dDataSetWithFunction[trainingSet3DList[[1]],
 pureSvm3dFunction,labels]
```

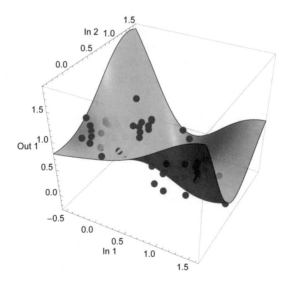

shows its similarity to a simple plane in the data region. From this latter (crucial) point of view the CR based strategy was no real improvement beyond mere cosmetics because the random selection based strategy already achieved the optimum reasonable outcome of the classification procedure. But the more uniform behavior of training and test set may be crucial for more challenging machine learning tasks. The successful outlined validation procedure by partitioning data sets in training and test sets will be explored more thoroughly next.

4.8 Training and test set partitioning

The validation strategy introduced in the last section with a partitioning of a data set into a training and a test set rises two questions:

- *Question 1*: How should a single I/O pair for each set be selected?
- *Question 2*: How many I/O pairs should each set contain after partitioning?

General guidelines that address these questions may read as follows:

- *Guideline 1*: Both sets should cover a similar input's space, i.e. possess a similar spatial diversity of inputs.
- *Guideline 2*: The training set should be kept as small as possible but allow for a high overall predictivity.

Guideline 1 may be taken into account by using a cluster based approach to get representatives for the training set (compare chapter 3). Then the issue remains which individual cluster member is to be taken as the representative for the training set. Unfortunately this leads to an extremely difficult optimization task: Think about a small data set with just 100 I/O pairs that is partitioned in a training set of 25 I/O pairs and a test set of 75 I/O pairs respectively. If a cluster based approach is used the I/O pair's inputs are split into 25 groups with 4 I/O pairs on average. If now one member of each cluster is chosen for the training set this evaluates to at least one quadrillion

```
ScientificForm[4.0^25,1]
```

$1. \times 10^{15}$

possible different training and test sets just for this small training fraction of 0.25 (in detail: There are 4 choices for a member of the first cluster. Each of these choices may be combined with the 4 choices of the next cluster which evaluates to $4 \times 4 = 16$ combination possibilities for two clusters. Thus 25 clusters evaluate to $4^{25} \approx 1.000.000.000.000.000$ possible different training sets). There is no practically feasible way to evaluate the optimum training set within this vast number of possibilities. Even an evolutionary algorithm based strategy would be computationally far too demanding in most cases. Therefore only heuristic partitioning strategies and optimization approaches may be applied which by no means guarantee to achieve an even tolerable selection. Heuristic strategies are guided by apparently reasonable ideas for selecting or optimizing representatives. The latter usually involves only a few trial steps for optimization. The heuristic strategies of the previous section were a straightforward "50:50/random strategy" at first (i.e. the whole data set was split into a training and test set of equal size where each I/O pair was randomly chosen to belong to the training or test set respectively) and a more elaborate "50:50/cluster representatives (CR) strategy" afterwards (which used cluster representatives for I/O pair selection). More intricate heuristics are outlined throughout this section. As a start the general superiority of a CR based selection is illustrated next with a more difficult classification example.

4.8.1 Cluster representatives based selection

```
Clear["Global`*"];
<<CIP `CalculatedData`
<<CIP `Utility`
<<CIP `Graphics`
<<CIP `Cluster`
<<CIP `SVM`
```

Consider the following inputs with their corresponding color classes:

```
centroid1={0.3,0.7};
standardDeviation=0.05;
numberOfCloudInputs=60;
cloudDefinition1={centroid1,numberOfCloudInputs,standardDeviation};
inputs1=CIP `CalculatedData`GetDefinedGaussianCloud[
 cloudDefinition1];
points2DWithPlotStyle1={inputs1,{PointSize[0.02],Black}};
centroid2={0.7,0.3};cloudDefinition2={centroid2,numberOfCloudInputs,
 standardDeviation};
inputs2=CIP `CalculatedData`GetDefinedGaussianCloud[
 cloudDefinition2];
points2DWithPlotStyle2={inputs2,{PointSize[0.02],Blue}};
centroid3={0.5,0.5};
standardDeviation=0.05;
numberOfCloudInputs=10;
cloudDefinition3={centroid3,numberOfCloudInputs,standardDeviation};
inputs3=CIP `CalculatedData`GetDefinedGaussianCloud[
 cloudDefinition3];
points2DWithPlotStyle3={inputs3,{PointSize[0.02],Orange}};
centroid4={0.8,0.8};
cloudDefinition4={centroid4,numberOfCloudInputs,standardDeviation};
inputs4=CIP `CalculatedData`GetDefinedGaussianCloud[
 cloudDefinition4];
points2DWithPlotStyle4={inputs4,{PointSize[0.02],Yellow}};
centroid5={0.2,0.2};
cloudDefinition5={centroid5,numberOfCloudInputs,standardDeviation};
inputs5=CIP `CalculatedData`GetDefinedGaussianCloud[
 cloudDefinition5];
points2DWithPlotStyle5={inputs5,{PointSize[0.02],Pink}};
points2DWithPlotStyleList={points2DWithPlotStyle1,
 points2DWithPlotStyle2,points2DWithPlotStyle3,
 points2DWithPlotStyle4,points2DWithPlotStyle5};
```

```
labels={"x","y","Inputs with corresponding color classes"};
CIP`Graphics`PlotMultiple2dPoints[points2DWithPlotStyleList,labels]
```

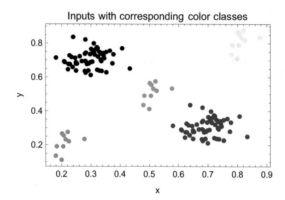

By visual inspection it is expected that a successful machine learning approach should yield a 100% classification success since the clouds can be unambiguously separated. Since a 100% success rate can (almost) always be achieved by a non-linear method as shown in the previous section a validation procedure is crucial to separate 100% overfitting from 100% predictivity. After generation of the corresponding classification data set

```
cloudDefinitions={cloudDefinition1,cloudDefinition2,
  cloudDefinition3,cloudDefinition4,cloudDefinition5};
classificationDataSet=
CIP`CalculatedData`GetGaussianCloudsDataSet[
  cloudDefinitions];
```

and selection of a comparatively small randomly chosen training set with only 25% of the original data set's I/O pairs

```
trainingFraction=0.25;
trainingAndTestSet=CIP`Cluster`GetRandomTrainingAndTestSet[
  classificationDataSet,trainingFraction];
trainingSet=trainingAndTestSet[[1]];
testSet=trainingAndTestSet[[2]];
```

it becomes visible that the input's space if no longer satisfactorily covered (with the training inputs in green/light gray and the test inputs in red/dark gray):

```
inputsOfTrainingSet=CIP`Utility`GetInputsOfDataSet[trainingSet];
inputsOfTestSet=CIP`Utility`GetInputsOfDataSet[testSet];
trainingPoints2DWithPlotStyle={inputsOfTrainingSet,
  {PointSize[0.02],Green}};
```

```
testPoints2DWithPlotStyle={inputsOfTestSet,{PointSize[0.02],Red}};
points2DWithPlotStyleList={testPoints2DWithPlotStyle,
 trainingPoints2DWithPlotStyle};
labels={"x","y","Training inputs (green), Test inputs (red)"};
inputsGraphics=CIP'Graphics'PlotMultiple2dPoints[
 points2DWithPlotStyleList,labels]
```

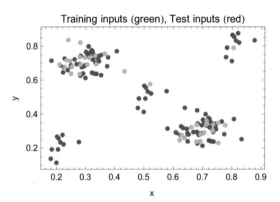

The training inputs are concentrated in the two bigger clusters whereas the inputs of the three small clusters are under-represented. A systematic exploration of kernel function settings for a SVM based machine learning approach as in the previous section (with the training result in green/light gray and the test result in red/dark gray)

```
kernelFunctionList=Table[{"Wavelet",kernelParameter},
 {kernelParameter,0.05,1.0,0.05}];
svmInfoList=CIP'SVM'FitSvmSeries[trainingSet,kernelFunctionList];
svmSeriesClassificationResult=
 CIP'SVM'GetSvmSeriesClassificationResult[trainingAndTestSet,
 svmInfoList];
CIP'SVM'ShowSvmSeriesClassificationResult[
 svmSeriesClassificationResult]
```

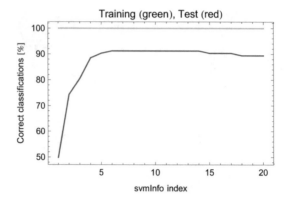

Best test set classification with svmInfo index = $\{6,7,8,9,10,11,12,13,14\}$

shows a distinct difference between the training and test set's results: Whereas the training set always yields the expected 100% success rate (no matter overfitted or not) the test set never comes close: The best predictivity is around 90% only. Note that this can not be attributed to a deficiency of the machine learning method: The learning procedure itself relies on the training data. If parts of the input's space are not covered and thus not trained they are simply unknown to the decision surfaces so they yield an arbitrary result for corresponding inputs. A training and test set selection based on cluster representatives (CRs) with the same training fraction

```
trainingFraction=0.25;
trainingAndTestSet=
 CIP`Cluster`GetClusterBasedTrainingAndTestSet[
 classificationDataSet,trainingFraction];
trainingSet=trainingAndTestSet[[1]];
testSet=trainingAndTestSet[[2]];
inputsOfTrainingSet=CIP`Utility`GetInputsOfDataSet[trainingSet];
inputsOfTestSet=CIP`Utility`GetInputsOfDataSet[testSet];
trainingPoints2DWithPlotStyle={inputsOfTrainingSet,
 {PointSize[0.02],Green}};
testPoints2DWithPlotStyle={inputsOfTestSet,{PointSize[0.02],Red}};
points2DWithPlotStyleList={testPoints2DWithPlotStyle,
 trainingPoints2DWithPlotStyle};
labels={"x","y","Training inputs (green), Test inputs (red)"};
inputsGraphics=CIP`Graphics`PlotMultiple2dPoints[
 points2DWithPlotStyleList,labels]
```

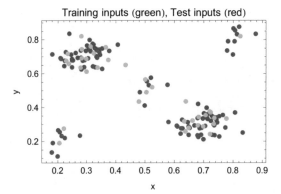

reduces this problem as shown: The input's space coverage is obviously improved. A repeated exploration with the same kernel function settings

```
kernelFunctionList=Table[{"Wavelet",kernelParameter},
 {kernelParameter,0.05,1.0,0.05}];
svmInfoList=CIP`SVM`FitSvmSeries[trainingSet,kernelFunctionList];
svmSeriesClassificationResult=
 CIP`SVM`GetSvmSeriesClassificationResult[trainingAndTestSet,
 svmInfoList];
CIP`SVM`ShowSvmSeriesClassificationResult[
 svmSeriesClassificationResult]
```

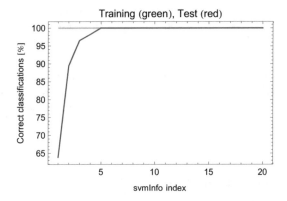

Best test set classification with svmInfo index = {5,6,7,8,9,10,11,12,13,14,15,16,17,18,19,20}

now leads to a satisfactory 100% success rate for the training as well as the test set for quite a number of different kernel functions after some overfitting settings at the beginning. Finally the necessary minimum size of the training set can be explored by use of a good SVM with an appropriate kernel function

```
kernelFunction={"Wavelet",0.55};
trainingFractionList=Table[trainingFraction,
 {trainingFraction,0.05,0.50,0.05}];
svmClassificationScan=
 CIP`SVM`ScanClassTrainingWithSvm[
 classificationDataSet,kernelFunction,trainingFractionList];
CIP`SVM`ShowSvmClassificationScan[
 svmClassificationScan]
```

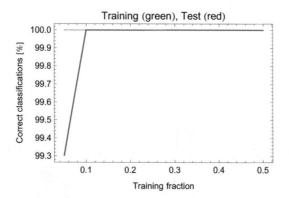

Best test set classification with index = {2,3,4,5,6,7,8,9,10}

The index mentioned in "Best test set classification ... " refers to svmClassificationScan, i.e. svmClassification-
Scan[[2]] corresponds to trainingFractionList[[2]] with a value of 0.1 (= 10%).

where it is found that a CR based training set size of only 10% of the original
data set is necessary to lead to a 100% success rate for the training as well as the
test set classifications. This result addresses the second question above about the
necessary minimum size of a training set with the highest overall predictivity for
this classification task.

4.8.2 Iris flower classification revisited

```
Clear["Global`*"];
<<CIP`ExperimentalData`
<<CIP`Cluster`
<<CIP`MLR`
<<CIP`Perceptron`
```

The prediction of iris flower species from their sepal and petal size data

```
classificationDataSet=
  CIP'ExperimentalData'GetIrisFlowerClassificationDataSet[];
```

was already discussed in chapter 3 on the basis of unsupervised learning. There it was shown that a purely clustering-based class predictor

```
clusterInfo=CIP'Cluster'FitCluster[classificationDataSet];
```

leads to an overall success rate of about 90%

```
CIP'Cluster'ShowClusterSingleClassification[
  {"CorrectClassification"},classificationDataSet,clusterInfo]
```

89.3% correct classifications

with distinctly different success rates for the three iris flower species (classes):

```
CIP'Cluster'ShowClusterSingleClassification[
  {"CorrectClassificationPerClass"},classificationDataSet,
  clusterInfo]
```

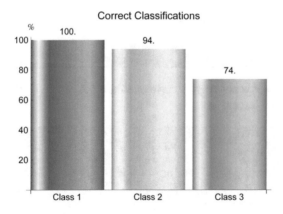

This result is not overall satisfactory so a supervised learning approach may be worth an attempt. Since a MLR based classification is not prone to overfitting it may be directly tried for a large variety of CR based training and test set sizes:

```
trainingFractionList=Table[trainingFraction,
  {trainingFraction,0.01,0.90,0.01}];
mlrClassificationScan=
  CIP'MLR'ScanClassTrainingWithMlr[
```

```
classificationDataSet,trainingFractionList];
CIP`MLR`ShowMlrClassificationScan[
 mlrClassificationScan]
```

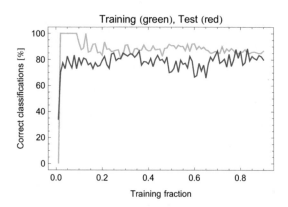

Best test set classification with index = {83}

The result is not promising and reveals an even reduced predictivity in comparison to the purely clustering-based class predictor before. Therefore a non-linear method is clearly indicated. A three-layer perceptron with only two hidden neurons

```
numberOfHiddenNeurons=2;
```

is chosen since it is not very prone to overfitting. A scan with different small CR based training and test set sizes

```
trainingFractionList=Table[trainingFraction,
 {trainingFraction,0.01,0.30,0.01}];
perceptronClassificationScan=
 CIP`Perceptron`ScanClassTrainingWithPerceptron[
 classificationDataSet,numberOfHiddenNeurons,trainingFractionList];
CIP`Perceptron`ShowPerceptronClassificationScan[
 perceptronClassificationScan]
```

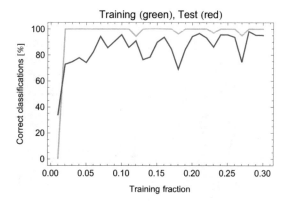

Best test set classification with index = {28}

reveals a training fraction of 0.28 (i.e. a training set with 28% of all I/O pairs of the data set) to be exceptionally predictive:

```
index=28;
trainingAndTestSetsInfo=
 perceptronClassificationScan[[1]];
trainingAndTestSet=trainingAndTestSetsInfo[[index,1]];
perceptronInfo=trainingAndTestSetsInfo[[index,2]];
```

The overall prediction success rate arrives at

```
CIP`Perceptron`ShowPerceptronClassificationResult[
  {"CorrectClassification"},trainingAndTestSet,perceptronInfo]
```

Training Set:

100.% correct classifications

Test Set:

98.1% correct classifications

with iris flower species (class) related success rates of

```
CIP`Perceptron`ShowPerceptronClassificationResult[
  {"CorrectClassificationPerClass","WrongClassificationPairs"},
  trainingAndTestSet,perceptronInfo]
```

Training Set:

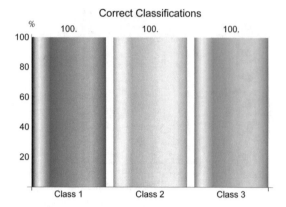

All I/O pairs are correctly classified

Test Set:

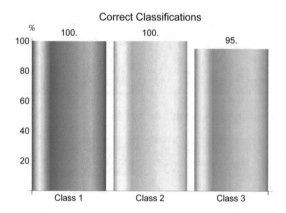

Wrong I/O pair index = 69; input = {63., 28., 51., 15.}; class desired/machine = 3 / 2
Wrong I/O pair index = 82; input = {61., 26., 56., 14.}; class desired/machine = 3 / 2

This result is a distinct advantage to the pure clustering-based class predictor be-fore. A small training set with less than a third of the total I/O pairs leads to a high test set predictivity with only two classification errors. Note that both misclassified inputs are closely neighbored in the input's space. With the obtained supervised learning based class predictor an acceptable solution to the classification seems to be found. The only dissatisfying incidence of the latter approach is the apparent hop-ping behavior of the test set predictivity obvious in the training set size scan above. If training fraction 0.27 is chosen instead of 0.28 with nearly the same training set size

```
index=27;
trainingAndTestSetsInfo=
  perceptronClassificationScan[[1]];
trainingAndTestSet=trainingAndTestSetsInfo[[index,1]];
perceptronInfo=trainingAndTestSetsInfo[[index,2]];
```

the overall predictivity

```
CIP`Perceptron`ShowPerceptronClassificationResult[
  {"CorrectClassification"},trainingAndTestSet,perceptronInfo]
```

Training Set:

95.% correct classifications

Test Set:

74.5% correct classifications

as well as the class related predictivity

```
CIP`Perceptron`ShowPerceptronClassificationResult[
  {"CorrectClassificationPerClass"},trainingAndTestSet,
  perceptronInfo]
```

Training Set:

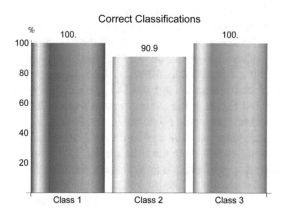

Test Set:

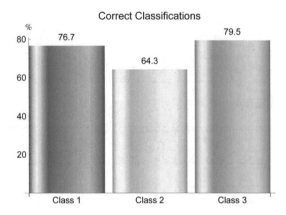

drops even below the predictivity level of the purely clustering-based class predictor. Thus the CR based training set selection led to a very unfavorable training set for training fraction 0.27 but to a highly predictive one for training fraction 0.28. This kind of sensitivity of a machine learning result to comparatively small parameter changes is a true burden of the methods in question and quite often encountered. In the current case the cluster representatives for the training set cover the input's space appropriately but are not the best choices for a good predictivity. A heuristic strategy to improve could be the following: I/O pairs are exchanged between the training and the test set since a swap of I/O pairs leaves the training and test set sizes unchanged. Candidates for this swapping procedure would be test set I/O pairs whose outputs are only poorly predicted with training set I/O pairs in use. After an exchange the machine learning fit is repeated and the deviations of the I/O pairs of the test set are re-evaluated. With this updated information at hand the next swap can be prepared etc. But a repeated unconstrained exchange could easily decrease the spatial diversity of the training and test sets I/O pair's inputs, e.g. the test set could shrink to only a small region of the I/O pairs input's space: Thus a high predictivity on the optimized test set would by no means imply a good general predictivity. To avoid or at least reduce possible spatial diversity losses an exchange could be confined to I/O pairs that belong to the same cluster. CIP provides a small number of these spatial diversity preserving training set optimization strategies. The default heuristics is abbreviated SingleGlobalMax: The single global test set I/O pair with the maximum deviation between its output and the machine prediction is chosen and exchanged with the current training set I/O pair of its cluster. Then the machine learning process is repeated and re-evaluated. This iteration is performed for a specified number of optimization steps. If the SingleGlobalMax training set optimization strategy is applied to the poor training fraction of 0.27 with the CR based training set from above as the first step

```
trainingFraction=0.27;
numberOfTrainingSetOptimizationSteps=20;
perceptronTrainOptimization=
 CIP`Perceptron`GetPerceptronTrainOptimization[
```

```
classificationDataSet,numberOfHiddenNeurons,trainingFraction,
numberOfTrainingSetOptimizationSteps];
CIP`Perceptron`ShowPerceptronTrainOptimization[
perceptronTrainOptimization]
```

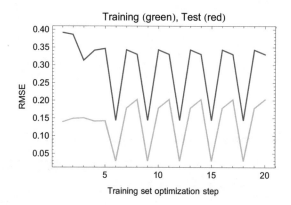

an improvement due to a decreased RMSE is obvious - but another problem occurred: The training set optimization is trapped in oscillations after step 6. This problem may be largely suppressed by a so-called blacklist modification of the heuristics
where an I/O pair is not allowed to be exchanged twice for a specified number of
optimization steps (in this case equal to the blacklist length):

```
numberOfTrainingSetOptimizationSteps=20;
blackListLength=20;
perceptronTrainOptimization=
 CIP`Perceptron`GetPerceptronTrainOptimization[
 classificationDataSet,numberOfHiddenNeurons,trainingFraction,
 numberOfTrainingSetOptimizationSteps,
 UtilityOptionBlackListLength -> blackListLength];
CIP`Perceptron`ShowPerceptronTrainOptimization[
 perceptronTrainOptimization]
```

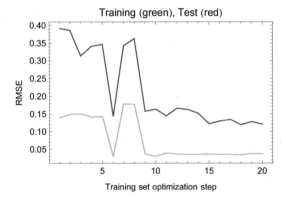

Now a considerable improvement is obtained and the best training set

```
bestIndex=
 CIP'Perceptron'GetBestPerceptronClassOptimization[
 perceptronTrainOptimization];
trainingAndTestSet=
 perceptronTrainOptimization[[3,bestIndex]];
perceptronInfo=
 perceptronTrainOptimization[[4,bestIndex]];
CIP'Perceptron'ShowPerceptronClassificationResult[
 {"CorrectClassification"},trainingAndTestSet,perceptronInfo]
```

Training Set:

100.% correct classifications

Test Set:

99.1% correct classifications

leads to class predictions with a satisfactory success rate. If the described training
and test set optimization is applied to the whole training fraction scan before

```
trainingFractionList=Table[trainingFraction,
 {trainingFraction,0.01,0.30,0.01}];
numberOfTrainingSetOptimizationSteps=20;
blackListLength=20;
perceptronClassificationScan=
 CIP'Perceptron'ScanClassTrainingWithPerceptron[
 classificationDataSet,numberOfHiddenNeurons,trainingFractionList,
 UtilityOptionOptimizationSteps ->
 numberOfTrainingSetOptimizationSteps,
 UtilityOptionBlackListLength -> blackListLength];
CIP'Perceptron'ShowPerceptronClassificationScan[
 perceptronClassificationScan]
```

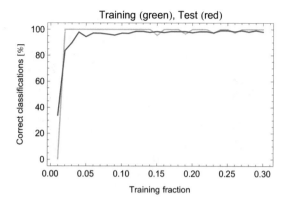

Best test set classification with index = {24, 25, 27, 29}

a dramatically improved picture is the result with smooth and high correct class prediction rates. The minimum training fraction with the highest detected test set prediction rate

```
bestIndex=24;
trainingFractionList[[bestIndex]]
```

0.24

performs excellent on the whole

```
trainingAndTestSetsInfo=
  perceptronClassificationScan[[1]];
trainingAndTestSet=trainingAndTestSetsInfo[[bestIndex,1]];
perceptronInfo=trainingAndTestSetsInfo[[bestIndex,2]];
CIP`Perceptron`ShowPerceptronClassificationResult[
  {"CorrectClassification"},trainingAndTestSet,perceptronInfo]
```

Training Set:

100.% correct classifications

Test Set:

99.1% correct classifications

as well as class specific

```
CIP`Perceptron`ShowPerceptronClassificationResult[
  {"CorrectClassificationPerClass","WrongClassificationPairs"},
  trainingAndTestSet,perceptronInfo]
```

Training Set:

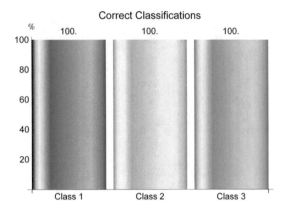

All I/O pairs are correctly classified

Test Set:

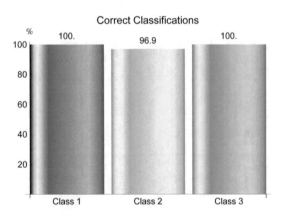

Wrong I/O pair index = 105; input = {60., 27., 51., 16.}; class desired/machine = 2 / 3

and is even smaller than the earlier top training fraction of 0.28. It is worth to note that the successful sketched training set optimization strategy must not be successful at all in general: There is no guarantee to improve - it is just heuristics. But again: A systematic training and test set enumeration and evaluation would inevitably fail due to the practically infinite number of combinations. A training rate of 0.24 for 150 iris flower I/O pairs means 36 clusters with 4 to 5 members each. If one member of each cluster is chosen for the training set this evaluates to at least

```
ScientificForm[4.0^36,1]
```

$5. \times 10^{21}$

possible different training and test sets just for this training fraction. Short cut heuristics are the only promising alternatives - and also the latter training fraction scan with 30 training fractions - each with 20 training set optimization steps and 3 perceptron fits for the 3 class output components - required already 1800 perceptron fits. For difficult machine learning problems one single fit may require hours up to days so the generation of the sketched training fraction scan would require months of CPU time on a single computer. On the other hand this kind of computation is particularly eligible for parallel architectures since there is mostly only a loose dependence of the individual perceptron fits (see Appendix A for corresponding parallelized calculations with CIP). This is why multicore architectures and grid computing play a major role in professional machine learning setups. Finally the ruled out MLR approach from the beginning of this subsection may be considered again and also enhanced by training set optimization:

```
trainingFractionList=Table[trainingFraction,
  {trainingFraction,0.01,0.90,0.01}];
numberOfTrainingSetOptimizationSteps=20;
blackListLength=20;
mlrClassificationScan=
  CIP `MLR`ScanClassTrainingWithMlr[
  classificationDataSet,trainingFractionList,
  UtilityOptionOptimizationSteps ->
  numberOfTrainingSetOptimizationSteps,
  UtilityOptionBlackListLength -> blackListLength];
CIP `MLR`ShowMlrClassificationScan[
mlrClassificationScan]
```

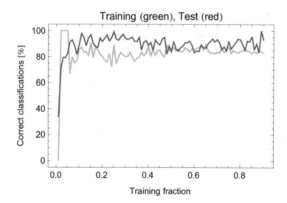

Best test set classification with index = {89}

But although the training fraction scan is clearly improved the linear MLR results are still significantly inferior to the non-linear perceptron predictions.

4.8.3 Adhesive kinetics regression revisited

```
Clear["Global`*"];
<< CIP`ExperimentalData`
<< CIP`Graphics`
<< CIP`MPR`
```

The adhesive kinetics data have already been successfully modelled by non-linear machine learning techniques. A specific validation procedure was not necessary for an assessment of the quality of the fit results since the experimental errors were known in advance and could be compared to the machine errors. In addition a visual 3D inspection was possible with a subset of I/O pairs. On the other hand it may be interesting to explore the minimum training set for this regression task. This is expected to be difficult since the data set

```
dataSet = CIP`ExperimentalData`GetAdhesiveKineticsDataSet[];
```

is only small

```
CIP`Graphics`ShowDataSetInfo[{"IoPairs"}, dataSet]
```

Number of IO pairs = 73

and the outputs' errors are reported to be in the order of 10 to 20%, i.e. the output values are at the borderline to be only semi-quantitative. So any alteration in an optimization procedure may cause drastic effects and may be hard to be judged. As shown before a quadratic-log MPR approach to the whole data set

```
polynomialDegree = 2;
dataTransformationMode = "Log";
quadraticLogMprInfo =
 CIP`MPR`FitMpr[dataSet, polynomialDegree,
  MprOptionDataTransformationMode -> dataTransformationMode];
CIP`MPR`ShowMprSingleRegression[
 {"RMSE"}, dataSet, quadraticLogMprInfo];
```

Root mean squared error (RMSE) = 2.607×10^2

leads to a satisfying fitting result with a RMSE of about 260 and acceptable residuals distribution. A scan of different CR based splits of the whole data set into a training and a test set with a quadratic-log MPR training

```
trainingFractionList =
 Table[trainingFraction, {trainingFraction, 0.3, 0.8, 0.05}];
quadraticLogMprRegressionScan =
 CIP`MPR`ScanRegressTrainingWithMpr[dataSet, polynomialDegree,
```

```
    trainingFractionList,
    MprOptionDataTransformationMode -> dataTransformationMode];
CIP`MPR`ShowMprRegressionScan[quadraticLogMprRegressionScan]
```

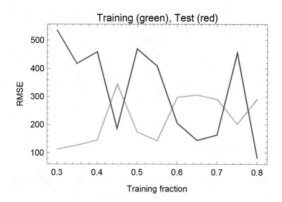

Best test set regression with index = {11}

which may be improved with a representative selection from each cluster by an
addition of several training set optimization steps (again with the SingleGlobalMax
heuristics and adequate blacklisting) for each training fraction of the quadratic-log
MPR scan

```
numberOfTrainingSetOptimizationSteps = 20;
blackListLength = 15;
quadraticLogMprRegressionScan =
  CIP`MPR`ScanRegressTrainingWithMpr[dataSet, polynomialDegree,
    trainingFractionList,
    MprOptionDataTransformationMode -> dataTransformationMode,
    UtilityOptionOptimizationSteps ->
    numberOfTrainingSetOptimizationSteps,
    UtilityOptionBlackListLength -> blackListLength];
CIP`MPR`ShowMprRegressionScan[quadraticLogMprRegressionScan]
```

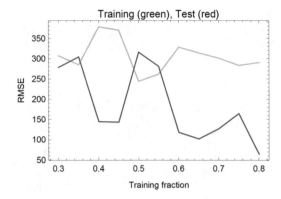

Best test set regression with index = {11}

indicates that already a small training fraction seems to be sufficient for successful model generation, e.g. a training fraction of only 0.4

```
index = 3;
trainingFractionList[[index]]
```

0.4

leads to a model with satisfactory residuals distribution for the whole (!) data set

```
trainingAndTestSetsInfo = quadraticLogMprRegressionScan[[1]];
trainingAndTestSet = trainingAndTestSetsInfo[[index, 1]];
quadraticLogMprFractionInfo = trainingAndTestSetsInfo[[index, 2]];
CIP`MPR`ShowMprSingleRegression[{"ModelVsDataPlot",
  "RelativeSortedResidualsPlot", "RelativeResidualsStatistics",
  "RMSE"}, dataSet, quadraticLogMprFractionInfo];
```

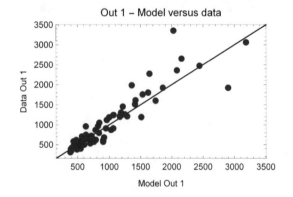

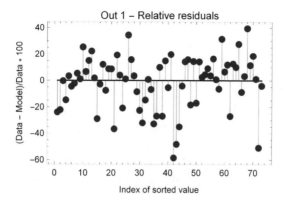

Out 1 : Relative residuals (100*(Data - Model)/Data):

Mean/Median/Maximum Value in % = 1.54×10^1 / 1.27×10^1 / 5.85×10^1

Root mean squared error (RMSE) = 2.637×10^2

which is very similar to the result of the quadratic-log MPR fit of the whole data set. This similarity may be confirmed by visual inspection of the overlay of both models:

```
polymerMassRatio = "80";
dataSet3D =
 CIP`ExperimentalData`GetAdhesiveKinetics3dDataSet[polymerMassRatio];
indexOfInput1 = 2;
indexOfInput2 = 3;
indexOfOutput = 1;
input = {80.0, 0.0, 0.0};
labels = {"In 2", "In 3", "Out 1"};
pureQuadraticLogMpr3dFunction =
 Function[{x, y},
 CIP`MPR`CalculateMpr3dValue[x, y, indexOfInput1, indexOfInput2,
  indexOfOutput, input, quadraticLogMprInfo]];
quadraticLogMpr3dPlot =
 CIP`Graphics`Plot3dDataSetWithFunction[dataSet3D,
  pureQuadraticLogMpr3dFunction, labels];
pureQuadraticLogMprFraction3dFunction =
 Function[{x, y},
 CIP`MPR`CalculateMpr3dValue[x, y, indexOfInput1, indexOfInput2,
  indexOfOutput, input, quadraticLogMprFractionInfo]];
quadraticLogMprFraction3dPlot =
 CIP`Graphics`Plot3dDataSetWithFunction[dataSet3D,
  pureQuadraticLogMprFraction3dFunction, labels];
Show[{quadraticLogMpr3dPlot, quadraticLogMprFraction3dPlot}]
```

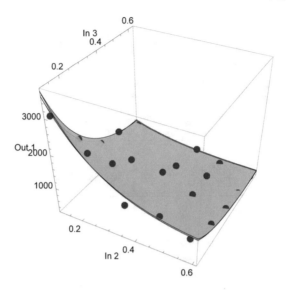

In summary this finding means that less than half of the adhesive kinetics data are in fact necessary to extract a satisfying model function (i.e. more than 50% of the lab work was not necessary in principle) - but such a clear assessment would be hard to make without visual inspection and an a priori good fit as a guideline.

4.8.4 Design of experiment

```
Clear["Global`*"];
<<CIP`CalculatedData`
<<CIP`Graphics`
<<CIP`SVM`
<<CIP`Cluster`
```

Scientific lab work usually generates data on the basis of a design of experiment (DoE), i.e. the setups and conditions for measurement are not chosen randomly but follow theoretical considerations or a specific systematics, e.g. derived from mathematical statistics. The DoE may be utilized to extract an optimum training set from the full data by an adequate procedure. For an example the inputs' locations for the chapter's introductory regression task may be chosen to form a two-dimensional grid, e.g. a 19×19 grid

```
pureOriginalFunction=Function[{x,y},
  1.9*(1.35+Exp[x]*Sin[13.0*(x-0.6)^2]*Exp[-y]* Sin[7.0*y])];
xRange={0.0,1.0};
yRange={0.0,1.0};
numberOfDataPerDimension=19;
```

```
standardDeviationRange={0.1,0.1};
dataSet3D=CIP`CalculatedData`Get3dFunctionBasedDataSet[
 pureOriginalFunction,xRange,yRange,numberOfDataPerDimension,
 standardDeviationRange];
labels={"x","y","z"};
viewPoint3D={0.0,0.0,3.0};
CIP`Graphics`Plot3dDataSet[dataSet3D,labels,
 GraphicsOptionViewPoint3D -> viewPoint3D]
```

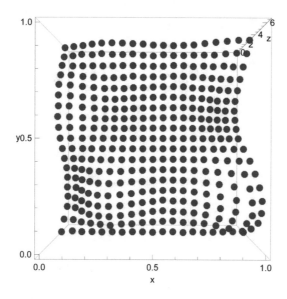

where the grid character becomes obvious by 3D inspection of the data set from above. As a rational choice for an adequate training set the extraction of a sub grid with a sufficient density is near at hand, e.g. a 10×10 sub grid:

```
numberOfPointsPerDimension=19;
trainingSet={};
testSet={};
Do[
 Do[
  If[OddQ[i] && OddQ[k],
  AppendTo[trainingSet,
   dataSet3D[[(i-1)*numberOfPointsPerDimension+k]]],
  AppendTo[testSet,
   dataSet3D[[(i-1)*numberOfPointsPerDimension+k]]],
  ];
  dataSet3D,
  {k,numberOfPointsPerDimension}
 ],
 {i,numberOfPointsPerDimension}
];
viewPoint3D={0.0,0.0,3.0};
CIP`Graphics`Plot3dDataSet[trainingSet,labels,
 GraphicsOptionViewPoint3D -> viewPoint3D]
```

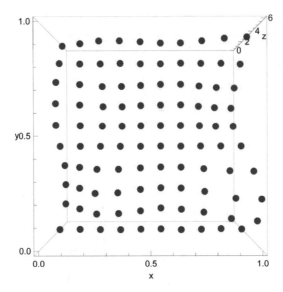

The remaining I/O pairs are accordingly assigned to the test set:

```
CIP `Graphics`Plot3dDataSet[testSet,labels,
  GraphicsOptionViewPoint3D -> viewPoint3D]
```

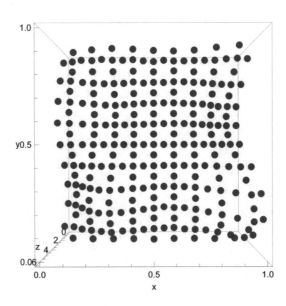

This choice seems to be an optimum with regard to similar spatial diversity and also satisfies intuitive aesthetic aspects. A SVM fit of the training set with the adequate kernel function

```
kernelFunction={"Wavelet",0.3};
svmInfo=CIP`SVM`FitSvm[trainingSet,kernelFunction];
```

yields satisfactory predictions for the training as well as the test set

```
trainingAndTestSet={trainingSet,testSet};
CIP`SVM`ShowSvmRegressionResult[{"ModelVsDataPlot",
 "AbsoluteSortedResidualsPlot","RMSE"},trainingAndTestSet,svmInfo]
```

Training Set:

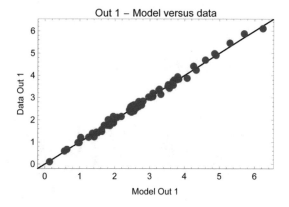

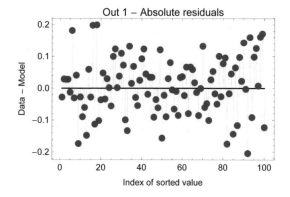

Root mean squared error (RMSE) = 8.904×10^{-2}

Test Set:

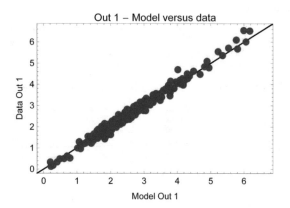

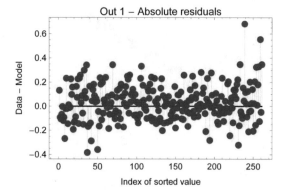

Root mean squared error (RMSE) = 1.508×10^{-1}

The RMSE values again correspond satisfactorily to the value of 0.1 used as the standard deviation for the calculation of the normally distributed I/O pairs' outputs. It may be interesting to compare the DoE-based partitioning to different training and test set splitting heuristics which completely neglect the DoE knowledge. This kind of exploration might lead to some insight about success or failure of these partitioning optimization strategies. The first attempt is a pure CR based training and test set partitioning with a training fraction equal to the sub-grid partitioning one. With the current sizes of the training set

```
Length[trainingSet]
```

100

and the test set

```
Length[testSet]
```

261

the training fraction is evaluated to be:

```
trainingFraction=
  N[Length[trainingSet]/(Length[trainingSet]+Length[testSet])]
```

0.277008

The CR based training set for this training fraction

```
trainingAndTestSet=CIP`Cluster`GetClusterBasedTrainingAndTestSet[
  dataSet3D,trainingFraction];
trainingSet=trainingAndTestSet[[1]];
testSet=trainingAndTestSet[[2]];
```

can be visually inspected to demonstrate a good spatial diversity but with inputs clearly different to those of the sub-grid before:

```
CIP`Graphics`Plot3dDataSet[trainingSet,labels,
  GraphicsOptionViewPoint3D -> viewPoint3D]
```

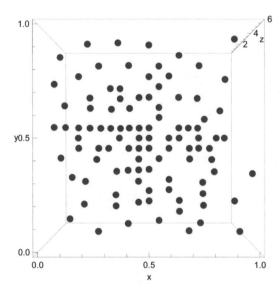

The SVM fit with the same kernel function

```
svmInfo=CIP'SVM'FitSvm[trainingSet,kernelFunction];
```

produces RMSE values for training and test set

```
CIP'SVM'ShowSvmRegressionResult[{"RMSE"},trainingAndTestSet,svmInfo]
```

Training Set:

Root mean squared error (RMSE) = 9.118×10^{-2}

Test Set:

Root mean squared error (RMSE) = 1.975×10^{-1}

which are a little inferior to those of the DoE-based fit before. So a further refinement with additional training set optimization steps on the basis of a distinct I/O pair exchange heuristics could lead to improvements. Note that this assumption is a kind of pure hope since none of all possible heuristics guarantees improvement. The AllClusterMax training set optimization strategy generates a training set in every iteration step which consists of the single maximum deviating I/O pairs of each cluster. So each optimization step may form a completely new training set. The application of this heuristics to the current regression task

```
numberOfTrainingSetOptimizationSteps=20;
deviationCalculationMethod="AllClusterMax";
svmTrainOptimization=
 CIP'SVM'GetSvmTrainOptimization[dataSet3D,
 kernelFunction,trainingFraction,
```

```
numberOfTrainingSetOptimizationSteps,
UtilityOptionDeviationCalculation ->
deviationCalculationMethod];
CIP`SVM`ShowSvmTrainOptimization[
svmTrainOptimization]
```

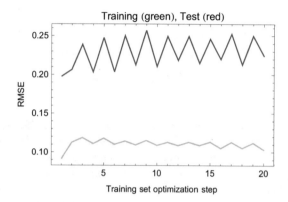

does not lead to any improvement to the pure CR based training set of step one. As a variant the AllClusterMean strategy can be examined where the training set in every iteration is formed by the clusters' I/O pairs which are nearest their corresponding cluster-specific mean deviation (and thus do not simply correspond to the cluster's maximum deviation). This more balancing global strategy

```
numberOfTrainingSetOptimizationSteps=20;
deviationCalculationMethod="AllClusterMean";
svmTrainOptimization=
  CIP`SVM`GetSvmTrainOptimization[dataSet3D,
  kernelFunction,trainingFraction,
  numberOfTrainingSetOptimizationSteps,
  UtilityOptionDeviationCalculation ->
  deviationCalculationMethod];
CIP`SVM`ShowSvmTrainOptimization[
  svmTrainOptimization]
```

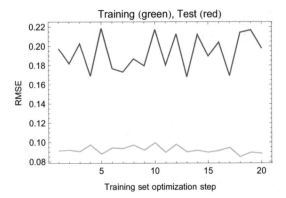

is able to achieve improvements in comparison to the pure CR based training set
of step one. More reserved heuristics do only change one single I/O pair in every
training set optimization step and not the training set as a whole. The SingleGlob-
alMean strategy evaluates the cluster with the maximum deviating I/O pair and replaces this cluster's training set member with the cluster's test set I/O pair which is
nearest its cluster-specific mean deviation. The application of the strategy

```
numberOfTrainingSetOptimizationSteps=20;
deviationCalculationMethod="SingleGlobalMean";
svmTrainOptimization=
 CIP'SVM'GetSvmTrainOptimization[dataSet3D,
 kernelFunction,trainingFraction,
 numberOfTrainingSetOptimizationSteps,
 UtilityOptionDeviationCalculation ->
 deviationCalculationMethod];
CIP'SVM'ShowSvmTrainOptimization[
 svmTrainOptimization]
```

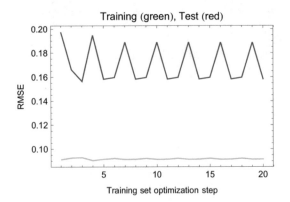

shows improvements but is trapped in an oscillation after a few steps. Single I/O pair exchange strategies are known to be prone to this kind of behavior as already discussed above. Therefore these strategies are enhanced by blacklisting to suppress oscillations: A single I/O pair is not allowed to be exchanged twice for a number of optimization steps (which may be defined by the blacklist length where the blacklist is removed after it arrived its maximum length). If the blacklist length is chosen to be equal to the number of optimization steps, i.e. no I/O pair is allowed to be exchanged twice,

```
numberOfTrainingSetOptimizationSteps=20;
deviationCalculationMethod="SingleGlobalMean";
blackListLength=20;
svmTrainOptimization=
 CIP`SVM`GetSvmTrainOptimization[dataSet3D,
 kernelFunction,trainingFraction,
 numberOfTrainingSetOptimizationSteps,
 UtilityOptionDeviationCalculation ->
 deviationCalculationMethod,
 UtilityOptionBlackListLength -> blackListLength];
CIP`SVM`ShowSvmTrainOptimization[
 svmTrainOptimization]
```

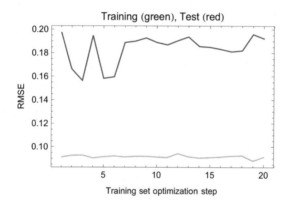

Training (green), Test (red)

the oscillation is successfully suppressed but only an early improvement is achieved after three optimization steps since all further optimization trials fail. A reduced blacklist length

```
numberOfTrainingSetOptimizationSteps=25;
deviationCalculationMethod="SingleGlobalMean";
blackListLength=10;
svmTrainOptimization=
 CIP`SVM`GetSvmTrainOptimization[dataSet3D,
 kernelFunction,trainingFraction,
 numberOfTrainingSetOptimizationSteps,
 UtilityOptionDeviationCalculation ->
 deviationCalculationMethod,
```

```
 UtilityOptionBlackListLength -> blackListLength];
CIP`SVM`ShowSvmTrainOptimization[
 svmTrainOptimization]
```

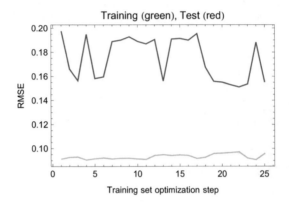

Training (green), Test (red)

allows a re-entry of already exchanged I/O pairs and leads to the best improvement in step 17 found so far. Finally the SingleGlobalMax strategy is explored (the CIP default strategy) which evaluates the cluster with the maximum deviating I/O pair and replaces this cluster's training set member with this maximum deviating I/O pair. A non-blacklisted application of this strategy

```
numberOfTrainingSetOptimizationSteps=20;
deviationCalculationMethod="SingleGlobalMax";
svmTrainOptimization=
 CIP`SVM`GetSvmTrainOptimization[
 dataSet3D,kernelFunction,trainingFraction,
 numberOfTrainingSetOptimizationSteps,
 UtilityOptionDeviationCalculation ->
 deviationCalculationMethod];
CIP`SVM`ShowSvmTrainOptimization[
 svmTrainOptimization]
```

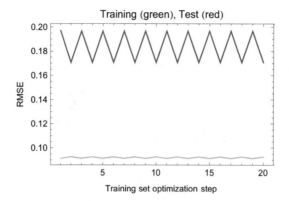

leads to only a minor improvement in step 2 and oscillations afterwards. A full blacklisting

```
numberOfTrainingSetOptimizationSteps=20;
deviationCalculationMethod="SingleGlobalMax";
blackListLength=20;
svmTrainOptimization=
 CIP`SVM`GetSvmTrainOptimization[dataSet3D,
 kernelFunction,trainingFraction,
 numberOfTrainingSetOptimizationSteps,
 UtilityOptionDeviationCalculation ->
 deviationCalculationMethod,
 UtilityOptionBlackListLength -> blackListLength];
CIP`SVM`ShowSvmTrainOptimization[
 svmTrainOptimization]
```

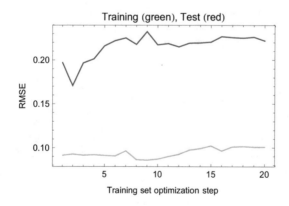

does not improve further. So a reduced blacklist length again

```
numberOfTrainingSetOptimizationSteps=20;
```

```
deviationCalculationMethod="SingleGlobalMax";
blackListLength=10;
svmTrainOptimization=
 CIP`SVM`GetSvmTrainOptimization[dataSet3D,
 kernelFunction,trainingFraction,
 numberOfTrainingSetOptimizationSteps,
 UtilityOptionDeviationCalculation ->
 deviationCalculationMethod,
 UtilityOptionBlackListLength -> blackListLength];
CIP`SVM`ShowSvmTrainOptimization[
 svmTrainOptimization]
```

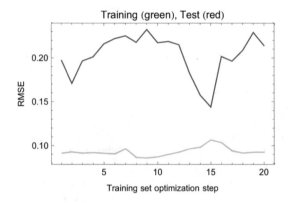

leads to the best result - in this case the best training/test set combination found so far for the current regression task. The latter optimization result reveals a nice "per aspera ad astra" (through difficulties to the stars) result: The small improvement in step 2 is followed by a number of apparently unfavorable steps with increased test RMSE values but after the blacklist reset in step 11 a long RMSE value drop down is found which arrives at an improved RMSE minimum. This best training and test set partitioning

```
index=CIP`SVM`GetBestSvmRegressOptimization[
 svmTrainOptimization];
trainingAndTestSetList=svmTrainOptimization[[3]];
svmInfoList=svmTrainOptimization[[4]];
trainingAndTestSet=trainingAndTestSetList[[index]];
trainingSet=trainingAndTestSet[[1]];
testSet=trainingAndTestSet[[2]];
svmInfo=svmInfoList[[index]];
```

still contains an absolutely satisfying spatial diverse training set

```
CIP`Graphics`Plot3dDataSet[trainingSet,labels,
 GraphicsOptionViewPoint3D -> viewPoint3D]
```

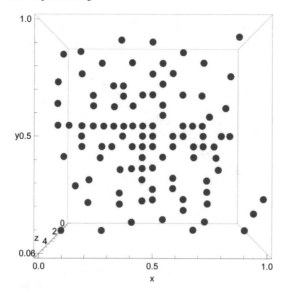

with RMSE values

```
CIP`SVM`ShowSvmRegressionResult[{"RMSE"},trainingAndTestSet,svmInfo]
```

Training Set:

Root mean squared error (RMSE) = 1.065×10^{-1}

Test Set:

Root mean squared error (RMSE) = 1.438×10^{-1}

and deviation plots for training and test set

```
CIP`SVM`ShowSvmRegressionResult[{"ModelVsDataPlot",
  "AbsoluteSortedResidualsPlot","RMSE"},trainingAndTestSet,svmInfo]
```

Training Set:

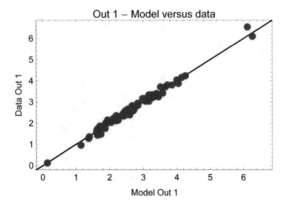

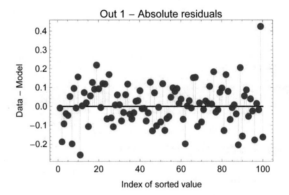

Root mean squared error (RMSE) = 1.065×10^{-1}

Test Set:

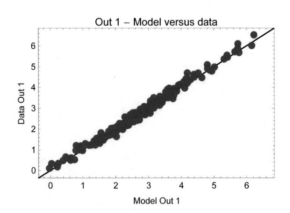

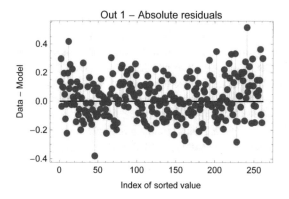

Root mean squared error (RMSE) = 1.438×10^{-1}

that are even slightly superior to the DoE-based sub grid result above. But note that it was a kind of educated good luck that led to this excellent final result. In summary a training and test set partitioning could be achieved by mere CR based partitioning with some following optimization steps which is comparable to the DoE-based sub grid approach. But the latter has the crucial advantage of being computationally far less expensive: That is why experimental scientists and data analysts should closely collaborate at very early project stages to design experiments together. So at best the machine learning process(es) can already accompany and support experimental work during its performance. This becomes even more important for modern miniaturized, parallelized and automated lab processes by intensive use of robotics. Otherwise the traditional "experiments first-data analysis second" procedure may give away an awful lot.

4.8.5 Concluding remarks

The partitioning of a data set into a training and a test set is a crucial step for successful machine learning: If a small training set possesses a high predictivity for a test set whose inputs cover a comparable input space a convincing result is usually obtained in agreement with the two guidelines sketched at the beginning of this section. But in general machine learning results should be treated with caution. There is a wide range for educated cheating: Almost always something can be learned and then predicted and thus machine learning results are often dubious. Proper validation is essential and far more demanding than mere data fitting. It is interesting to note that commercial machine learning applications are far more trustworthy than many academic reports since the first are inevitably assessed in practice with a painful penalty for deceits where the latter are often produced by harassed students under pressure with a fire and forget mentality.

With the discussed partitioning of data into a training and a test set the domain of cross validation was touched: This scientific field uses more elaborate partitioning schemes and operations upon them to validate data but faces the unfortunate drawback that these methods often require astronomic computational resources or time periods beyond all bearings. So in most cases the sketched or similar validation heuristics underlie practically feasible validation procedures - and as shown they may lead to helpful results for the practitioner despite their non-optimal and preliminary character. Nevertheless their deficiencies have to be kept in mind and a report of machine learning results should always point them out in clear cut words.

4.9 Comparative machine learning

```
Clear["Global`*"];
<<CIP`CalculatedData`
<<CIP`Graphics`
<<CIP`MLR`
<<CIP`MPR`
<<CIP`SVM`
<<CIP`Perceptron`
```

When machine learning is to be applied to a set of data a decision has to be made about the method to use. In general the best method for the data in question can not a priori be determined by theoretical considerations as already pointed out earlier. Thus experience, educated guesses and a lot of trial and error are necessary to proceed. And there is more than one way to skin a cat. Several machine construction processes may be performed and compared afterwards. This approach may be sketched in an illustrative manner with the chapter's introductory 3D data set generated on the basis of a true function:

```
pureOriginalFunction=Function[{x,y},
 1.9*(1.35+Exp[x]*Sin[13.0*(x-0.6)^2]*Exp[-y]* Sin[7.0*y])];
xRange={0.0,1.0};
yRange={0.0,1.0};
labels={"x","y","z"};
numberOfDataPerDimension=10;
standardDeviationRange={0.1,0.1};
dataSet3D=CIP`CalculatedData`Get3dFunctionBasedDataSet[
 pureOriginalFunction,xRange,yRange,numberOfDataPerDimension,
 standardDeviationRange];
trainingSet=dataSet3D;
testSet={};
trainingAndTestSet={trainingSet,testSet};
CIP`Graphics`Plot3dDataSetWithFunction[dataSet3D,
 pureOriginalFunction,labels]
```

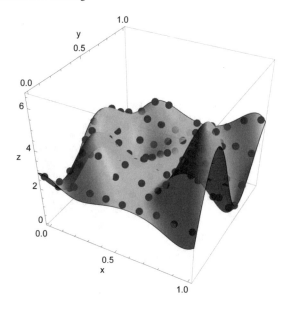

An initial linear MLR trial

```
mlrInfo=CIP`MLR`FitMlr[dataSet3D];
```

can be directly ruled out by visual inspection of the overlay of the approximated linear model function and the true function

```
originalFunctionGraphics3D=CIP`Graphics`Plot3dFunction[
  pureOriginalFunction,xRange,yRange,labels];
pureMlr3dFunction=Function[{x,y},
  CalculateMlr3dValue[x,y,mlrInfo]];
approximatedMlrFunctionGraphics3D=CIP`Graphics`Plot3dFunction[
  pureMlr3dFunction,xRange,yRange,labels];
Show[originalFunctionGraphics3D,approximatedMlrFunctionGraphics3D]
```

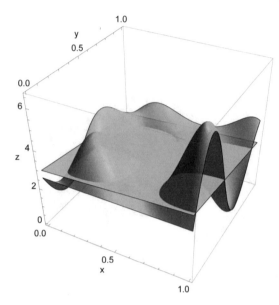

since the machine learning problem is strongly non-linear: So non-linear methods are clearly advised. MPR needs an a priori specification of the polynomial degree as its structural hyperparameter: Since an adequate polynomial degree is usually not known in advance (given that one exists at all for the problem in question) a reasonable range of values is scanned:

```
polynomialDegreeList=
  Table[polynomialDegree,{polynomialDegree,1,15}]
```

$\{1,2,3,4,5,6,7,8,9,10,11,12,13,14,15\}$

```
mprInfoList=CIP`MPR`FitMprSeries[dataSet3D, polynomialDegreeList];
mprSeriesRmse=
  CIP`MPR`GetMprSeriesRmse[trainingAndTestSet,mprInfoList];
CIP`MPR`ShowMprSeriesRmse[mprSeriesRmse]
```

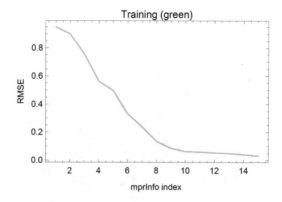

Best training set regression with mprInfo index = {15}

The continuously decreasing RMSE with increasing polynomial degree shows that a polynomial degree of 9 (which corresponds to index 9 in polynomialDegreeList and mprInfoList)

```
mprInfoIndex=9;
polynomialDegreeList[[mprInfoIndex]]
```

9

```
mprInfo=mprInfoList[[mprInfoIndex]];
CIP`MPR`ShowMprSingleRegression[{"RMSE"},dataSet3D,mprInfo];
```

Root mean squared error (RMSE) = 8.53×10^{-2}

leads to a model with a RMSE value near 0.1 (the value that was used for the data generation around the true function). The resulting approximated MPR model function adequately approximates the true function which is again demonstrated by their overlay:

```
pureMpr3dFunction=
  Function[{x,y},CalculateMpr3dValue[x,y,mprInfo]];
approximatedMprFunctionGraphics3D=
  CIP`Graphics`Plot3dFunction[pureMpr3dFunction,xRange,yRange,
   labels];
Show[originalFunctionGraphics3D,approximatedMprFunctionGraphics3D]
```

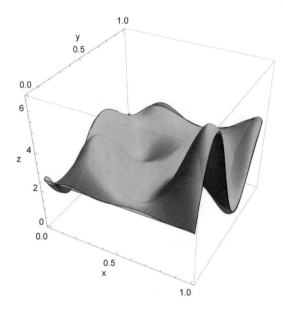

Support vector regression needs an a priori specification of the kernel function which is not known in general. If a type of kernel function is arbitrarily chosen its parameters are in question: These have to be scanned over a range of reasonable values to construct an adequate kernel. The wavelet kernel was already shown to be successful at the introduction of this chapter. A systematic scan of its kernel parameter in a range between 0.1 and 1.0 in steps of 0.1

```
kernelFunctionList=Table[{"Wavelet",kernelParameter},
  {kernelParameter,0.1,1.0,0.1}]
```

{{Wavelet, 0.1}, {Wavelet, 0.2}, {Wavelet, 0.3}, {Wavelet, 0.4}, {Wavelet, 0.5}, {Wavelet, 0.6},
{Wavelet, 0.7}, {Wavelet, 0.8}, {Wavelet, 0.9}, {Wavelet, 1.}}

```
svmInfoList=CIP`SVM`FitSvmSeries[dataSet3D,kernelFunctionList];
svmSeriesRegressionResult=
  CIP`SVM`GetSvmSeriesRmse[trainingAndTestSet,svmInfoList];
CIP`SVM`ShowSvmSeriesRmse[svmSeriesRegressionResult]
```

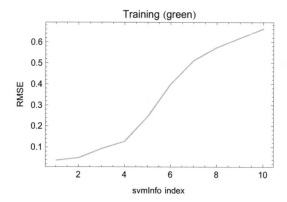

Best training set regression with svmInfo index = {1}

shows a continuously increasing RMSE value from values below 0.1 (which indicates overfitting since the data are better approximated than they should be according to the standard deviation of 0.1 used for their generation) to values considerably above 0.1 (which means an only poor approximation of the true function). A good machine learning result with a RMSE around 0.1 for a wavelet parameter of 0.3

```
svmInfoIndex=3;
svmInfo=svmInfoList[[svmInfoIndex]];
CIP`SVM`GetKernelFunction[svmInfo]
```

{Wavelet, 0.3}

thus determines an adequate kernel function (a kind of bump, compare above):

```
a=0.3;
x=0.0;
pureWaveletKernel2D=Function[y,CIP`SVM`KernelWavelet[{x},{y},a]];
argumentRange={-1.0,1.0};
functionValueRange={-0.4,1.1};
labels={"x","z","Wavelet kernel function"};
CIP`Graphics`Plot2dFunction[pureWaveletKernel2D,argumentRange,
  functionValueRange,labels]
```

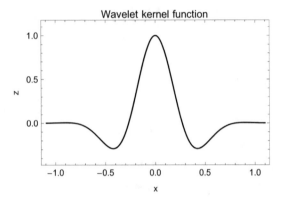

The overlay of the true function with the approximated model function shows the convincing result for this kernel:

```
pureSvm3dFunction=Function[{x,y},
 CIP `SVM`CalculateSvm3dValue[x,y,svmInfo]];
approximatedFunctionGraphics3D=CIP `Graphics`Plot3dFunction[
 pureSvm3dFunction,xRange,yRange,labels];
Show[originalFunctionGraphics3D,approximatedFunctionGraphics3D]
```

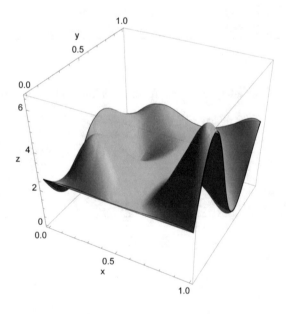

The analog procedure may be performed for a Gaussian radial basis function (RBF) kernel. A scan of its parameter

```
kernelFunctionList=Table[
  kernelFunction={"GaussianRBF",kernelParameter},
  {kernelParameter,1.0,15.0,1.0}]
```

{{GaussianRBF, 1.}, {GaussianRBF, 2.}, {GaussianRBF, 3.}, {GaussianRBF, 4.}, {GaussianRBF, 5.},
{GaussianRBF, 6.}, {GaussianRBF, 7.}, {GaussianRBF, 8.}, {GaussianRBF, 9.}, {GaussianRBF, 10.},
{GaussianRBF, 11.}, {GaussianRBF, 12.}, {GaussianRBF, 13.}, {GaussianRBF, 14.}, {GaussianRBF, 15.}}

```
svmInfoList=CIP`SVM`FitSvmSeries[dataSet3D,kernelFunctionList];
svmSeriesRegressionResult=
  CIP`SVM`GetSvmSeriesRmse[trainingAndTestSet,svmInfoList];
CIP`SVM`ShowSvmSeriesRmse[svmSeriesRegressionResult]
```

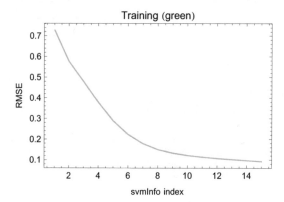

Best training set regression with svmInfo index = {15}

recommends a value of 14.0 for a RMSE value around 0.1

```
svmInfoIndex=14;
svmInfo=svmInfoList[[svmInfoIndex]];
CIP`SVM`GetKernelFunction[svmInfo]
```

{GaussianRBF, 14.}

to form an adequate Gaussian RBF kernel function:

```
beta=14.0;
x=0.0;
pureWaveletKernel2D=Function[y,
  CIP`SVM`KernelGaussianRbf[{x},{y},beta]];
argumentRange={-1.0,1.0};
functionValueRange={-0.1,1.1};
labels={"x","z","Gaussian RBF kernel function"};
```

```
CIP 'Graphics 'Plot2dFunction[pureWaveletKernel2D,argumentRange,
  functionValueRange,labels]
```

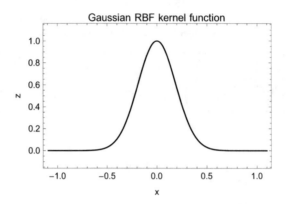

The overlay is again convincing:

```
pureSvm3dFunction=Function[{x,y},
  CIP 'SVM 'CalculateSvm3dValue[x,y,svmInfo]];
approximatedFunctionGraphics3D=CIP 'Graphics 'Plot3dFunction[
  pureSvm3dFunction,xRange,yRange,labels];
Show[originalFunctionGraphics3D,approximatedFunctionGraphics3D]
```

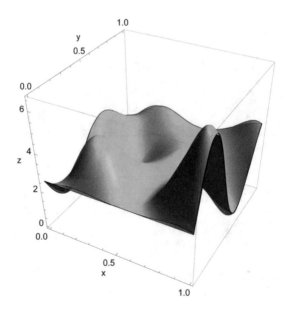

As a third alternative a universal Fourier kernel function may be chosen and its parameter scanned:

```
kernelFunctionList=Table[
 kernelFunction={"UniversalFourier",kernelParameter},
 {kernelParameter,0.1,0.9,0.1}]
```

{{UniversalFourier,0.1},{UniversalFourier,0.2},{UniversalFourier,0.3},{UniversalFourier,0.4},
{UniversalFourier,0.5},{UniversalFourier,0.6},{UniversalFourier,0.7},{UniversalFourier,0.8},
{UniversalFourier,0.9}}

```
svmInfoList=CIP`SVM`FitSvmSeries[dataSet3D,kernelFunctionList];
svmSeriesRegressionResult=
 CIP`SVM`GetSvmSeriesRmse[trainingAndTestSet,svmInfoList];
CIP`SVM`ShowSvmSeriesRmse[svmSeriesRegressionResult]
```

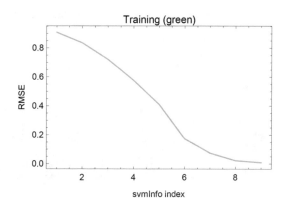

Best training set regression with svmInfo index = {9}

A value of 0.7 for a RMSE value around 0.1

```
svmInfoIndex=7;
svmInfo=svmInfoList[[svmInfoIndex]];
CIP`SVM`GetKernelFunction[svmInfo]
```

{UniversalFourier,0.7}

now leads to a successful kernel function

```
q=0.7;
x=0.0;
pureWaveletKernel2D=Function[y,
 CIP`SVM`KernelUniversalFourier[{x},{y},q]];
```

```
argumentRange={-1.0,1.0};
functionValueRange={0.0,3.0};
labels={"x","z","Universal Fourier kernel function"};
CIP'Graphics'Plot2dFunction[pureWaveletKernel2D,argumentRange,
  functionValueRange,labels]
```

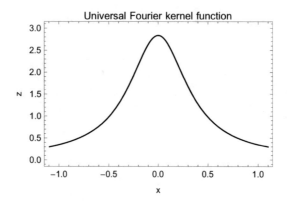

with an excellent overlay result:

```
pureSvm3dFunction=Function[{x,y},
  CIP'SVM'CalculateSvm3dValue[x,y,svmInfo]];
approximatedFunctionGraphics3D=CIP'Graphics'Plot3dFunction[
  pureSvm3dFunction,xRange,yRange,labels];
Show[originalFunctionGraphics3D,approximatedFunctionGraphics3D]
```

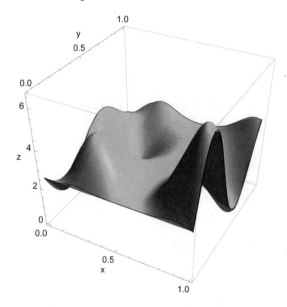

If the three adequate kernel functions are compared they are similar in principal (i.e. bump-like) but in detail notably different. Nevertheless they lead to approximated model functions of comparable quality. A perceptron based approach raises the question for the necessary number of hidden neurons to build a sufficient number of adequate bumps. Thus a scan of perceptron fits with different numbers of hidden neurons

```
numberOfHiddenNeuronsList=Table[numberOfHiddenNeurons,
 {numberOfHiddenNeurons,2,20}]
```

$\{2,3,4,5,6,7,8,9,10,11,12,13,14,15,16,17,18,19,20\}$

```
perceptronInfoList=CIP`Perceptron`FitPerceptronSeries[
 dataSet3D,numberOfHiddenNeuronsList];
perceptronSeriesRmse=CIP`Perceptron`GetPerceptronSeriesRmse[
 trainingAndTestSet,perceptronInfoList];
CIP`Perceptron`ShowPerceptronSeriesRmse[perceptronSeriesRmse]
```

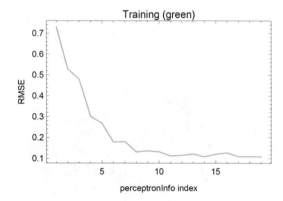

Best training set regression with perceptronInfo index = {14}

suggests at least 12 hidden neurons

```
perceptronInfoIndex=11;
perceptronInfo=perceptronInfoList[[perceptronInfoIndex]];
CIP'Perceptron'GetNumberOfHiddenNeurons[perceptronInfo]
```

12

to be a minimum sufficient number (but compare the discussion in section 4.12).
The overlay result

```
purePerceptron3dFunction=Function[{x,y},
 CIP'Perceptron'CalculatePerceptron3dValue[x,y,
 perceptronInfo]];
approximatedFunctionGraphics3D=CIP'Graphics'Plot3dFunction[
 purePerceptron3dFunction,xRange,yRange,labels];
Show[originalFunctionGraphics3D,approximatedFunctionGraphics3D]
```

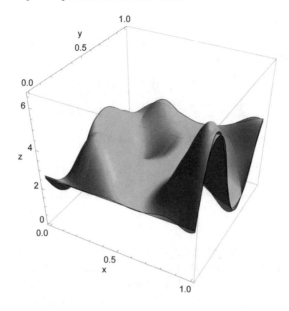

shows a satisfying approximation quality which is apparently a bit inferior to the successful SVM based results before. The initial statement that there is more than one way to skin a cat finally becomes obvious: For the current example a number of explored non-linear methods performed comparably well and it is just a matter of taste which one to choose (where the fast MPR approach would in practice be preferred due to its superior fitting performance).

4.10 Relevance of input components and minimal models

```
Clear["Global`*"];
<<CIP`ExperimentalData`
<<CIP`Perceptron`
<<CIP`Graphics`
<<CIP`DataTransformation`
```

Adequate inputs are a necessary basis for all machine learning tasks since something is to be learned for specific inputs. An input itself is a mathematical vector that codes information in an appropriate way for the machine learning task. From a practitioner's point of view an input should contain all useful information, i.e. the input's components should each have a precise scientific meaning. But if a single input component is really meaningful to the machine learning task is hard to judge in advance. This raises the question of the relevance of input components. A minimal predictive model with an optimum set of inputs should only contain those components which are necessary to successfully perform the machine learning task and omit possible redundant or irrelevant components which only boost an input's length but do not contribute to improved learning. As an example the well-defined iris flower inputs are examined.

```
classificationDataSet=
 CIP`ExperimentalData`GetIrisFlowerClassificationDataSet[];
```

An iris flower input consists of four components

```
CIP`Graphics`ShowDataSetInfo[{"InputComponents"},
 classificationDataSet];
```

Number of input components = 4

which are the sepal length (component 1), the sepal width (component 2), the petal length (component 3) and the petal width (component 4). These four quantities are meaningful to the biologist and thus the natural basis for a classification effort. It was shown above that a small perceptron with only two hidden neurons

```
numberOfHiddenNeurons=2;
```

leads to a satisfying classification result:

```
perceptronInfo=
 CIP`Perceptron`FitPerceptron[classificationDataSet,
  numberOfHiddenNeurons];
trainingAndTestSet={classificationDataSet,{}};
CIP`Perceptron`ShowPerceptronClassificationResult[
 {"CorrectClassification"},trainingAndTestSet,perceptronInfo]
```

Training Set:

99.3% correct classifications

The relevance of a specific input component may be determined by omission of this component with a re-evaluation of the machine learning task. If this is performed for all components (with a successive leave-one-out strategy) the least relevant component becomes obvious which is the component that leads to a minimum change for the worse (e.g. for a RMSE decrease) in comparison to the earlier machine learning result. After removal of this minimum impact component this process may be repeated for the second component etc. until the one single most meaningful component is reached. If this strategy is applied to the iris flower classification

```
perceptronInputRelevanceClass=
 CIP`Perceptron`GetPerceptronInputRelevanceClass[trainingAndTestSet,
  numberOfHiddenNeurons];
CIP`Perceptron`ShowPerceptronInputRelevanceClass[
 perceptronInputRelevanceClass]
```

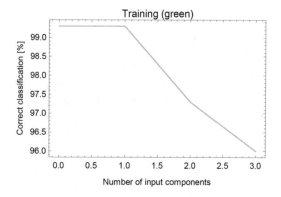

Input component list = {1,2,3}

the result is a relevance ranking of the four input components where the first component of the "Input component list" is the minimum impact component with lowest relevance etc. and the one component that survived the successive removal process (not shown in the "Input component list") is the most relevant and meaning-ful component (which is component 4 - the petal width - in this case), in a nutshell: Relevance of component $4 > 3 > 2 > 1$. A second finding is the insignificance of component 1 (the sepal length) for the classification task (since the classification success is unchanged after removal of this component) and the comparatively small impact of all 3 minor impact components of the "Input component list" on the pre-diction performance (see the diagram above): The maximum impact component 4 alone already leads to 96% correct classifications. The heuristic successive leave-one-out strategy for the input components relevance determination may be reversed to a successive component-inclusion strategy: In a first step all single input compo-nents are scanned for the one that leads to the best classification result with maxi-mum predictivity. Then this single best input component is fixed and the remaining input components are scanned for the next most relevant component so that a pair of two relevant input components is achieved. This procedure is continued to get a relevant triple etc.

```
perceptronInputComponentRelevanceListForClassification=
 CIP `Perceptron `GetPerceptronInputInclusionClass[trainingAndTestSet,
     numberOfHiddenNeurons];
 CIP `Perceptron `ShowPerceptronInputRelevanceClass[
 perceptronInputComponentRelevanceListForClassification]
```

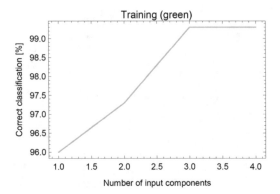

Input component list = $\{4,3,2,1\}$

For the iris flower classification task the resulting relevance ranking given in the "Input component list" is identical to the one obtained before by the leave-one-out strategy (both strategies perfectly mirror each other). But note that in general different heuristics of systematic input relevance determination do not necessarily coincide and above all do not lead to optimum subsets of relevant input components (i.e. those with the highest predictability) but may be regarded as a sensible (fast) heuristic choice only. A true optimum input component subset search (with e.g. an evolutionary search strategy) would be computationally far more demanding. Finally the high predictivity of component 4 (the petal width) may be visualized. First the petal width values for the three iris flower species (classes) are obtained for the classification data set

```
classIndex=1;
class1SubSet=
 CIP `DataTransformation`GetSpecificClassDataSubSet[
  classificationDataSet,classIndex];
class1SubSetInputs=class1SubSet[[All,1]];
class1SubSetPetalWidths=class1SubSetInputs[[All,4]];

classIndex=2;
class2SubSet=
 CIP `DataTransformation`GetSpecificClassDataSubSet[
  classificationDataSet,classIndex];
class2SubSetInputs=class2SubSet[[All,1]];
class2SubSetPetalWidths=class2SubSetInputs[[All,4]];

classIndex=3;
class3SubSet =
 CIP `DataTransformation`GetSpecificClassDataSubSet[
  classificationDataSet,classIndex];
class3SubSetInputs=class3SubSet[[All,1]];
class3SubSetPetalWidths=class3SubSetInputs[[All,4]];
```

and visualized as colored points

```
class1SubSetPetalWidthPoints=
 Table[{class1SubSetPetalWidths[[i]],1.2},{i,
  Length[class1SubSetPetalWidths]}];
class2SubSetPetalWidthPoints=
 Table[{class2SubSetPetalWidths[[i]],1.2},{i,
  Length[class2SubSetPetalWidths]}];
class3SubSetPetalWidthPoints=
 Table[{class3SubSetPetalWidths[[i]],1.2},{i,
  Length[class3SubSetPetalWidths]}];

class1SubSetPoints2DWithPlotStyle=
 {class1SubSetPetalWidthPoints,
  {PointSize[0.03],Opacity[0.3,Black]}};
class2SubSetPoints2DWithPlotStyle=
 {class2SubSetPetalWidthPoints,
  {PointSize[0.03],Opacity[0.3,Blue]}};
class3SubSetPoints2DWithPlotStyle=
 {class3SubSetPetalWidthPoints,
  {PointSize[0.03],Opacity[0.3,Red]}};

points2DWithPlotStyleList={class1SubSetPoints2DWithPlotStyle,
 class2SubSetPoints2DWithPlotStyle,
 class3SubSetPoints2DWithPlotStyle};
labels={"Petal width [mm]","Output",
 "Class 1 (black), 2 (blue), 3 (red)"};
argumentRange={0.0,26.0};
functionValueRange={-0.1,1.3};
pointGraphics=
 CIP'Graphics'PlotMultiple2dPoints[points2DWithPlotStyleList,
  labels,
  GraphicsOptionArgumentRange2D -> argumentRange,
  GraphicsOptionFunctionValueRange2D -> functionValueRange]
```

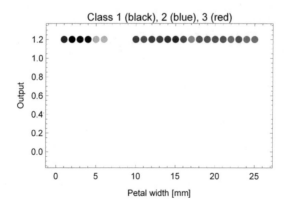

at an output *y*-value of 1.2 above. The points are displayed with a high level of transparency so a more intense color means more iris flower data with this distinct petal width value. Mixed colors indicate class overlaps that prevent simple and clear separations. The perceptron fit with the training set with the removed input components 1 to 3

```
inputComponentsToBeRemoved={1,2,3};
reducedTrainingSet=
 CIP`DataTransformation`RemoveInputComponentsOfDataSet[
   classificationDataSet,inputComponentsToBeRemoved];
reducedTrainingAndTestSet={reducedTrainingSet,{}};
perceptronInfo=
 CIP`Perceptron`FitPerceptron[reducedTrainingSet,
   numberOfHiddenNeurons];
CIP`Perceptron`ShowPerceptronClassificationResult[
 {"CorrectClassification"},reducedTrainingAndTestSet,
 perceptronInfo]
```

Training Set:

96.% correct classifications

yields a high prediction rate of 96% as already shown above. Since there is now only one input component (the petal width) and three output components (one for each class) the decision lines of each output component may be displayed in addition where each decision line is colored according to its corresponding class (i.e. iris flower species):

```
reducedTrainingSetList=
 CIP`DataTransformation`TransformDataSetToMultipleDataSet[
   reducedTrainingSet];

perceptronInfo=
 CIP`Perceptron`FitPerceptron[reducedTrainingSetList[[1]],
   numberOfHiddenNeurons];
pureClass1Function=
 Function[x,
  CIP`Perceptron`CalculatePerceptron2dValue[x,perceptronInfo]];
plotStyle={Thick,Black};
class1Graphics=
 CIP`Graphics`Plot2dFunction[pureClass1Function,argumentRange,
   functionValueRange,labels,
   GraphicsOptionLinePlotStyle -> plotStyle];

perceptronInfo=
 CIP`Perceptron`FitPerceptron[reducedTrainingSetList[[2]],
   numberOfHiddenNeurons];
pureClass2Function=
 Function[x,
  CIP`Perceptron`CalculatePerceptron2dValue[x,perceptronInfo]];
plotStyle={Thick,Blue};
class2Graphics=
 CIP`Graphics`Plot2dFunction[pureClass1Function,argumentRange,
   functionValueRange,labels,
   GraphicsOptionLinePlotStyle -> plotStyle];

perceptronInfo=
 CIP`Perceptron`FitPerceptron[reducedTrainingSetList[[3]],
   numberOfHiddenNeurons];
pureClass3Function=
 Function[x,
  CIP`Perceptron`CalculatePerceptron2dValue[x,perceptronInfo]];
plotStyle={Thick,Red};
class3Graphics =
 CIP`Graphics`Plot2dFunction[pureClass3Function,argumentRange,
   functionValueRange,labels,
   GraphicsOptionLinePlotStyle -> plotStyle];
```

```
Show[pointGraphics, class1Graphics, class2Graphics, class3Graphics]
```

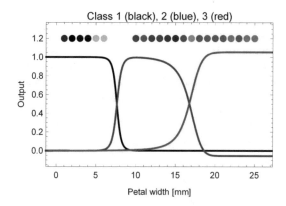

Note that for each petal width value the winner output is the decision line with the highest output value. Thus it becomes obvious that a petal width of less than about 6 mm is attributed to the class 1 (iris setosa, black), values between about 6 and 17 mm are assigned to the class 2 (iris versicolor, blue) and values above 17 mm belong to the class 3 (iris virginica, red). Since the petal width allows for this simple partitioning scheme its success is explained. The information of components 1 to 3 may be regarded as a refinement of this rough picture which leads to an improved predictivity.

The sketched strategies for minimal model construction by input component relevance analysis may in general provide valuable insights into dependencies of scientific quantities which in turn may substantially contribute to a scientist's understanding of a problem as well as motivate further investigations. If a possible input component is already known to be more or less redundant (i.e. its information is already contained in other input components) it should of course be omitted from the very beginning: The computational effort is increased with every additional input component and should be kept as small as possible. Also the proneness to overfitting increases with an increased length of the inputs and a minimal predictive model is a virtue on its own (compare the appendix of this chapter for examples and a more detailed discussion).

4.11 Pattern recognition

```
Clear["Global`*"];
<<CIP`ExperimentalData`
```

```
<<CIP`Utility`
<<CIP`Cluster`
<<CIP`DataTransformation`
<<CIP`MLR`
```

The recognition of patterns belongs to the most prominent and most demanding applications of machine learning, e.g. the classification of biological tissues or medical images. For simplicity the discussion in this section is confined to the recognition of digital grayscale images with different face types. Digital images are composed of pixels (picture elements) in a rectangular arrangement with a defined width and height (in pixels): A 640×480 digital image consists of

```
640*480
```

```
307200
```

307200 pixels arranged in 480 rows with 640 pixels in each row. Each pixel contains a specific color information. A grayscale pixel may contain 256 possible shades of gray ranging from black (with value 0) to white (with value 255). Thus a digital image can be represented as a matrix of numbers. Pattern recognition is demonstrated in the following for the intuitive problem of face detection (i.e. classification) with grayscale images of cat, dog and human faces. An image classification data set is obtained from the CIP ExperimentalData package (see Appendix A):

```
imageClassificationDataSet1=
  CIP`ExperimentalData`GetFacesWhiteImageDataSet[];
```

The data set contains

```
imageInputs1=CIP`Utility`GetInputsOfDataSet[
  imageClassificationDataSet1];
Length[imageInputs1]
```

```
18
```

18 I/O pairs: The faces of 6 cats (class 1), 6 dogs (class 2) and 6 humans (class 3)

```
GraphicsGrid[
  Table[
    Image[imageInputs1[[(i-1)*6+j]],"Byte"],
    {i,3},{j,6}
  ],
  ImageSize->300
]
```

Note that an above display pixel is not equal to an image pixel since the images are automatically scaled for better visibility.

with each class shown in one row. Each input is a 30×30 grayscale image

```
Dimensions[imageInputs1[[1]]]
```

{30,30}

which contains $30 \times 30 = 900$ pixels where each pixel contains a specific shade of gray (out of 256 possible values). Thus each input may be coded as a vector with 900 components where each component contains the grayscale value of its corresponding pixel. If the image classification data set is transformed in this way to a classification data set (with the rows of the rectangular pixel matrix structure concatenated to form a mere vector)

```
classificationDataSet1=
 CIP `DataTransformation`ConvertImageDataSet[
 imageClassificationDataSet1];
inputs1=CIP `Utility`GetInputsOfDataSet[classificationDataSet1];
```

clustering and machine learning tasks may be performed. An initial clustering of the inputs into 3 classes (the natural choice)

```
numberOfClusters=3;
clusterInfo=CIP `Cluster`GetFixedNumberOfClusters[inputs1,
 numberOfClusters];
```

yields 3 asymmetric classes

```
CIP `Cluster`ShowClusterResult[
 {"EuclideanDistanceDiagram","ClusterStatistics"},clusterInfo]
```

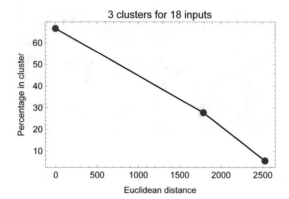

Cluster 1 : 12 members (66.6667%) with distance = 0.

Cluster 2 : 5 members (27.7778%) with distance = 1787.3

Cluster 3 : 1 members (5.55556%) with distance = 2525.37

where the data set's class assignments do not correspond to the detected clusters:

```
sortResult=CIP`DataTransformation`SortClassificationDataSet[
  classificationDataSet1];
classIndexMinMaxList=sortResult[[2]]
```

{{1,6},{7,12},{13,18}}

```
clusterOccupancies=CIP`Cluster`GetClusterOccupancies[
  classIndexMinMaxList,clusterInfo];
CIP`Cluster`ShowClusterOccupancies[clusterOccupancies]
```

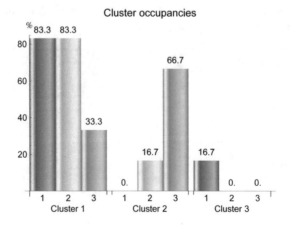

Cluster 1 contains the majority of cats and dogs with two humans. Cluster 2 is human dominated with just one dog and cluster 3 consists of only one cat. Cats and dogs are detected to be more similar since they are predominantly joined in one cluster. It becomes clear that a mere clustering of image inputs can not satisfactorily group the three types of faces. Thus supervised learning is advised. A linear MLR approach

```
mlrInfo1=CIP`MLR`FitMlr[classificationDataSet1];
CIP`MLR`ShowMlrSingleClassification[
  {"CorrectClassification"},classificationDataSet1,mlrInfo1]
```

100.% correct classifications

already yields a perfect 100% correct face detection (note that a MLR approach is not prone to overfitting thus a partitioning in training and test set is not necessary. But the success of a linear method is an exception chosen for simplicity: In general non-linear machine learning is necessary with all the difficulties discussed in previous sections). To dig a little deeper into the subtleties of pattern recognition consider the following image classification data set:

```
imageClassificationDataSet2=
 CIP`ExperimentalData`GetFacesGrayImageDataSet[];
imageInputs2=CIP`Utility`GetInputsOfDataSet[
 imageClassificationDataSet2];
GraphicsGrid[
 Table[
  Image[imageInputs2[[(i-1)*6+j]],"Byte"],
  {i,3},{j,6}
 ],
 ImageSize->300
 ]
```

It is identical to the one before with the difference that now the background of each image is not white but gray. If this data set is classified with the MLR predictor achieved before

```
classificationDataSet2=CIP`DataTransformation`ConvertImageDataSet[
 imageClassificationDataSet2];
CIP`MLR`ShowMlrSingleClassification[
 {"CorrectClassification"},classificationDataSet2,mlrInfo1]
```

33.3% correct classifications

a very poor predictivity of only 33.3% results. This seems puzzling since the faces are exactly the same but it is quite simple to understand: Humans automatically separate the background from a body of interest, a difficult operation they are usually not aware of. Since the machine learning process is expected to work human-like (the machine is anthropomorphized) it is expected to recognize the faces it learned also in another context. But the machine did not learn faces: It learned "faces on a white background" since this was the information which was presented in the training. And therefore a "face on a gray background" is in general unknown to the predictor which inevitably leads to a poor prediction performance. If the two image classification data sets are joined

```
joinedImageClassificationDataSets12=
 Join[imageClassificationDataSet1,imageClassificationDataSet2];
joinedImageInputs12=CIP`Utility`GetInputsOfDataSet[
 joinedImageClassificationDataSets12];
GraphicsGrid[
 Table[
  Image[joinedImageInputs12[[(i-1)*6+j]],"Byte"],
  {i,6},{j,6}
 ],
 ImageSize->300
]
```

and used to train the (MLR) machine

```
joinedClassificationDataSets12=
 CIP 'DataTransformation 'ConvertImageDataSet[
 joinedImageClassificationDataSets12];
 mlrInfo12=CIP 'MLR 'FitMlr[joinedClassificationDataSets12];
 CIP 'MLR 'ShowMlrSingleClassification[
 {"CorrectClassification"},joinedClassificationDataSets12,mlrInfo12]
```

100.% correct classifications

a 100% correct face detection is achieved for the joined data. A third image classification data set

```
imageClassificationDataSet3=
 CIP 'ExperimentalData 'GetFacesBlackImageDataSet[];
imageInputs3=CIP 'Utility 'GetInputsOfDataSet[
 imageClassificationDataSet3];
GraphicsGrid[
 Table[
  Image[imageInputs3[[(i-1)*6+j]],"Byte"],
  {i,3},{j,6}
 ],
 ImageSize->300
]
```

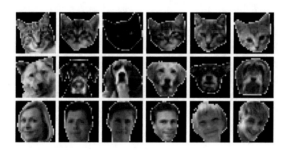

with the known faces but a black background is now directly a 100% correctly recognized

```
classificationDataSet3=CIP 'DataTransformation 'ConvertImageDataSet[
 imageClassificationDataSet3];
CIP 'MLR 'ShowMlrSingleClassification[
 {"CorrectClassification"},classificationDataSet3,mlrInfo12]
```

100.% correct classifications

by the MLR predictor based on the white and gray background data sets: The different backgrounds in the training data taught the machine that the background is not that important for face detection. Thus the machine becomes more tolerant to different backgrounds and more predictive on that score. A prediction on the basis of the first MLR predictor based only on the white background data set

```
CIP'MLR'ShowMlrSingleClassification[
 {"CorrectClassification"},classificationDataSet3,mlrInfo1]
```

33.3% correct classifications

yields again the expected poor result as before for the gray background data set. If all three image classification data sets are joined

```
joinedImageClassificationDataSets123=Join[
 imageClassificationDataSet1,imageClassificationDataSet2,
 imageClassificationDataSet3];
joinedImageInputs123=CIP'Utility'GetInputsOfDataSet[
 joinedImageClassificationDataSets123];
GraphicsGrid[
 Table[
  Image[joinedImageInputs123[[(i-1)*6+j]],"Byte"],
  {i,9},{j,6}
 ],
 ImageSize->300
]
```

and trained

```
joinedClassificationDataSets123=
 CIP`DataTransformation`ConvertImageDataSet[
 joinedImageClassificationDataSets123];
mlrInfo123=CIP`MLR`FitMlr[joinedClassificationDataSets123];
CIP`MLR`ShowMlrSingleClassification[
 {"CorrectClassification"},joinedClassificationDataSets123,
 mlrInfo123]
```

100.% correct classifications

a background independent face type MLR predictor with a 100% correct training detection rate is achieved. Its predictivity may be further explored with blurred versions of the images used for the training

```
blurredJoinedImageClassificationDataSets123=
 CIP`DataTransformation`BlurImageDataSet[
 joinedImageClassificationDataSets123];
blurredJoinedImageInputs123=CIP`Utility`GetInputsOfDataSet[
 blurredJoinedImageClassificationDataSets123];
```

```
GraphicsGrid[
 Table[
  Image[blurredJoinedImageInputs123[[(i-1)*6+j]],"Byte"],
  {i,9},{j,6}
 ],
 ImageSize->300
]
```

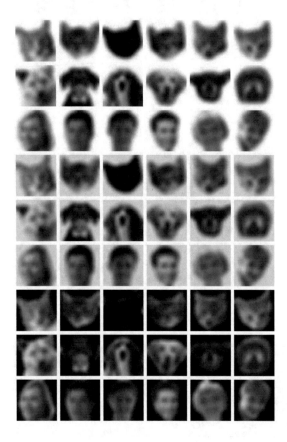

up to a degree where humans would still be able to recognize the face type without failure. A test with the blurred images

```
blurredJoinedClassificationDataSets123=
 CIP`DataTransformation`ConvertImageDataSet[
 blurredJoinedImageClassificationDataSets123];
CIP`MLR`ShowMlrSingleClassification[
 {"CorrectClassification","CorrectClassificationPerClass"},
 blurredJoinedClassificationDataSets123,mlrInfo123]
```

96.3% correct classifications

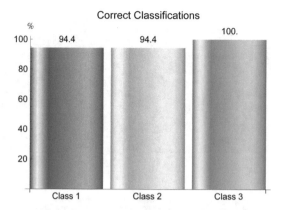

yields a notable correct detection result over 95% where the two misclassifications only affect the more similar cats and dogs (with their similarity already shown by the initial clustering result above). Thus a machine trained with specific faces is also able to recognize more abstract faces. It may be interesting to reverse the procedure: The machine is trained with the abstract blurred faces

```
mlrInfoBlurred123=CIP`MLR`FitMlr[
 blurredJoinedClassificationDataSets123];
CIP`MLR`ShowMlrSingleClassification[
 {"CorrectClassification"},blurredJoinedClassificationDataSets123,
 mlrInfoBlurred123]
```

100.% correct classifications

and then tested with the specific faces:

```
CIP`MLR`ShowMlrSingleClassification[
 {"CorrectClassification"},joinedClassificationDataSets123,
 mlrInfoBlurred123]
```

100.% correct classifications

In contrast to the result before now a 100% correct detection rate for both data sets is achieved: The blurred faces taught the machine to learn more generalizable face intrinsic characteristics which then led to an improved prediction result for the specific ones. Pattern recognition always needs an optimum level of abstraction to be most predictive. This finding also indicates that the image size could be reduced to still allow for a successful face recognition. If the image size is reduced from 30×30 images to 20×20 images with a scaling factor of 2/3 for the joined white and gray background images as a training set

```
scaleFactor=2/3;
```

```
reducedJoinedImageClassificationDataSets12=
 CIP`DataTransformation`ScaleSizeOfImageDataSet[
 joinedImageClassificationDataSets12,scaleFactor];
reducedJoinedImageInputs12=CIP`Utility`GetInputsOfDataSet[
 reducedJoinedImageClassificationDataSets12];
GraphicsGrid[
 Table[
  Image[reducedJoinedImageInputs12[[(i-1)*6+j]],"Byte"],
  {i,6},{j,6}
 ],
 ImageSize->300
]
```

Note that the display size of the images did not change. The reduced image size appears as a decreased and more coarse-grained resolution.

```
Dimensions[reducedJoinedImageInputs12[[1]]]
```

{20,20}

and the black background images as a test set

```
reducedImageClassificationDataSet3=
 CIP`DataTransformation`ScaleSizeOfImageDataSet[
 imageClassificationDataSet3,scaleFactor];
reducedImageClassificationInputs3=CIP`Utility`GetInputsOfDataSet[
 reducedImageClassificationDataSet3];
GraphicsGrid[
```

```
Table[
 Image[reducedImageClassificationInputs3[[(i-1)*6+j]],"Byte"],
 {i,3},{j,6}
 ],
 ImageSize->300
]
```

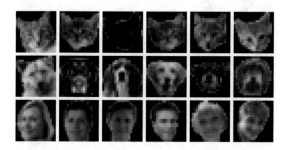

the face detection

```
reducedJoinedClassificationDataSets12=
 CIP`DataTransformation`ConvertImageDataSet[
 reducedJoinedImageClassificationDataSets12];
reducedClassificationDataSet3=
 CIP`DataTransformation`ConvertImageDataSet[
 reducedImageClassificationDataSet3];
mlrInfoReduced12=CIP`MLR`FitMlr[
 reducedJoinedClassificationDataSets12];
trainingAndTestSet={reducedJoinedClassificationDataSets12,
 reducedClassificationDataSet3};
CIP`MLR`ShowMlrClassificationResult[{"CorrectClassification"},
 trainingAndTestSet,mlrInfoReduced12]
```

Training Set:

100.% correct classifications

Test Set:

100.% correct classifications

keeps being perfect. Even a further reduction to 10×10 images for training

```
scaleFactor=1/3;
reducedJoinedImageClassificationDataSets12=
 CIP`DataTransformation`ScaleSizeOfImageDataSet[
 joinedImageClassificationDataSets12,scaleFactor];
reducedJoinedImageInputs12=CIP`Utility`GetInputsOfDataSet[
 reducedJoinedImageClassificationDataSets12];
GraphicsGrid[
 Table[
  Image[reducedJoinedImageInputs12[[(i-1)*6+j]],"Byte"],
  {i,6},{j,6}
  ],
 ImageSize->300
]
```

```
Dimensions[reducedJoinedImageInputs12[[1]]]
```

$\{10,10\}$

and test

```
reducedImageClassificationDataSet3=
 CIP`DataTransformation`ScaleSizeOfImageDataSet[
 imageClassificationDataSet3,scaleFactor];
reducedImageClassificationInputs3=CIP`Utility`GetInputsOfDataSet[
 reducedImageClassificationDataSet3];
GraphicsGrid[
 Table[
  Image[reducedImageClassificationInputs3[[(i-1)*6+j]],"Byte"],
  {i,3},{j,6}
  ],
 ImageSize->300
 ]
```

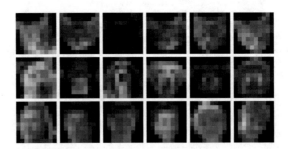

is still successful with the (MLR) machine

```
reducedJoinedClassificationDataSets12=
  CIP`DataTransformation`ConvertImageDataSet[
  reducedJoinedImageClassificationDataSets12];
reducedClassificationDataSet3=
  CIP`DataTransformation`ConvertImageDataSet[
  reducedImageClassificationDataSet3];
mlrInfoReduced12=CIP`MLR`FitMlr[
  reducedJoinedClassificationDataSets12];
trainingAndTestSet={reducedJoinedClassificationDataSets12,
  reducedClassificationDataSet3};
CIP`MLR`ShowMlrClassificationResult[{"CorrectClassification"},
  trainingAndTestSet,mlrInfoReduced12]
```

Training Set:

100.% correct classifications

Test Set:

100.% correct classifications

(Humans start having some recognition problems at this state of image resolution: On average they make three classification mistakes as a result of a small personal survey). Note that a size reduction from 30×30 images to 10×10 images means a dramatic inputs' size decrease from inputs with 900 components to inputs with only 100 components. But also the 100 pixels of the 10×10 images may be further reduced with a relevance determination of each pixel (i.e. each input component, compare the previous section above):

```
mlrInputRelevanceClass=
  CIP`MLR`GetMlrInputRelevanceClass[
  trainingAndTestSet];
CIP`MLR`ShowMlrInputRelevanceClass[
  mlrInputRelevanceClass]
```

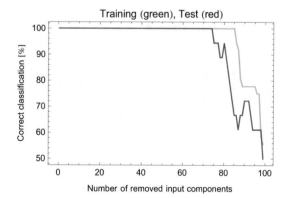

Removed input component list = {1,2,3,4,5,6,7,8,9,10,11,12,13,14,15,16,17,18,19,20,
21,22,23,24,25,26,27,28,29,30,31,32,33,34,35,36,37,38,39,40,41,42,43,44,45,46,47,48,49,50,
51,52,53,55,56,57,58,59,60,61,62,66,67,69,72,70,65,68,71,74,79,81,85,86,89,84,91,76,98,94,88,
78,90,92,54,75,93,83,82,99,80,77,95,63,100,87,97,64,96}

It becomes obvious that the majority of pixels can be discarded without any loss of predictive success: Only after elimination of more than 70% of the pixels the predictivity decreases, i.e. about the right 30 pixels are enough. There is of course a lower border for size reductions, e.g. if the size is reduced to 6×6 images for training

```
scaleFactor=1/5;
reducedJoinedImageClassificationDataSets12=
 CIP'DataTransformation'ScaleSizeOfImageDataSet[
 joinedImageClassificationDataSets12,scaleFactor];
reducedJoinedImageInputs12=CIP'Utility'GetInputsOfDataSet[
 reducedJoinedImageClassificationDataSets12];
GraphicsGrid[
 Table[
  Image[reducedJoinedImageInputs12[[(i-1)*6+j]],"Byte"],
  {i,6},{j,6}
 ],
 ImageSize->300
]
```

```
Dimensions[reducedJoinedImageInputs12[[1]]]
```

{6,6}

and test

```
reducedImageClassificationDataSet3=
 CIP`DataTransformation`ScaleSizeOfImageDataSet[
 imageClassificationDataSet3,scaleFactor];
reducedImageClassificationInputs3=CIP`Utility`GetInputsOfDataSet[
 reducedImageClassificationDataSet3];
GraphicsGrid[
 Table[
  Image[reducedImageClassificationInputs3[[(i-1)*6+j]],"Byte"],
  {i,3},{j,6}
 ],
 ImageSize->300
]
```

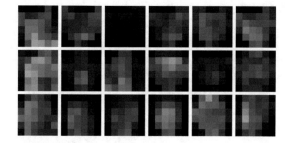

the machine starts to fail:

```
reducedJoinedClassificationDataSets12=
 CIP`DataTransformation`ConvertImageDataSet[
 reducedJoinedImageClassificationDataSets12];
reducedClassificationDataSet3=
 CIP`DataTransformation`ConvertImageDataSet[
 reducedImageClassificationDataSet3];
mlrInfoReduced12=CIP`MLR`FitMlr[
 reducedJoinedClassificationDataSets12];
trainingAndTestSet={reducedJoinedClassificationDataSets12,
 reducedClassificationDataSet3};
CIP`MLR`ShowMlrClassificationResult[
 {"CorrectClassification","CorrectClassificationPerClass"},
 trainingAndTestSet,mlrInfoReduced12]
```

Training Set:

100.% correct classifications

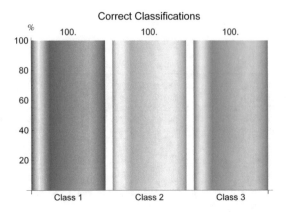

Test Set:

66.7% correct classifications

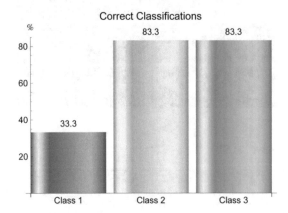

In this case it is the detection of cats which collapses first. Note that the image size reduction led to an unfavorable 36 pixels (the size reduction eliminated necessary information for face detection) whereas the right 30 pixels before were enough for a 100% prediction success. In general successful machine learning based pattern recognition needs adequate image preprocessing steps like a segmentation of relevant objects, rotations and translations, contrast enhancements, filtering techniques, noise suppression, image size reductions, color standardizations, wavelet transformations, spectral analysis or numerous others. Adequate image preprocessing towards a minimum number of input components is also mandatory to reduce the computational effort for the machine learning tasks since inputs with hundreds, thousands or even more input components for pure images complicate internal calculations and the whole optimization process considerably. Thus adequate preprocessing is crucial for successful machine learning. This does not only hold for pattern recognition: Well-prepared data are a virtue to the data analyst whereas input garbage only leads to output garbage no matter how sophisticated the machine learning method tries to be (the already mentioned GIGO - garbage-in/garbage-out - effect).

4.12 Technical optimization problems

```
Clear["Global`*"];
<<CIP`CalculatedData`
<<CIP`Graphics`
<<CIP`Perceptron`
```

All machine learning issues discussed so far did only address structural (hyper-parameter) problems of machine learning like the choice of a SVM kernel function or the number of hidden neurons of a perceptron approach. All technical parameters that guide the different machine learning methods were set to adequate default values and were therefore hidden by parameters which may optionally be changed but must not be specified in advance. In general these technical parameters must be properly adjusted for a specific machine learning task performed with a specific method. This means all sketched structural problems may be additionally spoiled by technical problems. It is this evil mixture of problems that often leads to a state of frustration when dealing with practically challenging machine learning tasks. To demonstrate the influence of technical parameters a fundamental parameter for every iterative procedure is chosen as an example: The maximum number of allowed iterations. This parameter is (or at least should be) essential since iterative procedures may run eternally under certain circumstances: An optimization procedure may get trapped in an oscillation around an optimum or run towards infinity forever (only stopped by an inevitable overflow error). Thus in each iterative step the current step number is compared to the maximum allowed number of iterations and the whole iterative process is stopped if this upper bound is exceeded. It is most often not desired for an iterative procedure to arrive at this upper bound: An optimization procedure should ideally stop before according to an a priori precision criterion (which is another technical parameter that guides the optimization process). For an illustration the chapter's introductory 3D data set (generated with a standard deviation of 0.1 on the basis of a true function) is used again

```
pureOriginalFunction=
 Function[{x,y},
  1.9*(1.35+Exp[x]*Sin[13.0*(x-0.6)^2]*Exp[-y]*Sin[7.0*y])];
xRange={0.0,1.0};
yRange={0.0,1.0};
labels={"x", "y", "z"};
numberOfDataPerDimension=10;
standardDeviationRange={0.1,0.1};
dataSet3D=
 CIP`CalculatedData`Get3dFunctionBasedDataSet[
```

```
    pureOriginalFunction,xRange,yRange,
    numberOfDataPerDimension,standardDeviationRange];
trainingSet=dataSet3D;
testSet={};
trainingAndTestSet={trainingSet,testSet};
CIP'Graphics'Plot3dDataSetWithFunction[
  dataSet3D,pureOriginalFunction,labels]
```

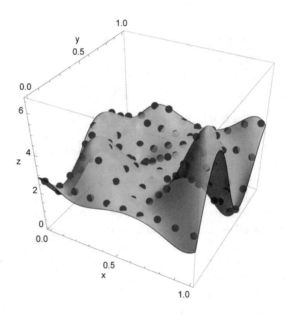

where a successful machine learning approach is expected to achieve a corresponding RMSE of about 0.1 that mirrors the data deviations. Above a perceptron based approach was analyzed to obtain an adequate number of hidden neurons to properly approximate the original surface with default settings for the minimization algorithm and all its technical optimization parameters where the default minimization is based on Mathematica's FindMinimum command with the (Polak-Ribiere variant of the) Conjugate-Gradient method (see [FitPerceptron])

```
OptionValue[CIP'Perceptron'PerceptronOptionsTraining,
  PerceptronOptionOptimizationMethod]
```

FindMinimum

and the default value for the maximum number of iterations is 10000:

```
OptionValue[CIP'Perceptron'PerceptronOptionsOptimization,
  PerceptronOptionMaximumIterations]
```

10000

In the following the hidden neuron scan is repeated

```
numberOfHiddenNeuronsList=
 Table[numberOfHiddenNeurons,{numberOfHiddenNeurons,1,15}]
```

$\{1,2,3,4,5,6,7,8,9,10,11,12,13,14,15\}$

but with a very small number of only 500 maximum iterations and an alternative Backpropagation plus Momentum (corrected steepest gradient descent) minimization algorithm (where parallelized calculation is used to accelerate the scan of mutually independent perceptrons, see Appendix A):

```
CIP`Utility`SetNumberOfParallelKernels[0];
maximumNumberOfIterations=500;
optimizationMethod="BackpropagationPlusMomentum";
perceptronInfoList=
 CIP`Perceptron`FitPerceptronSeries[dataSet3D,
  numberOfHiddenNeuronsList,
  PerceptronOptionMaximumIterations -> maximumNumberOfIterations,
  PerceptronOptionOptimizationMethod -> optimizationMethod,
  UtilityOptionCalculationMode -> "ParallelCalculation"];
perceptronSeriesRmse=
 CIP`Perceptron`GetPerceptronSeriesRmse[trainingAndTestSet,
  perceptronInfoList];
CIP`Perceptron`ShowPerceptronSeriesRmse[perceptronSeriesRmse]
```

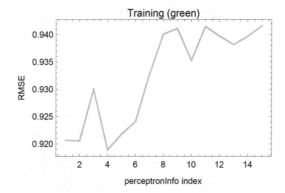

Best training set regression with perceptronInfo index = $\{4\}$

The RMSE of the trained perceptrons is not only almost an order of magnitude above the desired value of 0.1 but also an increased number of hidden neurons leads to poorer learning results (where the converse behaviour is expected, i.e. a drop of the RMSE with an increasing number of hidden neurons since the more bumps the more adequate the model function should be constructed). Thus it can be deduced that a maximum number of only 500 iterations is simply inadequate for the machine learning task and stops the minimzation process far too early somewhere over the rainbow. A tenfold increase to 5000 maximum iterations

```
maximumNumberOfIterations=5000;
perceptronInfoList=
 CIP`Perceptron`FitPerceptronSeries[dataSet3D,
  numberOfHiddenNeuronsList,
  PerceptronOptionMaximumIterations -> maximumNumberOfIterations,
  PerceptronOptionOptimizationMethod -> optimizationMethod,
  UtilityOptionCalculationMode -> "ParallelCalculation"];
perceptronSeriesRmse=
 CIP`Perceptron`GetPerceptronSeriesRmse[trainingAndTestSet,
  perceptronInfoList];
CIP`Perceptron`ShowPerceptronSeriesRmse[perceptronSeriesRmse]
```

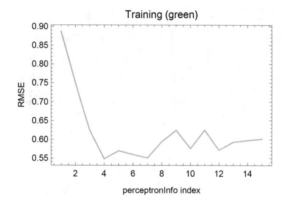

Best training set regression with perceptronInfo index = {4}

improves the learning result but is still not convincing for all but the smallest perceptrons. Another tenfold increase to 50000 maximum iterations

```
maximumNumberOfIterations=50000;
perceptronInfoList=
 CIP`Perceptron`FitPerceptronSeries[dataSet3D,
  numberOfHiddenNeuronsList,
  PerceptronOptionMaximumIterations -> maximumNumberOfIterations,
  PerceptronOptionOptimizationMethod -> optimizationMethod,
  UtilityOptionCalculationMode -> "ParallelCalculation"];
perceptronSeriesRmse=
 CIP`Perceptron`GetPerceptronSeriesRmse[trainingAndTestSet,
  perceptronInfoList];
CIP`Perceptron`ShowPerceptronSeriesRmse[perceptronSeriesRmse]
```

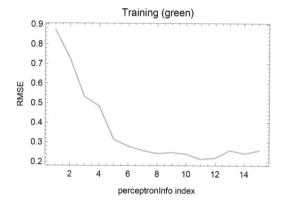

Training (green)

Best training set regression with perceptronInfo index = {11}

now leads to a more appealing scan which only seems to to have insufficient convergence for the biggest perceptrons with more than 11 hidden neurons. Despite the RMSE decrease to about 0.2 the perceptron approach may be interpreted to fail to achieve the desired value of about 0.1 thus a perceptron may not be judged as an adequate approach for the machine learning task in question. But if the same number of 50000 maximum iterations steps is used with the default Conjugate-Gradient minimization method (which is generally known to be superior to an even corrected simple steepest gradient descent)

```
perceptronInfoList=
 CIP`Perceptron`FitPerceptronSeries[dataSet3D,
   numberOfHiddenNeuronsList,
   PerceptronOptionMaximumIterations -> maximumNumberOfIterations,
   UtilityOptionCalculationMode -> "ParallelCalculation"];
perceptronSeriesRmse=
 CIP`Perceptron`GetPerceptronSeriesRmse[trainingAndTestSet,
   perceptronInfoList];
CIP`Perceptron`ShowPerceptronSeriesRmse[perceptronSeriesRmse]
```

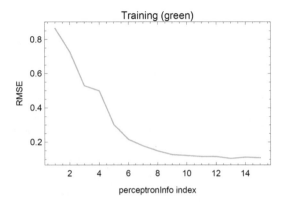

Best training set regression with perceptronInfo index = {13}

```
trainingPoints2D=perceptronSeriesRmse[[1]];
trainingPoints2D[[13]]
```

{13.,0.104956}

an overall convincing result is obtained with an RMSE of 0.1 for 13 hidden neurons. The Backpropagation plus Momentum algorithm simply failed to arrive at an adequate minimum with the set of technical optimization parameters chosen - where the latter statement is crucial: The algorithm is not only governed by the maximum number of iterations but also by a minimum and maximum learning rate (which determines the step size decrease during learning) as well as a momentum parameter (for correction of the steepest descent direction). These additional optional technical control parameters were set to their default values where alternative more adequate values could have improved the minimization result.

In summary technical problems may be properly detected and successfully fixed by adequate algorithm and parameter changes - but some evil potential remains: Among the worst consequences of technical problems are wrong decisions like an unfounded statement about the inapplicability of a specific machine learning approach or even the unnecessary termination of a research effort due to its apparent failure.

4.13 Cookbook recipes for machine learning

As in the earlier chapters the discussion of supervised machine learning is summarized in a number of cookbook recipes:

- **The data:** The quality of the data is essential for the success of a machine learning approach (also compare chapter 5). Since they may usually not be visually inspected due to their multiple dimensions specific care has to be taken. Outliers may play an evil role since they try to mask themselves even worse in comparison to curve fitting. Adequate data preprocessing may be crucial for a machine learning approach to be successful at all - the question of proper information encoding therefore is at heart of scientific disciplines like cheminformatics and bioinformatics. In practice the data generators (i.e. the lab scientists) and the data analyzers are often not identical. Therefore the latter should cooperate as closely as possible with the former to get a feeling about data quality (in fact it is this separation of scientists due to the inevitable division of labor and professions which is responsible for a lot of misinterpretation up to complete data analysis failure).

- **The linear and polynomial approach:** Nature is essentially non-linear. But a linear MLR machine learning approach is usually extremely fast (performed within seconds even for large data sets), not affected by technical or structural problems (if adequate software is used) and not prone to overfitting. Although a linear method will most often fail to get a successful result it may at least provide a feeling for the degree of non-linearity of the machine learning task under consideration. In practice many tasks may only be slightly non-linear and thus may be successfully tackled by the similarly fast polynomial MPR extension. Alternatively a more powerful non-linear method with structural hyperparameters that allow for near-linear regression or decision surfaces (e.g. realized with a bigger width-parameter for a wavelet kernel function or a small number of hidden neurons) also shows a faster performance. If a fast approach is already successful you are done! It is often astonishing that a powerful (but slow) non-linear method with all its subtleties and problems is applied where an extremely fast linear or polynomial approach would be equally successful.

- **Preparing training and test sets:** Unless there is a data set with known output errors for a regression task to assess the quality of a machine learning result the data must be partitioned into a training and test set. The concrete partitioning is a crucial step and an inadequate partitioning may spoil all further machine learning efforts. As outlined there is no ideal way available for a concrete partitioning and brute force strategies are not feasible. Thus this challenge has to be tackled by heuristic considerations in combination with related experience. As general rules of thumb a CR based training set selection is often superior to a purely random one and a training and test set of at least equal size are desirable. But these are only crude guidelines since every specific task requires its specific treatment.

- **The choice of method:** There are numerous machine learning methods and there is nothing like the single best choice for all purposes. On the other hand there are many ways to skin a cat: Usually a method is chosen on the basis of personal preferences or individual experiences. And if this method can be successfully applied there is no need to investigate alternatives. Only failure may motivate the evaluation of further methods.

- **Setting of structural and technical parameters:** Almost all machine learning procedures start with some default settings of its parameters which is known to be successful in similar tasks. At first an adjustment of the technical parameters is essential and afterwards the structural issues can be tackled with a proper technical setup. Unfortunately this sequential approach does not always work since there may be evil entanglements between technical and structural problems. Again there is no ideal way to solve these issues and in the end simple trial and error may be the road to success.

- **A proper estimation of computational effort:** Machine learning tasks may need considerable computational resources (multicore workstations up to grid computing) and long periods of time (ranging from hours over days and weeks up to months). Thus an adequate initial estimate of these requirements is indispensable. In practice estimates may be deduced from experience with similar tasks or preliminary investigations. In general machine learning efforts should not be initiated if there is no proper prospect for them to succeed (also compare chapter 5).
- **The interpretation of results:** Machine learning procedures may lead to results ranging from pure bullshit up to valuable and magically seeming insights. Due to their numerous parameters an awful lot can be tuned and problems can be subtly hidden. Thus there is a wide field for educated cheating already discussed in chapter 2. As a rule of thumb published machine learning results should always be regarded with care - again in general commercial applications are more trustworthy than academic claims. It is a scientifically venerable attitude to disclose the encountered and assumed problems for a specific machine learning task. All validation efforts should be outlined with care and thoroughness. Unfortunately a lot of publications lack these fundamentals which led machine learning to become somewhat dubious for many practitioners.

If properly applied supervised machine learning can be an extremely valuable tool to tackle complex and difficult scientific problems that otherwise could not be mastered. The sketched problems are remarkable but so are the possible benefits. Despite its wide range of applicability due to its universal character supervised machine learning has of course its limitations that are usually determined by the provided data: Machine learning is not able to extract something out of nothing - if it is not in the data it can not be modelled. The foundation of its magic remains the happenstance that a complex and non-trivial relationship (i.e. a non-linear model/decision function) may be created without further instructions or superior knowledge. The final next chapter discusses some of its consequences for the generation of new knowledge and the views of computational intelligence.

4.14 Appendix - Collecting the pieces

```
Clear["Global`*"];
<<CIP`Utility`
<<CIP`ExperimentalData`
<<CIP`DataTransformation`
<<CIP`Graphics`
```

```
<<CIP`Cluster`
<<CIP`MLR`
<<CIP`MPR`
<<CIP`Perceptron`
<<CIP`SVM`
```

In the following two complete machine learning approaches are outlined that comprise a number of topics already discussed in this and the previous chapter. This section contains no new material and is simply redundant from a scientific point of view. But redundancy is a virtue when trying to dig into a new discipline.

The first application chosen for demonstration from the field of medical decision support is easy to comprehend, an important area of research and it attracts a considerable attention (not only) from the machine learning community: The Wisconsin Diagnostic Breast Cancer (WDBC) data correlate features of cell nuclei extracted from breast tumor tissue (as input) with the tumor type after diagnosis (as output), i.e. they map the cell nuclei features onto a diagnosed benign (class 1) or malignant (class 2) tumor type (see Appendix A for details and [WDBC data] in the references). Thus the WDBC data may be used to construct a class predictor that supports the crucial benign/malignant decision in tumor diagnosis. The WDBC classification data set is available from the CIP ExperimentalData package:

```
classificationDataSet=
  CIP`ExperimentalData`GetWDBCClassificationDataSet[];
CIP`Graphics`ShowDataSetInfo[{"IoPairs","InputComponents",
    "OutputComponents","ClassCount"},classificationDataSet]
```

Number of IO pairs = 569

Number of input components = 30

Number of output components = 2

Class 1 with 357 members

Class 2 with 212 members

It consists of 569 I/O pairs (i.e. the diagnoses of 569 different patients with breast tumors) where each input consists of 30 components and each output of 2 components. Each input component describes a single quantity of a feature of a tumor tissue's cell nuclei and the two output components denote classes 1 (benign tumor, coded $\{1.0, 0.0\}$) and 2 (malignant tumor, coded $\{0.0, 1.0\}$). The number of benign tumor samples (357) exceeds the number of malignant samples (212) thus the classification data set is asymmetric with the benign class samples being overrepresented:

```
N[357/212]
```

1.68396

The first sensible step to construct a benign/malignant class predictor should be an unsupervised learning trial with a purely clustering-based class predictor. This predictor exploits only the spatial distribution of inputs in the input space to classify a single input (thus the predictor is unsupervised and not at all prone to overfit data): If a clustering-based class predictor exhibits a 100% predictive success rate this means that the inputs of benign and malignant tumors form clearly separated point clouds in the 30 dimensional input space. If the predictive success rate is less than a 100% the point clouds of benign and malignant tumors do overlap, i.e. they penetrate each other in some way. A clustering-based class predictor is constructed with the FitCluster method of the CIP Cluster package:

```
clusterInfo=CIP'Cluster'FitCluster[classificationDataSet];
CIP'Cluster'ShowClusterSingleClassification[{
  "CorrectClassification","CorrectClassificationPerClass",
  "WrongClassificationDistribution"},classificationDataSet,
  clusterInfo]
```

85.4% correct classifications

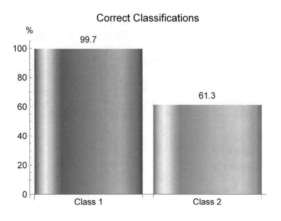

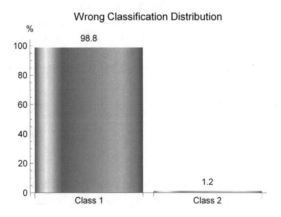

The overall predictive success rate of 85% shows that the benign and malignant point clouds in the input space are structured but they do not form clearly separated clusters. With a success rate of 99% the predictivity of class 1 benign tumors is nearly perfect but class 2 malignant predictions with only 61% success are rather poor (in medial practice this would mean that on average 39 out of 100 woman with malignant breast tumors would be diagnosed to have a benign tumor: A catastrophic result that would lead to a completely inadequate medical treatment. But if a tumor is predicted to be malignant this would be a comparatively reliable result since wrong class 2 predictions are rare with only 1% among all wrong predictions). In conclusion unsupervised learning does not seem to be able to successfully tackle the benign/malignant decision problem. Before turning to supervised machine learning methods (which will take the benign/malignant output diagnoses into account to control the learning process) a purely technical issue may be of interest. The FitCluster method above was called without any further technical parameters so it used its internal defaults (i.e. the default k-medoids clustering algorithm). If the clustering algorithm is changed to ART-2a an unexpected error occurs:

```
clusterMethod="ART2a";
clusterInfo=CIP`Cluster`FitCluster[classificationDataSet,
 ClusterOptionMethod -> clusterMethod];
```

Infinite expression 1/0 encountered.

If a computational method fails there are two possibilities: The method's code contains a bug or the method is not able to arrive at the desired result in principal. To analyze the ART-2a error it is necessary to have a look at the pure clustering of the inputs of the WDBC classification data set:

```
inputs=CIP`Utility`GetInputsOfDataSet[classificationDataSet];
```

Since ART-2a is fundamentally controlled by a vigilance parameter we set this parameter to a very small value to urge the method to construct only a few large clusters:

```
clusterMethod="ART2a";
vigilanceParameter=0.01;
clusterInfo = CIP`Cluster`GetClusters[inputs,
 ClusterOptionMethod -> clusterMethod,
 ClusterOptionVigilanceParameter -> vigilanceParameter];
CIP`Cluster`ShowClusterResult[{"NumberOfClusters"},clusterInfo]
```

Number of clusters = 9

With a very small vigilance parameter of 0.01 ART-2a detects 9 clusters. If the dependence of the detected number of clusters on the vigilance parameter is explored (first over the full range from 0 to 1

```
minimumVigilanceParameter=0.01;
maximumVigilanceParameter=0.99;
numberOfScanPoints=30;
art2aScanInfo = CIP`Cluster`GetVigilanceParameterScan[inputs,
 minimumVigilanceParameter,maximumVigilanceParameter,
 numberOfScanPoints];
CIP`Cluster`ShowVigilanceParameterScan[art2aScanInfo]
```

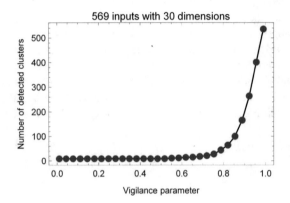

and then confined to the smaller range from 0 to 0.5)

```
minimumVigilanceParameter=0.01;
maximumVigilanceParameter=0.5;
numberOfScanPoints=30;
art2aScanInfo = CIP`Cluster`GetVigilanceParameterScan[inputs,
 minimumVigilanceParameter,maximumVigilanceParameter,
 numberOfScanPoints];
CIP`Cluster`ShowVigilanceParameterScan[art2aScanInfo]
```

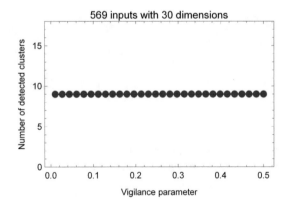

it becomes obvious that the 9-cluster result is the method's answer over a wide range of low vigilance. If we force the ART-2a algorithm to reduce the vigilance parameter towards zero to arrive at exactly 2 clusters (decreasing the vigilance parameter means fewer and larger clusters)

```
numberOfClusters=2;
clusterInfo = CIP'Cluster'GetFixedNumberOfClusters[inputs,
  numberOfClusters,ClusterOptionMethod -> clusterMethod];
```

Infinite expression 1/0 encountered.

the error occurs and that is exactly what happened during the predictor construction: Also in the limit of a vanishingly small vigilance parameter ART-2a is not able to detect the desired 2 classes - and in numerical computing with a limited number of digits a vanishing small value will inevitably be equated with zero and lead to problems. Thus ART-2a is simply inadequate to construct a clustering-based class predictor for the particular classification data set in question. This finding emphasizes the need for a growing tool box of comparable computational methods where the one will fail in a particular situation while another may possibly succeed.

When it comes to supervised machine learning it is always advised to start with a linear method. A Multiple Linear Regression (MLR) method performs very fast and is not prone to overfitting - but of course limited in principle due to its linear nature. If MLR is used to construct a benign/malignant class predictor on the basis of the complete classification data set

```
mlrInfo=CIP'MLR'FitMlr[classificationDataSet];
CIP'MLR'ShowMlrSingleClassification[{"CorrectClassification",
  "CorrectClassificationPerClass",
  "WrongClassificationDistribution"},classificationDataSet,mlrInfo]
```

96.5% correct classifications

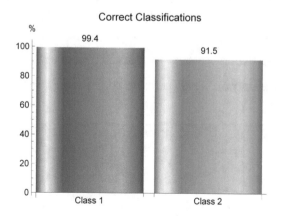

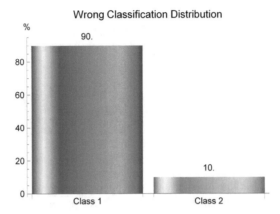

Wrong Classification Distribution

a remarkable overall predictive success rate of over 96% is achieved where the predictivity of benign class 1 tumors with a nearly perfect 99% success rate is again clearly superior to the malignant class 2 predictions with only 91%. But compared to the unsupervised learning approach before the supervised learning improves the malignant class 2 predictivity significantly from a 61% to a 91% success rate! (Again a malignant prediction is comparatively reliable due to only 10% wrong class 2 predictions among all wrong predictions - and this finding remains valid in the following but is skipped as a subtlety to ease the discussion.) The relative success of the linear approach suggests that the classification problem in question can be characterized as near-linear, i.e. it probably does not demand highly non-linear curved decision surfaces. This has to be taken into account when the non-linear methods will be applied. But before we can get even more from the MLR approach. In general it is desirable to construct a minimal predictive model, i.e. a model with the smallest subset of input components/features that is able to successfully predict the tumor type. Thus a relevance analysis of the input components is indicated. Since this analysis is ideally suited for parallelized computation (the relevance of an input component may be evaluated independently from those of other input components) the CIP parallel calculation option should be initialised (see Appendix A for details):

```
CIP`Utility`SetNumberOfParallelKernels[0];
```

If the relevance analysis is restricted to the six most significant input components only

```
trainingAndTestSet={classificationDataSet,{}};
numberOfInclusionsPerStepList={1,1,1,1,1,1};
mlrInputComponentRelevanceListForClassification=
  CIP`MLR`GetMlrInputInclusionClass[trainingAndTestSet,
    UtilityOptionInclusionsPerStep -> numberOfInclusionsPerStepList,
    UtilityOptionCalculationMode -> "ParallelCalculation"];
CIP`MLR`ShowMlrInputRelevanceClass[
  mlrInputComponentRelevanceListForClassification]
```

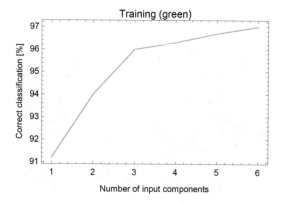

Input component list = $\{28, 21, 22, 24, 15, 29\}$

```
numberOfComponents=6;
inputComponentInclusionList=
 CIP`MLR`GetMlrClassRelevantComponents[
  mlrInputComponentRelevanceListForClassification,
  numberOfComponents]
```

$\{28, 21, 22, 24, 15, 29\}$

and the classification data set is reduced to these six input components

```
reducedDataSet=
 CIP`DataTransformation`IncludeInputComponentsOfDataSet[
  classificationDataSet,inputComponentInclusionList];
```

a MLR fit

```
mlrInfo=CIP`MLR`FitMlr[reducedDataSet];
CIP`MLR`ShowMlrSingleClassification[{"CorrectClassification",
 "CorrectClassificationPerClass",
 "WrongClassificationDistribution"},reducedDataSet,mlrInfo]
```

97.% correct classifications

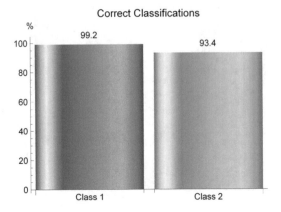

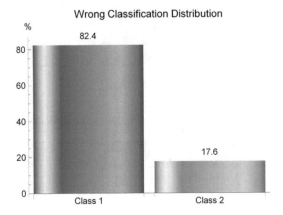

leads to a classification success comparable to the one with the complete classification data set (the correct classifications are even slightly improved - a subtlety which is caused by the fact that within CIP a classification task is internally coded as a corresponding regression task where a decreased RMSE does not necessarily lead to an improved classification because classification follows a winner-take-all principle, see chapter 1). For an additional assessment of the predictive power of the minimal model the reduced classification data set is split into equally sized and heuristically optimized training and test sets (by use of the default SingleGlobal-Max optimization strategy with blacklisting to avoid oscillations - to finally arrive at comparable training and test inputs that cover a similar input space with a similar spatial diversity, compare above and chapter 3):

```
trainingFraction=0.50;
numberOfTrainingSetOptimizationSteps=10;
blackListLength=5;
```

```
mlrTrainOptimization=
  CIP`MLR`GetMlrTrainOptimization[reducedDataSet,trainingFraction,
  numberOfTrainingSetOptimizationSteps,
  UtilityOptionBlackListLength -> blackListLength];
bestOptimization="MinimumDeviation";
bestClassificationStep=CIP`MLR`GetBestMlrClassOptimization[
  mlrTrainOptimization,
  UtilityOptionBestOptimization -> bestOptimization,
  UtilityOptionCalculationMode -> "ParallelCalculation"];
optimizedTrainingAndTestSetList=mlrTrainOptimization[[3]];
bestTrainingAndTestSet=
  optimizedTrainingAndTestSetList[[bestClassificationStep]];
```

The training set becomes enriched in the underrepresented malignant class 2 tumor samples

```
bestTrainingSet=bestTrainingAndTestSet[[1]];
CIP`Graphics`ShowDataSetInfo[{"ClassCount"},bestTrainingSet]
```

Class 1 with 166 members

Class 2 with 118 members

```
N[166/118]
```

1.40678

and the test set is reciprocally depleted (compare benign/malignant ratios with ratio 1.68 of the complete data set above):

```
bestTestSet=bestTrainingAndTestSet[[2]];
CIP`Graphics`ShowDataSetInfo[{"ClassCount"},bestTestSet]
```

Class 1 with 191 members

Class 2 with 94 members

```
N[191/94]
```

2.03191

The MLR classification success for both sets

```
optimizedMlrInfoList=mlrTrainOptimization[[4]];
bestMlrInfo=optimizedMlrInfoList[[bestClassificationStep]];
CIP`MLR`ShowMlrClassificationResult[{"CorrectClassification"},
  bestTrainingAndTestSet,bestMlrInfo]
```

Training Set:

96.8% correct classifications

Test Set:

97.9% correct classifications

is comparable and demonstrates a predictivity that corresponds to the findings for the complete classification data set. It is interestingly enough and not uncommon that the majority - 80% in this case - of the input components can be omitted without any significant loss of predictivity: The quantities used to describe the cell nuclei reveal a high mutual redundancy with respect to the classification task although this may not be obvious to the medicinal scientist - an insight that can motivate further investigations: A reduced number of necessary input components/features usually not only simplifies the computational machine learning process but alleviates the whole medical diagnosis procedure (but compare additional comments about method dependencies below).

To investigate possible further improvements the non-linear machine learning methods must be taken into account. But for non-linear methods overfitting will become a severe problem - again note that a non-linear method will almost always arrive at a perfect 100% predictor by creating a simply overfitted look-up table for the data. In addition the non-linear methods require the definition of structural hyperparameters for their operation. Since the WDBC data set classification task could be characterized above as near-linear a Multiple Polynomial Regression (MPR) approach with a polynomial degree of only 2

```
polynomialDegree=2;
```

may be regarded as the first adequate step into non-linearity. But also this humble MPR fit

```
mprInfo = CIP`MPR`FitMpr[classificationDataSet,polynomialDegree];
CIP`MPR`ShowMprSingleClassification[{"CorrectClassification",
  "CorrectClassificationPerClass"},classificationDataSet,mprInfo]
```

100.% correct classifications

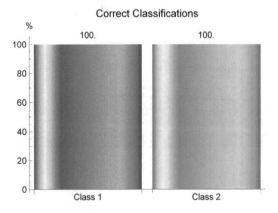

leads directly to a perfect 100% correct classifications which indicates an un-wanted overfitted model without predictivity. This result becomes intelligible if the number of parameters of the MPR model is investigated

```
CIP`MPR`GetMprNumberOfParameters[classificationDataSet,
  polynomialDegree]
```

496

which is similar to the number of I/O pairs of the complete classification data set. Thus the construction of a minimal predictive model by reduction of the input space is advised (with a corresponding reduction of MPR parameters) which may again be performed by a relevance analysis of the input components:

```
mprInputComponentRelevanceListForClassification=
 CIP`MPR`GetMprInputInclusionClass[trainingAndTestSet,
  polynomialDegree,
  UtilityOptionInclusionsPerStep -> numberOfInclusionsPerStepList,
UtilityOptionCalculationMode -> "ParallelCalculation"];
CIP`MPR`ShowMprInputRelevanceClass[
 mprInputComponentRelevanceListForClassification]
```

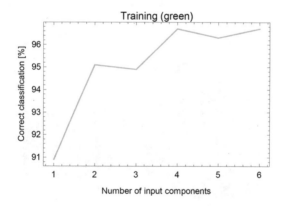

Input component list = {24, 28, 21, 22, 25, 11}

```
inputComponentInclusionList=
 CIP`MPR`GetMprClassRelevantComponents[
  mprInputComponentRelevanceListForClassification,
  numberOfComponents]
```

{24, 28, 21, 22, 25, 11}

Note that the detected most relevant input components do not coincide with those of the MLR approach before which means that the relevance of an input component in general depends on the machine learning method used and can not be regarded as an objective characteristic of the specific input feature for the machine learning task in question - a statement that pours some cold water on the probably too enthusiastic statements about the relevance of input components made above. There is an additional finding that the number of detected relevant MPR input components is smaller in comparison to the MLR approach (i.e. for MPR only 4 input components seem to be sufficient for a minimal predictive model whereas 6 input components were necessary for MLR - but for simplicity the discussion is continued with the same number of 6 input components). The reduced classification data set (with 6 input components)

```
reducedDataSet=
 CIP`DataTransformation`IncludeInputComponentsOfDataSet[
  classificationDataSet,inputComponentInclusionList];
```

leads to a MPR fit with a classification success,

```
mprInfo=CIP`MPR`FitMpr[reducedDataSet,polynomialDegree];
CIP`MPR`ShowMprSingleClassification[{"CorrectClassification",
 "CorrectClassificationPerClass",
 "WrongClassificationDistribution"},reducedDataSet,mprInfo]
```

96.7% correct classifications

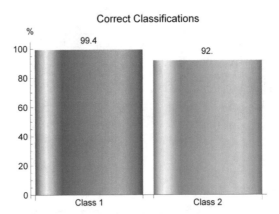

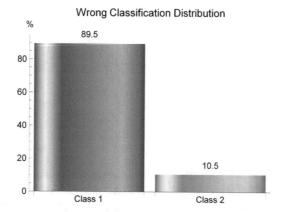

training/test set enrichments/depletions

```
mprTrainOptimization=
 CIP`MPR`GetMprTrainOptimization[reducedDataSet, polynomialDegree,
  trainingFraction,numberOfTrainingSetOptimizationSteps,
  UtilityOptionBlackListLength -> blackListLength];
bestClassificationStep=CIP`MPR`GetBestMprClassOptimization[
 mprTrainOptimization,
 UtilityOptionBestOptimization -> bestOptimization,
 UtilityOptionCalculationMode -> "ParallelCalculation"];
optimizedTrainingAndTestSetList=mprTrainOptimization[[3]];
bestTrainingAndTestSet=
 optimizedTrainingAndTestSetList[[bestClassificationStep]];
bestTrainingSet=bestTrainingAndTestSet[[1]];
CIP`Graphics`ShowDataSetInfo[{"ClassCount"},bestTrainingSet]
```

Class 1 with 166 members

Class 2 with 118 members

```
N[166/118]
```

1.40678

```
bestTestSet=bestTrainingAndTestSet[[2]];
CIP`Graphics`ShowDataSetInfo[{"ClassCount"},bestTestSet]
```

Class 1 with 191 members

Class 2 with 94 members

```
N[191/94]
```

2.03191

and a predictive power

```
optimizedMprInfoList=mprTrainOptimization[[4]];
bestMprInfo=optimizedMprInfoList[[bestClassificationStep]];
CIP`MPR`ShowMprClassificationResult[{"CorrectClassification"},
 bestTrainingAndTestSet,bestMprInfo]
```

Training Set:

97.5% correct classifications

Test Set:

97.5% correct classifications

which is comparable to the one of the MLR approach but no significant improvement. The MLR and MPR results may serve as a comparison to the prediction results of the highly non-linear methods which are now being explored. For three-layer perceptrons the crucial structural hyperparameter is the number of hidden neurons. For a near-linear problem a very small number of hidden neurons is advised so that the perceptron's ability to produce bumps is restricted and as a consequence its proneness to overfitting is reduced. If a minimum of 2 hidden neurons is used

```
numberOfHiddenNeurons=2;
```

again a suspicious classification success with possible overfitting appears:

```
perceptronInfo=CIP`Perceptron`FitPerceptron[classificationDataSet,
 numberOfHiddenNeurons,
 UtilityOptionCalculationMode -> "ParallelCalculation"];
CIP`Perceptron`ShowPerceptronSingleClassification[{
 "CorrectClassification","CorrectClassificationPerClass"},
 classificationDataSet,perceptronInfo]
```

99.6% correct classifications

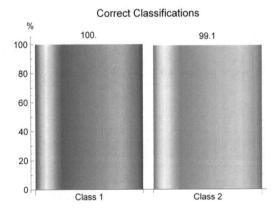

Relevance analysis of the input components

```
perceptronInputComponentRelevanceListForClassification=
 CIP 'Perceptron'GetPerceptronInputInclusionClass[trainingAndTestSet,
  numberOfHiddenNeurons,
  UtilityOptionInclusionsPerStep -> numberOfInclusionsPerStepList,
  UtilityOptionCalculationMode -> "ParallelCalculation"];
CIP 'Perceptron'ShowPerceptronInputRelevanceClass[
 perceptronInputComponentRelevanceListForClassification]
```

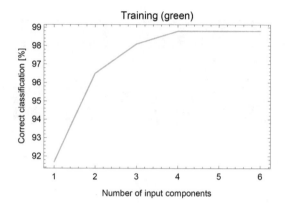

Input component list = {23, 25, 2, 4, 1, 7}

```
inputComponentInclusionList=
 CIP 'Perceptron'GetPerceptronClassRelevantComponents[
  perceptronInputComponentRelevanceListForClassification,
  numberOfComponents]
```

{23, 25, 2, 4, 1, 7}

with corresponding data set reduction

```
reducedDataSet =
 CIP 'DataTransformation'IncludeInputComponentsOfDataSet[
  classificationDataSet, inputComponentInclusionList];
```

and training/test splitting

```
perceptronTrainOptimization=
 CIP 'Perceptron'GetPerceptronTrainOptimization[reducedDataSet,
  numberOfHiddenNeurons, trainingFraction,
  numberOfTrainingSetOptimizationSteps,
  UtilityOptionBlackListLength -> blackListLength,
  UtilityOptionCalculationMode -> "ParallelCalculation"];
```

```
bestClassificationStep=
 CIP`Perceptron`GetBestPerceptronClassOptimization[
  perceptronTrainOptimization,
  UtilityOptionBestOptimization -> bestOptimization,
  UtilityOptionCalculationMode -> "ParallelCalculation"];
optimizedTrainingAndTestSetList=perceptronTrainOptimization[[3]];
bestTrainingAndTestSet=
 optimizedTrainingAndTestSetList[[bestClassificationStep]];
bestTrainingSet=bestTrainingAndTestSet[[1]];
CIP`Graphics`ShowDataSetInfo[{"ClassCount"},bestTrainingSet]
```

Class 1 with 169 members

Class 2 with 115 members

N[169/115]

1.46957

```
bestTestSet=bestTrainingAndTestSet[[2]];
CIP`Graphics`ShowDataSetInfo[{"ClassCount"},bestTestSet]
```

Class 1 with 188 members

Class 2 with 97 members

N[188/97]

1.93814

results in a convincing minimal model with excellent predictive success rates, very good generalization abilities and a slightly improved predictive power compared to the MLR/MPR approach:

```
optimizedPerceptronInfoList=perceptronTrainOptimization[[4]];
bestPerceptronInfo=
 optimizedPerceptronInfoList[[bestClassificationStep]];
CIP`Perceptron`ShowPerceptronClassificationResult[{
 "CorrectClassification"},bestTrainingAndTestSet,
 bestPerceptronInfo]
```

Training Set:

98.6% correct classifications

Test Set:

98.6% correct classifications

Again note that compared to MLR the number of detected relevant input components is decreased (i.e. for the perceptron approach only 4 input components seem to be sufficient for a minimal predictive model which was also found above for MPR with a polynomial degree of 2).

For sake of completeness a SVM approach is finally investigated where the crucial structural hyperparameter of a SVM - the kernel function - is in need. Again the near-linear classification problem characterization of the linear MLR approach is helpful - if we arbitrarily choose a wavelet kernel an adequate width parameter in a region that corresponds to wider bumps (which do not allow highly non-linear and very curved decision surfaces) is advised:

```
kernelFunction={"Wavelet",2.0};
```

The complete classification data set fit

```
svmInfo=CIP'SVM'FitSvm[classificationDataSet, kernelFunction,
 UtilityOptionCalculationMode -> "ParallelCalculation"];
CIP'SVM'ShowSvmSingleClassification[{"CorrectClassification",
 "CorrectClassificationPerClass"},classificationDataSet,svmInfo]
```

98.9% correct classifications

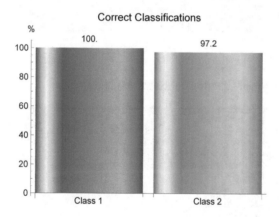

again seems to be too (overfitted) perfect. Input component relevance analysis

```
svmInputComponentRelevanceListForClassification=
 CIP'SVM'GetSvmInputInclusionClass[trainingAndTestSet,
  kernelFunction,
  UtilityOptionInclusionsPerStep -> numberOfInclusionsPerStepList,
  UtilityOptionCalculationMode -> "ParallelCalculation"];
CIP'SVM'ShowSvmInputRelevanceClass[
 svmInputComponentRelevanceListForClassification]
```

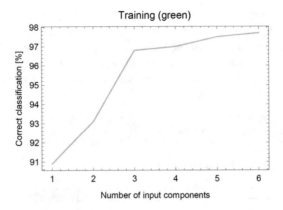

Input component list = $\{8, 24, 22, 28, 3, 21\}$

```
inputComponentInclusionList=
 CIP'SVM'GetSvmClassRelevantComponents[
  svmInputComponentRelevanceListForClassification,
  numberOfComponents]
```

$\{8, 24, 22, 28, 3, 21\}$

and the sequence of already established modelling steps

```
reducedDataSet=
 CIP'DataTransformation'IncludeInputComponentsOfDataSet[
  classificationDataSet, inputComponentInclusionList];
svmTrainOptimization=CIP'SVM'GetSvmTrainOptimization[
 reducedDataSet,kernelFunction,
 trainingFraction,numberOfTrainingSetOptimizationSteps,
 UtilityOptionBlackListLength -> blackListLength,
 UtilityOptionCalculationMode -> "ParallelCalculation"];
bestClassificationStep=CIP'SVM'GetBestSvmClassOptimization[
 svmTrainOptimization,
 UtilityOptionBestOptimization -> bestOptimization,
 UtilityOptionCalculationMode -> "ParallelCalculation"];
optimizedTrainingAndTestSetList=svmTrainOptimization[[3]];
bestTrainingAndTestSet=
 optimizedTrainingAndTestSetList[[bestClassificationStep]];
bestTrainingSet=bestTrainingAndTestSet[[1]];
CIP'Graphics'ShowDataSetInfo[{"ClassCount"},bestTrainingSet]
```

Class 1 with 169 members
Class 2 with 115 members

```
N[169/115]
```

1.46957

```
bestTestSet=bestTrainingAndTestSet[[2]];
CIP`Graphics`ShowDataSetInfo[{"ClassCount"},bestTestSet]
```

Class 1 with 188 members

Class 2 with 97 members

```
 N[188/97]
```

1.93814

finally leads to a true predictive minimal model which is comparable to the MLR/MPR approach but slightly inferior to the perceptron results before:

```
optimizedSvmInfoList=svmTrainOptimization[[4]];
bestSvmInfo=optimizedSvmInfoList[[bestClassificationStep]];
CIP`SVM`ShowSvmClassificationResult[{"CorrectClassification"},
 bestTrainingAndTestSet,bestSvmInfo]
```

Training Set:

97.2% correct classifications

Test Set:

97.9% correct classifications

In summary it can be concluded that the WDBC classification problem could be successfully tackled by supervised machine learning (further refinements like additional hyperparameter variations would be time-consuming and the room for improvement appears to be only small). It was not only possible to construct convincing minimal class predictors of sufficient quality that can successfully support diagnostic decisions in medical practice but also (unfortunately: method dependent) insights about the necessary features of a cell nucleus for successful benign/malignant classification could be stimulated.

As a second application the construction of a Quantitative Structure Property Relationship (QSPR) is briefly investigated. A QSPR model maps features of chemical structures (the input) to a property of interest (the output). The features of a chemical structure are so called (structural/molecular) descriptors, i.e. characteristic numbers which are calculated for the specific structure. The output quantity of interest is usually measured experimentally. The following QSPR data set

```
dataSet=CIP`ExperimentalData`GetQSPRDataSet01[];
CIP`Graphics`ShowDataSetInfo[{"IoPairs","InputComponents",
 "OutputComponents"},dataSet]
```

Number of IO pairs = 183

Number of input components = 155

Number of output components = 1

comprises 183 input/output (I/O) pairs that each represent a single chemical compound. Each input vector consists of 155 components where each component is a structural descriptor. Each corresponding output vector contains a single experimentally measured value for a compound-related physico-chemical quantity. Since the number of input components is nearly equal to the number of I/O pairs the usefulness of the whole data set is in doubt ("one point per axis"), i.e. also a linear MLR method is likely to be able to establish a perfect relationship between the molecular descriptors and the corresponding physico-chemical quantity

```
mlrInfo = CIP 'MLR 'FitMlr[dataSet];
CIP 'MLR 'ShowMlrSingleRegression[{"ModelVsDataPlot","RMSE"},
  dataSet,mlrInfo];
```

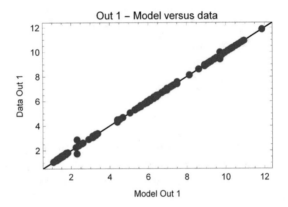

Root mean squared error (RMSE) = 7.145×10^{-2}

but without any predictive power: The model is a pure look-up table. This may be shown by partitioning of the whole data into a training and a test set of equal size (with a number of heuristic optimization trials of both sets already sketched above):

```
trainingFraction=0.5;
numberOfTrainingSetOptimizationSteps=10;
blackListLength=5;
mlrTrainOptimization=CIP 'MLR 'GetMlrTrainOptimization[dataSet,
  trainingFraction,numberOfTrainingSetOptimizationSteps,
  UtilityOptionBlackListLength -> blackListLength];
bestOptimization="MinimumDeviation";
bestRegressionStep=CIP 'MLR 'GetBestMlrRegressOptimization[
  mlrTrainOptimization,
  UtilityOptionBestOptimization -> bestOptimization];
trainingAndTestSetList = mlrTrainOptimization[[3]];
mlrInfoList=mlrTrainOptimization[[4]];
trainingAndTestSet=trainingAndTestSetList[[bestRegressionStep]];
bestMlrInfo=mlrInfoList[[bestRegressionStep]];
CIP 'MLR 'ShowMlrRegressionResult[{"ModelVsDataPlot","RMSE"},
```

```
trainingAndTestSet,bestMlrInfo]
```

Training Set:

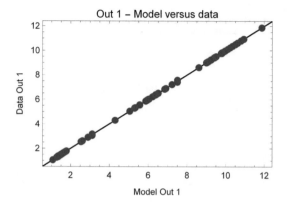

Root mean squared error (RMSE) = 1.618×10^{-2}

Test Set:

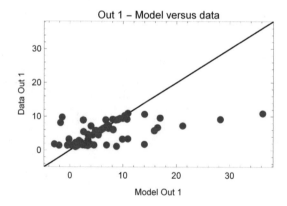

Root mean squared error (RMSE) = 5.075

The training data are perfectly "tabulated" but the prediction of the test data is extremely poor. In cheminformatics it is well known that structural descriptors are usually calculated with a do-everything-possible attitude in mind so that QSPR input spaces are often extremely oversized with mutually highly redundant descriptors. If the relevance of descriptors (input components) is successively evaluated with MLR

```
trainingAndTestSet={dataSet,{}};
numberOfInclusionsPerStepList=Table[1,{10}];
mlrInputComponentRelevanceListForRegression=
  CIP`MLR`GetMlrInputInclusionRegress[trainingAndTestSet,
   UtilityOptionInclusionsPerStep -> numberOfInclusionsPerStepList,
   UtilityOptionCalculationMode -> "ParallelCalculation"];
CIP`MLR`ShowMlrInputRelevanceRegress[
  mlrInputComponentRelevanceListForRegression]
```

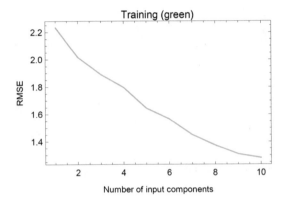

Input component list = $\{43, 102, 111, 128, 67, 100, 116, 97, 72, 14\}$

and MPR (with the most modest polynomial degree of 2)

```
polynomialDegree=2;
mprInputComponentRelevanceListForRegression=
  CIP`MPR`GetMprInputInclusionRegress[trainingAndTestSet,
   polynomialDegree,
   UtilityOptionInclusionsPerStep -> numberOfInclusionsPerStepList,
   UtilityOptionCalculationMode -> "ParallelCalculation"];
CIP`MPR`ShowMprInputRelevanceRegress[
  mprInputComponentRelevanceListForRegression]
```

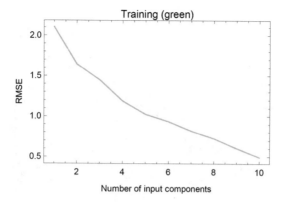

Input component list = {43, 102, 31, 91, 140, 42, 84, 120, 123, 113}

it becomes obvious that (for the MPR approach) only 5 input components (descriptors) out of 155

```
numberOfComponents=5;
inputComponentInclusionList =
 CIP`MPR`GetMprRegressRelevantComponents[
  mprInputComponentRelevanceListForRegression,numberOfComponents]
```

{43, 102, 31, 91, 140}

are necessary to arrive at an adequate RMSE of 1.0 for at least a semi-quantitative minimal model

```
reducedDataSet=
 CIP`DataTransformation`IncludeInputComponentsOfDataSet[dataSet,
  inputComponentInclusionList];
mprTrainOptimization=
 CIP`MPR`GetMprTrainOptimization[reducedDataSet,polynomialDegree,
  trainingFraction,numberOfTrainingSetOptimizationSteps,
  UtilityOptionBlackListLength -> blackListLength];
bestRegressionStep=
 CIP`MPR`GetBestMprRegressOptimization[mprTrainOptimization,
  UtilityOptionBestOptimization -> bestOptimization];
optimizedTrainingAndTestSetList=mprTrainOptimization[[3]];
mprInfoList=mprTrainOptimization[[4]];
bestTrainingAndTestSet=
 optimizedTrainingAndTestSetList[[bestRegressionStep]];
bestMprInfo=mprInfoList[[bestRegressionStep]];
CIP`MPR`ShowMprRegressionResult[{"ModelVsDataPlot","RMSE"},
 bestTrainingAndTestSet,bestMprInfo]
```

Training Set:

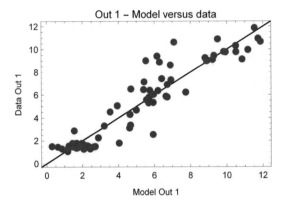

Root mean squared error (RMSE) = 1.071

Test Set:

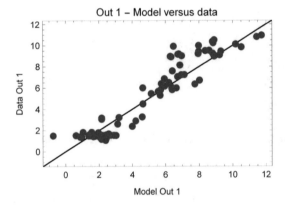

Root mean squared error (RMSE) = 1.072

which exhibits equal predictive power for training and test data and thus may be helpful in practice. But already a magnification of the input space with the 10 most relevant MPR descriptors out of 155

```
numberOfComponents=10;
inputComponentInclusionList=
 CIP`MPR`GetMprRegressRelevantComponents[
  mprInputComponentRelevanceListForRegression, numberOfComponents]
```

{43, 102, 31, 91, 140, 42, 84, 120, 123, 113}

again leads to overfitting with an overall loss of any predictivity:

```
reducedDataSet=
 CIP'DataTransformation'IncludeInputComponentsOfDataSet[dataSet,
  inputComponentInclusionList];
mprTrainOptimization=
 CIP'MPR'GetMprTrainOptimization[reducedDataSet,polynomialDegree,
  trainingFraction, numberOfTrainingSetOptimizationSteps,
  UtilityOptionBlackListLength -> blackListLength];
bestRegressionStep=
 CIP'MPR'GetBestMprRegressOptimization[mprTrainOptimization,
 UtilityOptionBestOptimization -> bestOptimization];
optimizedTrainingAndTestSetList=mprTrainOptimization[[3]];
mprInfoList=mprTrainOptimization[[4]];
bestTrainingAndTestSet=
 optimizedTrainingAndTestSetList[[bestRegressionStep]];
bestMprInfo=mprInfoList[[bestRegressionStep]];
CIP'MPR'ShowMprRegressionResult[{"ModelVsDataPlot",
 "RMSE"},bestTrainingAndTestSet,bestMprInfo]
```

Training Set:

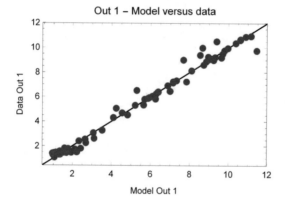

Root mean squared error (RMSE) = 4.194×10^{-1}
Test Set:

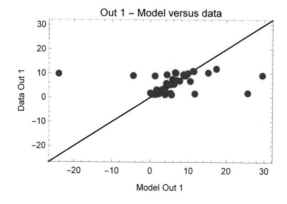

Root mean squared error (RMSE) = 5.432

Further (minor) improvements may be obtained with adequate hyperparameter-optimized perceptron or SVM approaches. As a conclusion it may be kept in mind that careful machine learning may also reveal helpful relationships for unfavorable data but of course can not overcome the GIGO (garbage-in/garbage-out) effect in general.

Chapter 5
Discussion

At the end of a tour from curve fitting to machine learning there are two kinds of questions that usually remain: The first kind is about the numerous details and side branches of the sketched topics that had to be omitted for the sake of readability and comprehensibility since limitations are inevitable and a bunch of important and interesting issues had to be skipped. The second kind of questions addresses the more abstract and general aspects that arise from the earlier discussions like the principal capabilities of machine learning. In this final chapter some so far neglected topics that belong to both kinds of questions are outlined.

First a crucial aspect of computation is discussed: Speed. A proper estimate of the time period necessary to perform a computational task is essential for almost all practical applications (section 5.1). After an initial fascination a deeper insight into machine learning often leads to a notion of disappointment about what can be expected from these methods in principal thus some basic possibilities and limits are discussed (section 5.2). The relations of the methods outlined on the road from curve fitting to machine learning to a possibly emerging computational intelligence are of general interest and thus briefly sketched (section 5.3). Final remarks close this chapter (section 5.4).

5.1 Computers are about speed

```
Clear["Global`*"];
<<CIP`CalculatedData`
<<CIP`Graphics`
<<CIP`MLR`
<<CIP`MPR`
<<CIP`Perceptron`
<<CIP`SVM`
<<CIP`CurveFit`
```

© Springer International Publishing Switzerland 2016   407
A. Zielesny, *From Curve Fitting to Machine Learning*, Intelligent Systems
Reference Library 109, DOI 10.1007/978-3-319-32545-3_5

Performance issues which are related to the road from curve fitting to machine learning were only marginally mentioned in the previous chapters. But they are of course at heart of any practical application: If there is not at least a vague estimate about the necessary computational resources and corresponding time periods to come to a successful result a research effort is immediately abandoned. As pointed out in chapter 1 all methods discussed so far can be mathematically traced back to optimization problems which can be tackled with particular adequate stepwise iterative procedures. A question about performance for a specific problem thus can be divided in two parts: How much time does a single optimization step need? And how many steps are necessary to arrive at a successful result? Whereas an approximate answer of the first question is feasible for most practical applications the second one can not be answered in general: There is no way of knowing the number of necessary iterations in advance for a non-linear optimization problem (linear optimization problems can almost always be tackled successfully in short time periods, see below). Putting both answers together the principal statement about any question of performance is not satisfying: We do not know! Fortunately there are a number of practical rules of thumb, a lot of experience with already performed similar problems as well as procedures of preliminary estimation which turn the sad general answer into a more optimistic version for many practically relevant situations. To put it short (and neglecting pathological cases): With today's computers curve fitting is usually performed on the fly (this means you can sit in front of your screen and wait for the result to emerge after a few seconds) whereas clustering and machine learning are typical batch tasks: They usually consume minutes (for very small problems like many of those discussed in the previous chapters) up to hours and days or even longer - they are started and performed in the background without being constantly monitored or waited for. An important characteristics of a method of choice concerning its necessary computational time consumption is its behavior for a varying problem size. This behavior can be experimentally deduced or derived from theoretical considerations. If for example a number of K data records is to be searched for a specific entry in a successive manner one after another (a so-called exact sequential search) the necessary maximum time period can be estimated to be "the time necessary to detect the entry for a single record" times K (where it is assumed that the entry detection for every single data record consumes the same time on average). What happens if the number of data records is doubled to $2K$? Then the maximum necessary time period simply doubles too. A sequential search is said to scale with $O(K)$ (read "order K" where "O" means order), i.e. the dependence between the data size and the search speed is linear. A sequential search is in fact a worst case scenario for exact data searching so there are more efficient algorithmic alternatives available like a binary tree search with $O(\log_2 K)$ or a hash-table search with $O(1)$. $O(\log_2 K)$ means that you can search 2 ($= 2^1$) data records in let's say 1 second, 4 ($= 2^2$) data records in 2 seconds, 8 ($= 2^3$) data records 3 seconds and 4.294.967.296 (about 4 billion) data records in 32 seconds:

```
2^32
```

4294967296

The search period increases only logarithmically with the data size. O(1) indicates that the search speed and the data size are decoupled: The search speed does no longer depend on the number of records (this is the holy grail of searching - in fact hashing in this context simply means the calculation of a position in a data table with a calculation time that does not depend of the number of the table's rows that corresponds to the data size). Note that the scaling behavior says nothing about the absolute time period necessary to search e.g. 12 data records with the different methods but it signals that a hash-table search will finally outperform its competitors for searching large data volumes. The thorough characterization of the interplay between data structures (like binary trees or hash-tables) and the algorithms that work upon them is at heart of computer science and every single choice for a practical application of a method is a compromise that (hopefully) best fits the specific needs: The selection of a method is usually based on its most attractive features (like maximum search speed) whereas its problems (e.g. additional memory consumption) must be tolerable. In general each computational method exhibits contradictory features (like speed versus memory) so that their comparison by adequate benchmarks is a difficult and professional science on its own with many traps and subtleties: A lot of published benchmark results simply compare apples and oranges. Thus great care is necessary to achieve trustable results that are able to meet the requirements of scientific validity. To demonstrate the estimation of necessary computational time periods the already discussed regression problem of chapter 4 may be utilized again: Normally distributed 3D points are generated at grid positions around a non-linear surface (here shown with a 10×10 grid with 100 I/O pairs)

```
pureOriginalFunction=Function[{x,y},
 1.9*(1.35+Exp[x]*Sin[13.0*(x-0.6)^2]*Exp[-y]* Sin[7.0*y])];
xRange={0.0,1.0};
yRange={0.0,1.0};
numberOfDataPerDimension=10;
standardDeviationRange={0.1,0.1};
dataSet3D=CIP`CalculatedData`Get3dFunctionBasedDataSet[
 pureOriginalFunction,xRange,yRange,numberOfDataPerDimension,
 standardDeviationRange];
labels={"x","y","z"};
CIP`Graphics`Plot3dDataSetWithFunction[dataSet3D,
 pureOriginalFunction,labels]
```

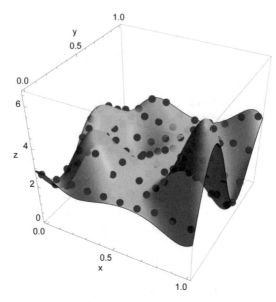

where the number of grid points (and thus the number of I/O pairs of the data set) is varied. How do the different machine learning methods scale with an increasing size of the data set? For the machine learning method implementations provided by the CIP package this may be analyzed by experiment: With Mathematica's AbsoluteTiming command the time period consumed by a specific Fit procedure can be measured and displayed in a number of I/O pairs versus training period diagram. For a Multiple Linear Regression (MLR) approach to the regression task the following result is obtained:

```
xyErrorData={};
rmsePoints2D={};
Do[
 dataSet3D=CIP `CalculatedData `Get3dFunctionBasedDataSet[
  pureOriginalFunction,xRange,yRange,numberOfDataPerDimension,
  standardDeviationRange];
 result=AbsoluteTiming[CIP `MLR `FitMlr[dataSet3D]];
 rainingPeriod=result[[1]];
 mlrInfo=result[[2]];
 numberOfIoPairs=numberOfDataPerDimension*numberOfDataPerDimension;
 AppendTo[xyErrorData,{numberOfIoPairs,trainingPeriod,1.0}];
 AppendTo[rmsePoints2D,{numberOfIoPairs,
  CIP `MLR `CalculateMlrDataSetRmse[dataSet3D,mlrInfo]}],
 {numberOfDataPerDimension,5,100,5}
];

minExponent=1.0;
maxExponent=4.0;
exponentStepSize=0.1;
exponentLabels={"Number of I/O pairs (K)","Training period [s]",
 "Training period = O(K^exponent)"};
CIP `CurveFit `ShowBestExponent[xyErrorData,minExponent,maxExponent,
 exponentStepSize,CurveFitOptionLabels -> exponentLabels];
```

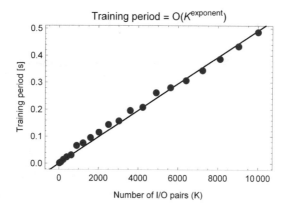

Best exponent = 1.

Over a wide range of data set sizes (from $K = 25$ I/O pairs up to $K = 10.000$ I/O pairs) the training period scales linear with an increasing number of I/O pairs K ($O(K)$ which corresponds to a "Best exponent" of 1.0). Each MLR fit is performed in fractions of a second (with a common notebook computer). This linear scaling behavior is the best we can expect for a machine learning method and confirms the general statement already mentioned above that linear methods are fast with today's computers. But of course a MLR approach is completely inadequate for a non-linear regression task which may be revealed by an inspection of the corresponding RMSE values of the regression results which should lie around 0.1 since the normally distributed data were generated with a standard deviation of 0.1 (see above):

```
qualityLabels={"Number of I/O pairs (K)","RMSE",
 "Quality of machine learning result"};
functionValueRange2D={0.0,1.2};
CIP`Graphics`Plot2dLineWithOptionalPoints[rmsePoints2D,
 rmsePoints2D,qualityLabels,
 GraphicsOptionFunctionValueRange2D -> functionValueRange2D]
```

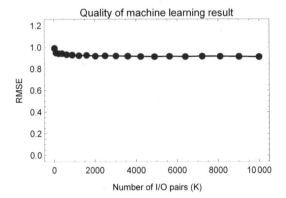

The determined RMSE values are an order of magnitude above expectation and thus MLR is clearly out of play. Multiple Polynomial Regression (MPR) with an adequate polynomial degree (compare chapter 4) exhibits a comparable fast performance with linear scaling (since the minimization task is also linear in its model parameters)

```
polynomialDegree=9;
xyErrorData={};
rmsePoints2D={};
Do[
 dataSet3D =
  CIP`CalculatedData`Get3dFunctionBasedDataSet[pureOriginalFunction,
   xRange,yRange,numberOfDataPerDimension,standardDeviationRange];
  result=AbsoluteTiming[CIP`MPR`FitMpr[dataSet3D,polynomialDegree]];
  trainingPeriod=result[[1]];
  mprInfo=result[[2]];
  numberOfIoPairs=numberOfDataPerDimension*numberOfDataPerDimension;
  AppendTo[xyErrorData,{numberOfIoPairs,trainingPeriod,1.0}];
  AppendTo[rmsePoints2D,{numberOfIoPairs,
   CIP`MPR`CalculateMprDataSetRmse[dataSet3D,mprInfo]}],
  {numberOfDataPerDimension,5,100,5}
];

CIP`CurveFit`ShowBestExponent[xyErrorData,minExponent,maxExponent,
 exponentStepSize,CurveFitOptionLabels -> exponentLabels];
```

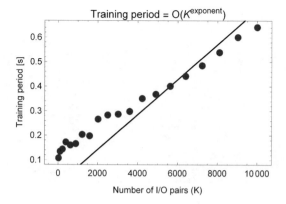

Best exponent = 1.

but with a far more satisfying result for the machine learning task in question (i.e. a desired RMSE of about 0.1 after some initial overfitting):

```
functionValueRange2D={0.0,0.2};
CIP'Graphics'Plot2dLineWithOptionalPoints[rmsePoints2D,
 rmsePoints2D,qualityLabels,
 GraphicsOptionFunctionValueRange2D -> functionValueRange2D]
```

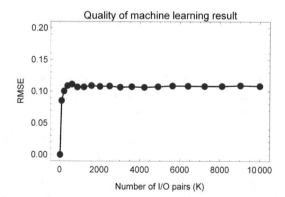

Next a Support Vector Machine (SVM) approach with an adequate kernel function and default settings may be explored (compare chapter 4):

```
kernelFunction={"Wavelet",0.3};
xyErrorData={};
rmsePoints2D={};
Do[
```

```
dataSet3D=CIP`CalculatedData`Get3dFunctionBasedDataSet[
  pureOriginalFunction,xRange,yRange,numberOfDataPerDimension,
  standardDeviationRange];
result=AbsoluteTiming[CIP`SVM`FitSvm[dataSet3D,kernelFunction]];
trainingPeriod=result[[1]];
svmInfo=result[[2]];
numberOfIoPairs=numberOfDataPerDimension*numberOfDataPerDimension;
AppendTo[xyErrorData,{numberOfIoPairs,trainingPeriod,1.0}];
AppendTo[rmsePoints2D,{numberOfIoPairs,
  CIP`SVM`CalculateSvmDataSetRmse[dataSet3D,svmInfo]}],
  {numberOfDataPerDimension,5,20}
];

CIP`CurveFit`ShowBestExponent[xyErrorData,minExponent,maxExponent,
  exponentStepSize,CurveFitOptionLabels -> exponentLabels];
```

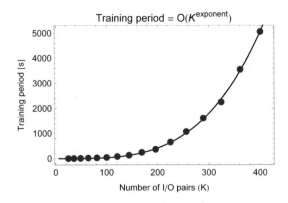

Best exponent = 3.6

For the data range from $K = 25$ I/O pairs to $K = 400$ I/O pairs the training period is well described to scale with $O(K^{3.6})$. Compared to a fast MLR or MPR fit which is performed in fractions of a second a SVM fit is terribly slow (5000 seconds correspond to roughly one and a half hour for a single fit with 400 I/O pairs). With such a polynomial scaling it may be deduced that a single SVM fit of 1.000 I/O pairs would require

```
factor=1000./400.
```

2.5

```
timePeriod=5000*factor^3.6
```

135380.

```
hours=timePeriod/(60*60)
```

37.6055

about one and a half day (38 hours) if an extrapolation is dared. The CIP default implementation of a SVM becomes prohibitive for larger data sets and is thus reasonably confined to machine learning tasks with only small data set sizes (compare Appendix A for CIP design goals). Note that these specific findings can not simply be generalized: There are far more efficient SVM implementations available! In general a SVM's training period scales between quadratically ($O(K^2)$) and cubically ($O(K^3)$) in the number of I/O pairs (see [Joachims 1999]) - nevertheless even with an efficient implementation and an improved polynomial scaling behavior SVMs are known to have a large data set problem currently addressed by many R&D efforts. A final look at the quality of the SVM fits shows their results

```
functionValueRange2D={0.0,0.2};
CIP`Graphics`Plot2dLineWithOptionalPoints[rmsePoints2D,
 rmsePoints2D,qualityLabels,
 GraphicsOptionFunctionValueRange2D -> functionValueRange2D]
```

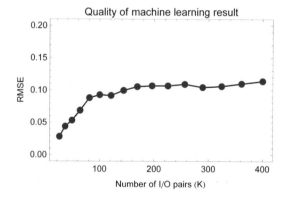

to be satisfying: Again, after some inevitable initial overfitting for the smallest data sets the RMSE values are distributed around the expected value of 0.1.

A similar perceptron analysis with an adequate number of hidden neurons and default settings (compare chapter 4)

```
numberOfHiddenNeurons=12;
xyErrorData={};
rmsePoints2D={};
Do[
 dataSet3D=CIP`CalculatedData`Get3dFunctionBasedDataSet[
  pureOriginalFunction,xRange,yRange,numberOfDataPerDimension,
```

```
    standardDeviationRange];
  result=AbsoluteTiming[CIP`Perceptron`FitPerceptron[dataSet3D,
    numberOfHiddenNeurons]];
  trainingPeriod=result[[1]];
  perceptronInfo=result[[2]];
  numberOfIoPairs=numberOfDataPerDimension*numberOfDataPerDimension;
  AppendTo[xyErrorData,{numberOfIoPairs,trainingPeriod,1.0}];
  AppendTo[rmsePoints2D,{numberOfIoPairs,
    CIP`Perceptron`CalculatePerceptronDataSetRmse[dataSet3D,
    perceptronInfo]}],
  {numberOfDataPerDimension,5,20}
];

CIP`CurveFit`ShowBestExponent[xyErrorData,minExponent,maxExponent,
  exponentStepSize,CurveFitOptionLabels -> exponentLabels];
```

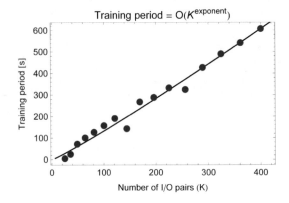

Best exponent = 1.1

leads to a near linear $O(K^{1.1})$ scaling. In addition a single (default CIP) perceptron training is to be about an order of magnitude faster than a (default CIP) SVM
one (again see Appendix A for comments on the CIP design goals). But an inspection of the regression results reveals that the perceptron fits did not quite yield the
expected RMSE values for most data set sizes:

```
functionValueRange2D={0.0,0.2};
CIP`Graphics`Plot2dLineWithOptionalPoints[rmsePoints2D,
  rmsePoints2D,qualityLabels,
  GraphicsOptionFunctionValueRange2D -> functionValueRange2D]
```

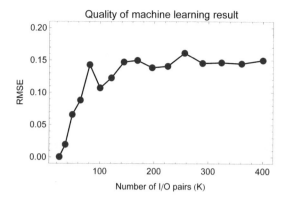

There is again an initial overfitting for very small data sets but then the RMSE values rise to about 0.15 which is a above the expected value of 0.1 and the SVM results (see chapter 4 for more adequate technical perceptron optimization parameters with improved learning results). Thus the perceptron training can not simply be compared to the more satisfying SVM training before - and this is exactly where the comparison of apples and oranges usually starts. In conclusion a fair comparison of methods has to take all these subtle differences (and many others like differences in memory consumption or the acceleration due to parallelized calculation setups) into account. As a final rule of thumb it is recommended to always perform initial investigations concerning speed, scaling behavior, memory consumption etc. before applying non-linear iterative optimization procedures to real-world problems in practice.

5.2 Isn't it just ...?

```
Clear["Global`*"];
<<CIP`CalculatedData`
<<CIP`DataTransformation`
<<CIP`Graphics`
<<CIP`Perceptron`
<<CIP`CurveFit`
<<CIP`SVM`

Off[FindMaximum::"lstol"];
```

The methods described on the road from curve fitting to machine learning are perceived by practitioners quite differently: Opinions range from "mere technical tools" (predominantly for curve fitting) up to an esteem of "intelligence" with a "touch of magic" (often attributed to successful machine learning results). This finding may be related to the fact that results of curve fitting can be directly inspected in a visual manner whereas the results of clustering and machine learning are less intuitive and more complex to grasp. In scientific education the same basic moods can be observed with students - especially an expectant curiosity about machine learning which is inspired by rumors regarding their magic capabilities.

5.2.1 ... optimization?

But after the background of the methods is outlined and traced back to mathematical optimization problems a swing in opinions occurs: "Isn't it just optimization?" is a common expression of disappointment. Machine learning seems to loose a lot of its initial fascination after an explanation of its basic machinery. This may be due to fact that an optimization task sounds easy: Just walk uphill (for maximization) or downhill (for minimization) - and your are done (in principal). It is quite common to the views about modern science that the level of sophistication necessary to tackle the details is undervalued after a principle understanding is achieved. But the progress of science is more and more absorbed by details while developing from its basic foundations in physics and chemistry to an understanding of complex systems in biology or ecology: "It's a mere detail that makes you dead or alive" as a friend summarizes his daily experience as a surgeon with the complex system homo sapiens. As pointed out in chapter 1 it is the details of optimization that lead to success of failure and there is no general way to avoid the latter. Optimization issues are deeply connected to the most challenging scientific problems like protein folding (where the correctly folded biologically active protein conformation is assumed to be a minimum energy state). Whereas optimization tasks sound easy they should be regarded with the necessary respect for an extremely difficult and active field of research.

5.2.2 ... data smoothing?

Another attitude often expressed by practitioners after getting acquainted with machine learning may be summarized with the question "Isn't it just data smoothing?" which implies that machine learning may be useful for a comprehensive description of data and possible interpolations of new values but not for any new insights. This issue is a bit more subtle and concerns the principle question: What can you get out of your data? Possibilities may be sketched with the following example: A relatively imprecise data set is generated around a 3D function

```
pureOriginalFunction=Function[{x,y},
  1.9*(1.35+Exp[x]*Sin[13.0*(x-0.6)^2]*Exp[-y]* Sin[7.0*y])];
xRange={0.0,0.5};
yRange={0.0,0.5};
numberOfDataPerDimension=10;
standardDeviationRange={0.75,0.75};
dataSet3D=CIP`CalculatedData`Get3dFunctionBasedDataSet[
```

```
 pureOriginalFunction,xRange,yRange,numberOfDataPerDimension,
 standardDeviationRange];
labels={"x","y","z"};
viewPoint3D={1.9, -3.0, 0.9};
CIP`Graphics`Plot3dDataSetWithFunction[dataSet3D,
 pureOriginalFunction,labels,
 GraphicsOptionViewPoint3D -> viewPoint3D]
```

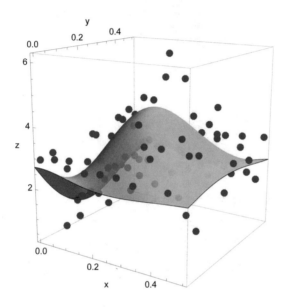

with a maximum in the data range:

```
FindMaximum[pureOriginalFunction[x,y],{x,0.2},{y,0.2}]
```

$\{4.55146, \{x \to 0.265291, y \to 0.204128\}\}$

At the position of the maximum a hole in the data set is generated

```
reducedDataSet3D={};
Do[
 If[!(dataSet3D[[i,1,1]]>0.1&& dataSet3D[[i,1,1]]<0.4 &&
  dataSet3D[[i,1,2]]>0.1 && dataSet3D[[i,1,2]]<0.4),
  AppendTo[reducedDataSet3D,dataSet3D[[i]]]
 ],
 {i,Length[dataSet3D]}
];
CIP`Graphics`Plot3dDataSetWithFunction[reducedDataSet3D,
 pureOriginalFunction,labels,
 GraphicsOptionViewPoint3D -> viewPoint3D]
```

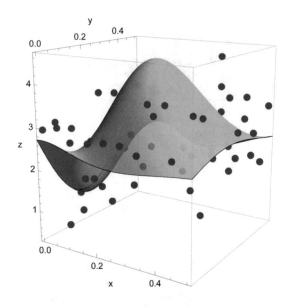

which becomes easily visible from above:

```
viewPoint3D={0.0,0.0,3.0};
CIP`Graphics`Plot3dDataSetWithFunction[reducedDataSet3D,
 pureOriginalFunction,labels,
 GraphicsOptionViewPoint3D -> viewPoint3D]
```

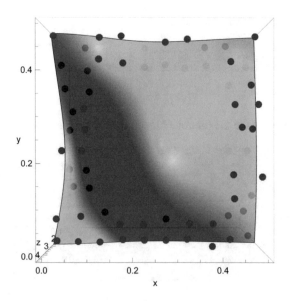

If an approximation of the original function is performed with a three-layer perceptron based on the data set with a hole

```
numberOfHiddenNeurons=3;
perceptronInfo=CIP`Perceptron`FitPerceptron[reducedDataSet3D,
 numberOfHiddenNeurons];
purePerceptron3dFunction=Function[{x,y},
 CIP`Perceptron`CalculatePerceptron3dValue[x,y,perceptronInfo]];
viewPoint3D={1.9, -3.0, 0.9};
CIP`Graphics`Plot3dDataSetWithFunction[reducedDataSet3D,
 purePerceptron3dFunction,labels,
 GraphicsOptionViewPoint3D -> viewPoint3D]
```

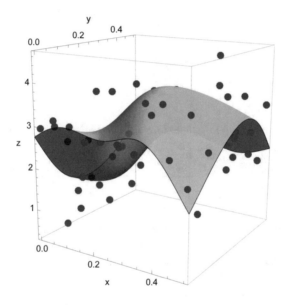

the approximated perceptron model function contains a maximum

```
FindMaximum[purePerceptron3dFunction[x,y],{x,0.2},{y,0.2}]
```

$\{3.96551, \{x \to 0.326515, y \to 0.170616\}\}$

near the true maximum of the original 3D function. If a SVM with an adequate kernel function is used for the regression task

```
kernelFunction={"Wavelet",0.5};
svmInfo=CIP`SVM`FitSvm[reducedDataSet3D,kernelFunction];
pureSvm3dFunction=Function[{x,y},
 CIP`SVM`CalculateSvm3dValue[x,y,svmInfo]];
CIP`Graphics`Plot3dDataSetWithFunction[reducedDataSet3D,
 pureSvm3dFunction,labels,GraphicsOptionViewPoint3D -> viewPoint3D]
```

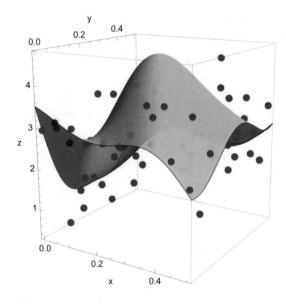

the approximation of the true maximum is even improved:

```
FindMaximum[pureSvm3dFunction[x,y],{x,0.2},{y,0.2}]
```

$\{4.72082, \{x \rightarrow 0.287884, y \rightarrow 0.190222\}\}$

This means that although the data are considerably error-biased and do not cover the interesting maximum region a machine learning method may be able to reveal a maximum which is indicated by the surrounding data. Since optima are often the primary targets of research and development a machine learning result as the one illustrated may lead to a true discovery (note that visual inspection is not possible in general but an exploration of the approximated model surface for optima may equally work in multiple dimensions). This promising feature has of course its limits: If the data hole is enlarged further

```
reducedDataSet3D={};
Do[
 If[!(dataSet3D[[i,1,1]]>0.05&& dataSet3D[[i,1,1]]<0.45 &&
  dataSet3D[[i,1,2]]>0.05 && dataSet3D[[i,1,2]]<0.45),
  AppendTo[reducedDataSet3D,dataSet3D[[i]]]
  ],
 {i,Length[dataSet3D]}
];
viewPoint3D={1.9, -3.0, 0.9};
CIP`Graphics`Plot3dDataSetWithFunction[reducedDataSet3D,
 pureOriginalFunction,labels,
 GraphicsOptionViewPoint3D -> viewPoint3D]
```

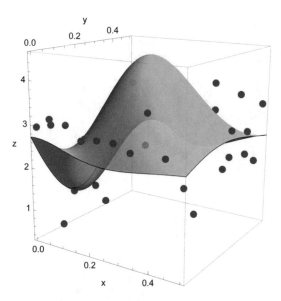

so that the true maximum is only framed by a few points

```
viewPoint3D={0.0,0.0,3.0};
CIP`Graphics`Plot3dDataSetWithFunction[reducedDataSet3D,
  pureOriginalFunction,labels,
  GraphicsOptionViewPoint3D -> viewPoint3D]
```

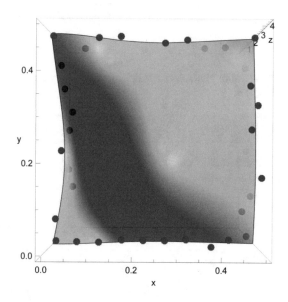

a machine learning approach is likely to completely fail. A perceptron model fit with an inadequate number of hidden neurons may lead to an approximated model function that does not describe the hole region at all,

```
numberOfHiddenNeurons=10;
perceptronInfo=CIP`Perceptron`FitPerceptron[reducedDataSet3D,
 numberOfHiddenNeurons];
purePerceptron3dFunction=Function[{x,y},
 CIP`Perceptron`CalculatePerceptron3dValue[x,y,perceptronInfo]];
CIP`Graphics`Plot3dDataSetWithFunction[reducedDataSet3D,
 purePerceptron3dFunction,labels]
```

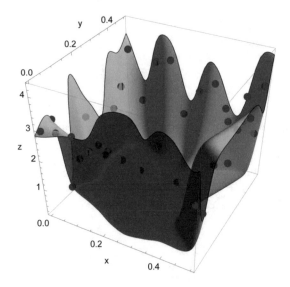

i.e. it leads to a flat valley with a zero gradient without any optimum. A maximum search must inevitably fail:

```
FindMaximum[purePerceptron3dFunction[x,y],{x,0.2},{y,0.2}]
```

Encountered a gradient that is effectively zero. The result returned may not be a maximum;

it may be a minimum or a saddle point.

$\{0.143304, \{x \to 0.2, y \to 0.2\}\}$

In this specific case a SVM approach with an adequate model function

```
kernelFunction={"Wavelet",0.5};
svmInfo=CIP`SVM`FitSvm[reducedDataSet3D,kernelFunction];
pureSvm3dFunction=Function[{x,y},
 CIP`SVM`CalculateSvm3dValue[x,y,svmInfo]];
```

```
viewPoint3D={1.9, -3.0, 0.9};
CIP 'Graphics 'Plot3dDataSetWithFunction[reducedDataSet3D,
  pureSvm3dFunction,labels,GraphicsOptionViewPoint3D -> viewPoint3D]
```

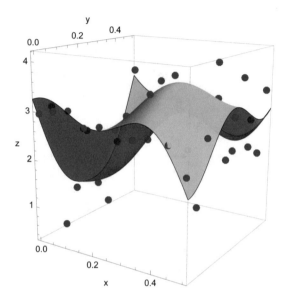

is still able to predict a maximum not too far from the true original one:

```
FindMaximum[pureSvm3dFunction[x,y],{x,0.2},{y,0.2}]
```

$\{3.68395,\{x \to 0.379893, y \to 0.15681\}\}$

But this is just good luck - and explains why machine learning performs astonishingly well for one problem and fails utterly for another. Although the problems and data sets may seem to be very similar it can be small and subtle differences in the data that lead to completely different outcomes. A final point may be a closer look of what was earlier described as "indicated by the surrounding data". Normally distributed data are generated around a 2D function

```
pureOriginalFunction=Function[x,0.5*x+3.0*Exp[-(x-4.0)^2]];
argumentRange={1.0,7.0};
numberOfData=100;
standardDeviationRange={0.1,0.1};
xyErrorData=CIP 'CalculatedData 'GetXyErrorData[pureOriginalFunction,
  argumentRange,numberOfData,standardDeviationRange];
labels={"x","y","Data above function"};
functionValueRange={0.0,5.5};
CIP 'Graphics 'PlotXyErrorDataAboveFunction[xyErrorData,
  pureOriginalFunction,labels,
```

```
GraphicsOptionFunctionValueRange2D -> functionValueRange]
```

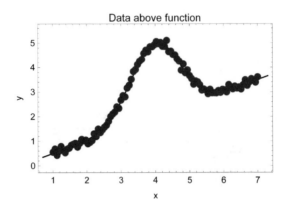

with a single maximum:

```
FindMaximum[pureOriginalFunction[x],{x,3.9}]
```

$\{5.02091,\{x \rightarrow 4.08392\}\}$

Again data are removed around the maximum

```
reducedXyErrorData={};
Do[
 If[!(xyErrorData[[i,1]]>3.0&&xyErrorData[[i,1]]<5.4),
  AppendTo[reducedXyErrorData,xyErrorData[[i]]]],
 {i,Length[xyErrorData]}
];
labels={"x","y","Reduced data above function"};
CIP`Graphics`PlotXyErrorDataAboveFunction[reducedXyErrorData,
 pureOriginalFunction,labels,
 GraphicsOptionFunctionValueRange2D -> functionValueRange]
```

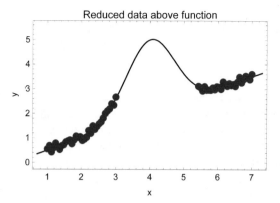

to produce a data hole. A perceptron fit to the data with an adequate number of hidden neurons

```
reducedDataSet=
 CIP 'DataTransformation'TransformXyErrorDataToDataSet[
 reducedXyErrorData];
numberOfHiddenNeurons=3;
perceptronInfo=CIP 'Perceptron'FitPerceptron[reducedDataSet,
 numberOfHiddenNeurons];
purePerceptron2dFunction=Function[x,
 CIP 'Perceptron'CalculatePerceptron2dValue[x,perceptronInfo]];
labels={"x","y","Reduced data above approximate model"};
CIP 'Graphics'PlotXyErrorDataAboveFunction[reducedXyErrorData,
 purePerceptron2dFunction,labels,
 GraphicsOptionFunctionValueRange2D -> functionValueRange]
```

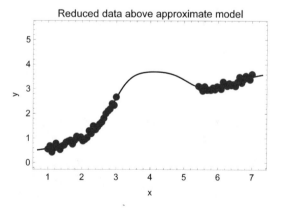

leads to an approximated model function with a maximum very close to the true original one:

```
FindMaximum[purePerceptron2dFunction[x],{x,3.9}]
```

$\{3.70255,\{x \to 4.11641\}\}$

It now becomes clear why the data can indicate a maximum: The data blocks on the left and on the right of the hole are best described by lines with positive curvature. For a continuous model function these lines must be connected by a line with negative curvature which inevitably produces a maximum in between. If the data blocks are further reduced

```
reducedXyErrorData={};
Do[
  If[!(xyErrorData[[i,1]]>2.0&&xyErrorData[[i,1]]<6.0),
   AppendTo[reducedXyErrorData,xyErrorData[[i]]]],
  {i,Length[xyErrorData]}
];
labels={"x","y","Reduced data above function"};
CIP`Graphics`PlotXyErrorDataAboveFunction[reducedXyErrorData,
  pureOriginalFunction,labels,
  GraphicsOptionFunctionValueRange2D -> functionValueRange]
```

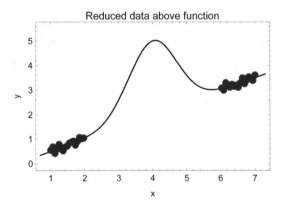

so that they no longer advise the maximum in between due to their curvature an approximated model function

```
reducedDataSet=
  CIP`DataTransformation`TransformXyErrorDataToDataSet[
  reducedXyErrorData];
numberOfHiddenNeurons=3;
perceptronInfo=CIP`Perceptron`FitPerceptron[reducedDataSet,
  numberOfHiddenNeurons];
purePerceptron2dFunction=Function[x,
  CIP`Perceptron`CalculatePerceptron2dValue[x,perceptronInfo]];
labels={"x","y","Reduced data above approximate model"};
CIP`Graphics`PlotXyErrorDataAboveFunction[reducedXyErrorData,
```

```
purePerceptron2dFunction, labels,
GraphicsOptionFunctionValueRange2D -> functionValueRange]
```

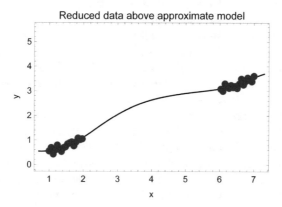

will most likely fail to suggest one (in this case a maximum is detected but far outside the data region without any meaning):

```
FindMaximum[purePerceptron2dFunction[x],{x,3.9}]
```

$\{3.78097, \{x \to 108.998\}\}$

It may be only a very few data that are crucial for success or failure and they are not known in advance. In summary data analysis may lead to a discovery of hidden optima which can not be easily derived from the pure data - but there is of course no guarantee of success.

5.3 Computational intelligence

Are computers intelligent? Or - more humble - is machine learning a first glimpse of computational intelligence? And by the way: What at all means to be intelligent? To put it short: There is no generally accepted definition of intelligence - it is an umbrella term that comprises an interplay of human behavioral properties like abstract thought, reasoning, planning, problem solving, communication or learning (see [Intelligence 2010]). As an intuitive definition one may suggest that intelligent behavior means to act as an average human being. This is the basis of a famous test for machine intelligence proposed by Alan Turing in 1950: If a machine's behavior may not be distinguished from human behavior then the machine must be attributed to be intelligent (see [Turing 1950]). But it may be comparatively easy to simulate human behavior as was early demonstrated by the definitely non-intelligent ELIZA computer program that mimicked a human psychotherapist (see [Weizenbaum 1966]) - and complex tests of human behavior which would be able to reveal fakes like ELIZA could easily have the catastrophic result to attribute non-humanity to living human beings. Thus the situation remains dodgy. In addition the introductory questions touch old and basic philosophical issues like the famous mind-body problem. The two major schools of thought that try to resolve this problem are dualism and monism: Dualism claims that mind and matter are different existing substances which interact in some way. Unfortunately nothing is said about the nature and the details of this mind-body interaction which makes this position quite obscure. Monism on the other hand regards mind as an emerging function of the dynamics of specifically organized matter so there is just matter as an ontological entity. Unfortunately monism is still not able to tell how this emergence of mind from matter is achieved in detail (this unsatisfying situation may be illustrated with a famous cartoon by Sydney Harris: There are two persons standing in front of a chalkboard which shows a mathematical derivation with an intermediate step paraphrased as *Then A Miracle Occurs*: One person points at this statement by telling the other *I think you should be more explicit here in step two*). In the 20th century the old philosophical problem was complemented with questions about a possible artificial intelligence which may be realized by computational devices as well as questions about intelligence acceleration as an effect of combining man and machine into cybernetic organisms (cyborgs). In a situation of vague definitions and fundamental unsolved problems it is not surprising that there is still a lively discussion of all these fascinating issues (e.g. see [Hofstadter 1981], [Penrose 1991], [Dreyfus 1992], [Penrose 1994], [Churchland 1996], [Kaku 1998], [Koch 2004], [Hawkins 2005], [Baggott 2005], [Kurzweil 2005], [Mitchell 2009], [Kurzweil 2012], [Barrat 2013] or [Ford 2015]).

From a scientific point of view biological systems in general and human brains in particular are just ordinary pieces of matter that obey the known laws of physics and chemistry. There is nothing special about them: Concepts like "supernormal vis vitalis" (specific forces of living systems beyond physics) or "immaterial souls" besides matter (essential for the dualist view, compare above) are outside any serious scientific discussion: There is simply no clue for their existence (that is why the majority of scientists nowadays tends to be monists and materialists with regarding matter as the only ontological entity - at least in their daily scientific work). But whereas the human brain is ordinary in a pure material sense it is an extremely intricate piece of matter in a structural sense with a hundred billion nerve cells (neurons) that form a hundred trillion connections (synapses) for mutual interaction: It is the most complex natural system of the known universe. And although there is an already broad and impressive knowledge about its neurobiological parts and their chemical and physical interactions it is still completely unknown how the whole system works and achieves its intelligent characteristics (this miracle is the monists' problem, compare above): All there is are appealing proposals (like the memory-prediction framework theory etc., see [Hawkins 2005], [Kurzweil 2012]) but no true "Theory of the Brain"- a challenge that is likely to become the foundational scientific feat of the 21th century with the aid of more developed computers. Today's computers are not able to simulate or even represent such a complex biological structure (see [Kurzweil 2005]): The human brain is able to update its hundred trillion (10^{14}) synapses about a 100 times a second (each neuron is a comparatively slow functional unit with a reset time of about 10 ms) which means that the brain performs about $10^{14} \times 100 = 10^{16}$ synaptic transactions per second. If a single synaptic transaction is formally described by about 1.000 computational calculations a digital computer would need to perform at least $10^{16} \times 1.000 = 10^{19}$ calculations per second to simulate a human brain (not too mention the necessary computational memory which would be at least $10^{14} \times 10 = 10^{15} = 1$petabyte $= 1.000$terabyte if every synapse is represented by only 10 bytes of memory). Compared to the fastest available computers with a performance of about 10^{15} calculations per second these (more than) rough estimates illustrate the existing complexity gap between the artificial in-silico machines and the evolutionary developed biological system. But within the next decades the computational devices may achieve similar (or even higher) levels of complexity as their biological predecessors - and early attempts to reverse-engineer the mammalian brain have already started (e.g. see [Blue Brain Project], [Human Brain Project]). So from a pure materialist's point of view today's computers can not be intelligent due to their insufficient degree of complexity compared to the human brain as a gold standard - where another interesting question remains currently unanswered whether human intelligence really needs the brain's complexity or may be realized by a simpler architecture. Also note that this view does not necessarily imply that powerful enough computers automatically produce intelligent behavior. But they are regarded as the necessary and sufficient tools to successfully tackle the mind-body problem at leasts from the monist's point of view.

Computers are often blamed because they are not able to understand simple daily-life situations. A brief example (inspired by Roger Schank, see [Schank 1977]) is the following message: *John went to a restaurant. He ordered lobster. He paid the bill and left.* A proper comprehension of this message could be tested with the question: "What did John eat?" A human being is extremely like to reply "Lobster, of course!" but the best a computer could respond is "I don't know!" - an answer that would be taken as an indication of failed understanding. But the computer's answer is correct since the message did not at all contain what John really ate. Humans do not consciously hear, smell, see or feel what they really hear, smell, see or feel: Every sensual input is automatically interpreted by the brain within a concrete situation. This interpretation is among others a function of the historical and sociocultural context in which the human being is living in. Concerning the above message an inhabitant of an American or European society knows that someone usually eats in a restaurant what he orders since it is normally not allowed to bring in own food etc. Thus the objectively missing contextual piece of information is automatically filled-in by the neural information processing. To let a computer act as a human being in daily-life situations an access to a human-comparable contextual memory would be necessary which is at present technically impossible to achieve (compare above, an approximation is the aim of the Cyc project, see [Cyc 2010]).

So the failure of the computer system is not a principle one: A human being would fail as well if it had to act within an unknown context since then the automatic fill-in could not work adequately. The related unsure feeling in an alien environment is often experienced by foreigners in their new home countries and this is why cross-cultural training becomes more and more popular in an increasingly globalized and flexible world.

The above arguments suggest that contemporary computers are not intelligent in comparison to an average culturally educated specimen of homo sapiens. But there is a justified hope to scientifically reveal the miracle of human intelligence and there is no principal objection against a future human-like machine intelligence with more developed computational devices.

And if the intelligence discussion is confined to the initial more humble question about a first glimpse of computational intelligence exerted by current machine learning the answer may turn from a simple "no" to a more optimistic outcome. It is inevitable for a corresponding line of thought to at least come to a preliminary working definition of what means intelligent and to compare its characteristics with the current power of machine learning. Besides the semantic ambiguities it has to be recognized that there is an additional well-known semantic shift of what is regarded to be intelligent which may be summarized by the following rule of thumb: If a machine is capable of performing an intelligent operation this operation is no longer called intelligent. Obviously homo sapiens likes its exceptional status on earth. For an illustration of emerging computational intelligence different ways of searching in data are sketched in the following. Searching is valued from being a dull task up to a challenge of the highest level of intellectual sophistication thus it may be worth to outline this ascent and its relation to the issue in question.

Let's start with an exact search: A data item is searched as a complete whole in a defined volume of data, e.g. a distinct name in a telephone book, a full chemical structure in a compound database or a complete biological sequence in a set of sequences. This type of search may be performed in a sequential manner (i.e. the data item is successively compared one after another to each single data item of the data volume) or with the data volume in a preprocessed state (e.g. as a binary tree or a hash-table) to increase the search speed. Although the details of the concrete computational implementation of an exact search may be very difficult and thus intellectually demanding an exact search is not regarded to be an intelligent operation. It is a dull task.

The next common step for a search with enhanced options is to soften the data item comparison to match only partially. For the name search in a telephone book this may be realized by the well-known wildcard characters like "*" at the beginning or the end of a name (left/right truncation, e.g. "May*" will find "Mayer", "Maybridge", "Mayfield" etc.). In chemical compound search this kind of structure retrieval is named substructure search: The structure query defines only a part of a full chemical structure (a substructure like a benzene ring or a functional chemical unit like a hydroxy group) and each full chemical structure of the database is checked for an occurrence of this substructure. Substructure searches are important for many areas of molecular research like chemical synthesis or drug development.

Similarly a biological sequence search will only look for the defined parts of the (base pair or amino acid) sequences of a gene or protein in the volume of sequences under investigation, e.g. to find highly conserved sequence regions in biological evolution. The technical implementation of a partial-match search may be very demanding (e.g. chemical substructure search is an active field of research for decades and the development of an adequate sub-graph isomorphism algorithm is a severe and an intellectually extremely challenging task) but also this kind of search is generally not attributed to be intelligent. Nonetheless a borderline of intelligence may come into sight: The latter kind of a defined softened search may retrieve data that were not primarily anticipated by a user (but of course are part of the user's defined solution space if the search performed in a correct manner). This means that emerging intelligence could be attributed to a computational method that yields results that are beyond simple anticipation of a human being: To perform the same task a human being would think that it has to apply its intelligent abilities rather than its brute labor. This is of course a very weak definition but may be at least operational for a division between (emerging) intelligent and (predominantly) non-intelligent computational methods: An exact search was characterized as a boring and dull task (too) often encountered in daily-life whereas for a partial-match search someone may feel an upcoming necessity to use his noddle.

If this line is followed then a true computationally intelligent search would be able to retrieve data that are related to the query in a non-trivial sophisticated manner. These results may only be similar or abstractly associated to what was searched for and they may reveal connections or insights that were not initially intended. In other words: Query (input) and answer (output) are related by a non-trivial input/output mapping function f. But the construction of these complex and non-trivial mapping functions f for an input/output pair obtained by output $= f$(input) or in a adequately coded mathematical notation $y = f(x)$ respectively is exactly the goal of machine learning as discussed in chapter 3 and 4. As a very basic example a name search could be fuzzy so much that only phonetically similarity is in question, i.e. query "Meyer" would yield the answers "Mayer", "Maher", "Mayr" etc. More advanced similarity or associative name searches could involve arising elements of the elaborate combined logic used by Sherlock Holmes and his successors to detect malefactors. A chemical structure search may try to retrieve structures that are in some way similar to a query structure. As far as topological similarity is in question such a query could of course yield the results of a corresponding substructure search but with an enhanced result set that consists of additional structures. The latter are not true substructures in a topological sense but resemble the query structure according to a defined measure of similarity. This measure may be defined quantitatively by an overlap (Tanimoto) coefficient that evaluates a similarity value in percent on the basis of individual bit vectors for the query and a test structure of the database: Each bit of a bit vector may encode the appearance (true) or absence (false) of a specific chemical group (like a benzene ring, a hydroxy group etc.) in a chemical structure. The Tanimoto coefficient then computes the ratio of the number of intersected "true"-bits (i.e. bits which are simultaneously true in both structures) to the number of united "true"-bits of query and test structure. If the ratio/coefficient is above a specific threshold (commonly 90%) then query and test structure are regarded to be similar and the test structure is transferred to the answer set. For more ambitious similarity searches the complexity of the similarity measure or the similarity detection function similarity value $= f$(test structure) will have to be readily increased up to an arbitrary difficult level, e.g. concerning the similarity in physico-chemical, environmental or pharmaceutical effects. A biological sequence related similarity search (a sequence alignment) makes use of scoring systems with (evolutionary derived) similarity matrices for monomer (base pair or amino acid) comparisons and specific penalties for alignment gaps. An optimum (local or global) sequence alignment between query and test sequence for a defined scoring system can be achieved with specific computational methods (dynamic programming): The sequence similarity search operation became so widespread and popular in molecular biology related research and development that the new verb "to blast" was established to denote its execution (named after the heuristic BLAST algorithm which is one of the most popular algorithms used for biological sequence alignment). As mentioned for chemical structures a biological sequence related similarity search may be extended to arbitrary levels of complexity, e.g. concerning the biological function of sequences or their specific expression under certain circumstances. Thus abstractly associated properties of chemical structures or biological sequences may be revealed by a trained computationally intelligent machine learning system which would be otherwise the results of attempts of well-educated human scientists (or simply impossible). And they may reveal completely new insights (the possible detection of hidden optima discussed above) that were not anticipated at all. The latest and globally recognized triumph of associative computational intelligence was achieved by IBM's Watson question answering computer system which excelled two human champions on the quiz show *Jeopardy!* in 2011 (see [IBM Watson]).

Computational intelligence which is modestly and preliminary characterized in the above manner is not yet ready to surpass its human predecessor in general but rather to accelerate and enhance human intelligence and creativity in many ways: It's the combination and hybridization of both man and machine that shapes the developments at the beginning of the 21th century. And a true man-machine interaction on a WYTIWYG (What You Think Is What You Get) basis comes into sight.

5.4 Final remark

In chapter 1 the motivation of this book was stated to show how specific situations of the interplay between data and models could be tackled: Firstly the situation was sketched where a model function f is known but not its parameters (denoted situation 2 in chapter 1) which was discussed on the basis of statistical curve fitting approaches. The second situation (situation 3 in chapter 1) contained the additional inconvenience that the model function f itself is unknown which led to attempts of unsupervised and supervised machine learning. The road from curve fitting to machine learning demonstrated how we can proceed from experimental data to models: A road that is often stony and full of perils and pitfalls. It was aimed to not only mention these difficulties but to outline how they can be successfully overcome. This implies the courage and the honesty to stop any analysis in the case of inadequate data to avoid the simple GIGO (garbage-in/garbage-out) effect. But appropriate data analysis and model construction reward all efforts with convincing results up to possible new insights that were not anticipated before. With a still exponentially increasing computational power the sketched methods will become more extensive and faster and thus more widespread and easier to use. Combination strategies and new heuristics for their application will emerge so that they become more and more ubiquitous tools (not only) in scientific research and development.

Appendix A
CIP -Computational Intelligence Packages

The Computational Intelligence Packages (CIP) are a high-level function library that is used for all demonstrations throughout this book (see [CIP]). It is built on top of the computing platform Mathematica (see [Mathematica]) to exploit its algorithmic and graphical capabilities. The CIP design goals were neither maximum speed nor minimum memory consumption but a largely unified and robust access to high-level functions necessary for demonstration purposes. Thus CIP is not an optimized and maximum efficient library for scientific application although it may be practically utilized in many operational areas (see comments below). Since CIP is open-source the library may be used as a starting point for customized and tailored extensions (for download location see [CIP]).

A.1 Basics

The unification goal of CIP design primarily addresses the optimization calculations: Data (with adequate data structures) are submitted to Fit methods provided by the CurveFit, Cluster, MLR, MPR, SVM or Perceptron packages to perform a corresponding optimization procedure. The result of the latter is a comprehensive info data structure, e.g. a curveFitInfo, clusterInfo, mlrInfo, mprInfo, svmInfo or perceptronInfo. This info data structure can then be passed to corresponding Show methods for multiple evaluation purposes like visual inspection of the goodness of fit or to Calculate methods for model related calculations. Similar operations of different packages are denoted in a similar manner to ease their use. Method signatures do mainly contain only structural hyperparameters where technical control parameters may be changed via options if necessary. CIP consists of the following packages:

- **Utility:** The Utility package is a basic package that collects several general methods used by other packages like GetMeanSquaredError which is used by all ma-

© Springer International Publishing Switzerland 2016

A. Zielesny, *From Curve Fitting to Machine Learning*, Intelligent Systems
Reference Library 109, DOI 10.1007/978-3-319-32545-3

chine learning related packages. Thus this package is used to decrease redundant code.

- **ExperimentalData:** The ExperimentalData package provides all data used throughout the book where details are provided below. This package makes use of the packages Utility, DataTransformation and CurveFit.

- **DataTransformation:** CIP performs many internal data transformations for different purposes, e.g. all data that are passed to a machine learning method are scaled before the operation (like ScaleDataMatrix) and re-scaled afterwards (like ScaleDataMatrixReverse). The DataTransformation package comprehends all these methods in a single package. It uses the Utility package.

- **Graphics:** The Graphics package tailors Mathematica's graphical functions for application throughout the book. It is used for all graphical representations and uses itself the Utility and DataTransformation packages.

- **CalculatedData:** The CalculatedData package complements the ExperimentalData package with methods for the generation of simulated data like normally distributed xy-error data around a function with GetXyErrorData. It uses methods from the Utility and DataTransformation packages.

The packages discussed so far complement and underlie the actual core packages of CIP. These core packages address curve fitting, clustering, multiple linear and polynomial regression, three-layer perceptrons and support vector machines:

- **CurveFit:** The CurveFit package tailors Mathematica's built in curve fitting method (NonlinearModelFit) for least-squares minimization and adds a smoothing cubic splines support. Since NonlinearModelFit is an algorithmic state-of-the-art implementation for curve fitting the CurveFit package is well-suited for professional data analysis purposes. It uses the Utility, Graphics, DataTransformation and CalculatedData packages.

- **Cluster:** The Cluster package tailors Mathematica's built in FindClusters method for clustering purposes and adds an adaptive resonance theory (ART-2a) support. FindClusters is an algorithmic state-of-the-art implementation for k-medoids clustering thus the Cluster package may be used for professional tasks (see [GetClusters]). It uses the Utility, Graphics and DataTransformation packages.

- **MLR and MPR:** The MLR and MPR packages tailor Mathematica's built in Fit method for multiple linear regression (MLR) and multiple polynomial regression (MPR). These algorithmic implementations are state-of-the-art so both packages may be used for professional applications (see [FitMlr] and [FitMpr]). They use the Utility, Graphics, DataTransformation and Cluster packages.

- **Perceptron:** The Perceptron package provides optimization algorithms for three-layer perceptrons. It utilizes Mathematica's FindMinimum (ConjugateGradient) or NMinimize (DifferentialEvolution) methods for minimization tasks (see [FindMinimum/FindMaximum] and [NMinimize/NMaximize]). The package also provides a Backpropagation plus Momentum minimization and a classical genetic algorithm based minimization. Although the quality of the minimization algorithms is state-of-the-art the specific calculation setup contains non-optimum redundancies that decrease performance and increase memory consumption.

Thus the usage of this package is confined to small data sets with about a (few) thousand I/O pairs for practical application. It uses the Utility, Graphics, Data-Transformation and Cluster packages.

- **SVM:** The SVM package provides constrained optimization algorithms for support vector machines (SVM). It utilizes Mathematica's FindMaximum (Interior-Point) or NMaximize (DifferentialEvolution) methods for constrained optimization tasks (see [FindMinimum/FindMaximum] and [NMinimize/NMaximize]). Although these algorithms are robust they do not exploit any specifics of the support vector objective function to increase optimization convergence speed etc. Therefore a practical application is advised to only very small data sets with less than a thousand I/O pairs. The package uses the Utility, Graphics, DataTransformation and Cluster packages.

A.2 Experimental data

```
Clear["Global`*"];
<<CIP`Utility`
<<CIP`DataTransformation`
<<CIP`ExperimentalData`
<<CIP`Graphics`
<<CIP`CurveFit`
```

The CIP package ExperimentalData provides mainly experimental and some theoretically calculated data which are used throughout the book. They are briefly sketched in the following.

A.2.1 Temperature dependence of the viscosity of water

The xy-error data describe the temperature dependence of the viscosity η of water (measured in centi-Poise which is the scientific unit of viscosity) in the temperature range from 293.15 to 323.15 K (20 to 50 degree Celsius) with a very small estimated experimental error of 0.0001 (10^{-4}) cP (see [Weast 1975] for reference):

```
xyErrorData=CIP`ExperimentalData`GetWaterViscosityXyErrorData[];
labels={"T [K]","\[Eta] [cP]","Viscosity of water"};
CIP`Graphics`PlotXyErrorData[xyErrorData,labels]
```

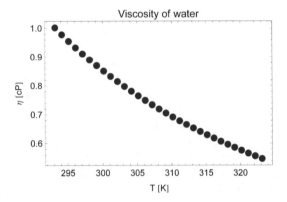

A.2.2 Potential energy surface of hydrogen fluoride

The *xy*-error data describe the potential energy of the hydrogen fluoride molecule (measured in Hartree) as a function of the interatomic distance (measured in Angstrom). A very small absolute error of 10^{-6} is assumed for the energy values. They were calculated with the ab-initio quantum-chemical software package Gaussian using a uMP4/6-311++g(3df, 3pd) model chemistry (see [Gaussian 2003]).

```
xyErrorData=
  CIP'ExperimentalData'GetHydrogenFluoridePESXyErrorData[];
labels={"H-F Distance [Angstrom]","Energy [Hartree]",
  "PES of hydrogen fluoride (HF)"};
CIP'Graphics'PlotXyErrorData[xyErrorData,labels]
```

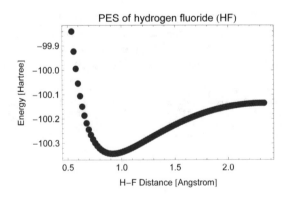

A.2.3 Kinetics data from time dependent IR spectra of the hydrolysis of acetanhydride

The hydrolysis of acetanhydride was monitored online by infrared (IR) spectra which were taken every 2 minutes (see [Meyer 2010]). The list of IR spectra

```
spectra=CIP`ExperimentalData`GetAcetanhydrideHydrolysisIRSpectra[];
Length[spectra]
```

22

ranges from spectrum 1 (at start = 0 minutes) to spectrum 22 (at the end = 44 minutes). Each spectrum is a list of 2D *xy*-points where the argument value (*x*) is the wave number in 1/cm and the dependent value (*y*) the corresponding absorption at this wave number. From the full spectrum

```
index=1;
time=(index-1)*2;
labels={"Wavenumber [1/cm]","Absorption",
  StringJoin["Full IR spectrum (",ToString[time]," min)"]};
pointSize=0.01;
CIP`Graphics`Plot2dPoints[spectra[[index]],labels,
  GraphicsOptionPointSize -> pointSize]
```

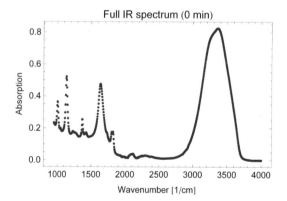

the acetanhydride absorption peak around 1140 1/cm is used for tracking the reduction of acetanhydride due to the hydrolysis reaction. The peak has a maximum height at the beginning of the reaction

```
partialSpectrum=Select[spectra[[index]],#[[1]]<1500&];
labels={"Wavenumber [1/cm]","Absorption",
  StringJoin["Zoomed IR spectrum (",ToString[time]," min)"]};
```

```
CIP 'Graphics 'Plot2dPoints[partialSpectrum, labels,
 GraphicsOptionPointSize -> pointSize]
```

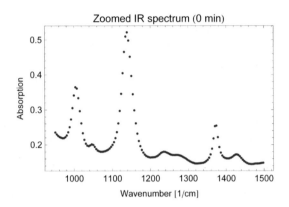

and then decreases with reaction progress

```
index=11;
time=(index-1)*2;
partialSpectrum=Select[spectra[[index]],#[[1]]<1500&];
labels={"Wavenumber [1/cm]","Absorption",
 StringJoin["Zoomed IR spectrum (",ToString[time]," min)"]};
CIP 'Graphics 'Plot2dPoints[partialSpectrum, labels,
 GraphicsOptionPointSize -> pointSize]
```

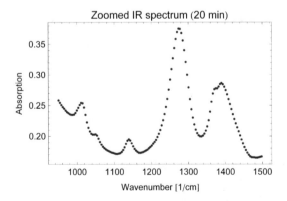

to an almost complete disappearance at the end of the monitored time period:

```
index=22;
time=(index-1)*2;
partialSpectrum=Select[spectra[[index]],#[[1]]<1500&];
labels={"Wavenumber [1/cm]","Absorption",
 StringJoin["Zoomed IR spectrum (",ToString[time]," min)"]};
CIP'Graphics'Plot2dPoints[partialSpectrum,labels,
 GraphicsOptionPointSize -> pointSize]
```

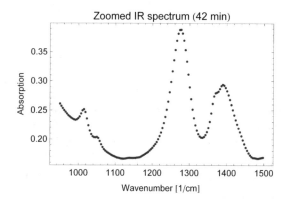

The height of the peak is linearly correlated to the chemical concentration of acetanhydride (Lambert-Beer law). Thus the most straightforward procedure to extract kinetics data from the spectra is to simply use the maximum peak absorption value at each time. This method is denoted "method 1" and executed with GetAcetanhydrideKineticsData1 of the CIP ExperimentalData package:

```
kineticsDataMethod1=GetAcetanhydrideKineticsData1[];
```

Note that the absorption at the end of the reaction with completely vanished acetanhydride is not zero as expected but has a baseline (background) value above zero:

```
labels={"Time [min]","Absorption",
 "Measured absorption maxima around 1140 1/cm"};
CIP'Graphics'Plot2dPoints[kineticsDataMethod1,labels]
```

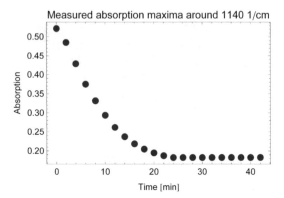

The non-zero baseline is an unlovely artifact of the measurement process. More elaborate methods of kinetics data extraction from the spectra try to take the unwanted baselines into account to correct the absorption for background independence. A possible more elaborated method (denoted "method 2") proceeds as follows. The peak of each spectrum is isolated in the wave number range 1060 to 1220 1/cm (shown here for the first spectrum at the beginning of the reaction):

```
index=1;
time=(index-1)*2;
labels={"Wavenumber [1/cm]","Absorption",
 StringJoin["IR spectrum (",ToString[time]," min)"]};
partialSpectrum=
 Select[spectra[[index]],(#[[1]]<1220 && #[[1]]>1060)& ];
CIP 'Graphics 'Plot2dPoints[partialSpectrum,labels]
```

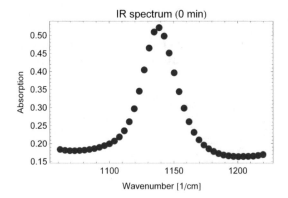

The absorption values are assumed to be very precise with a small absolute error of 0.0005 and the spectral data are transformed to xy-error data for curve fitting:

```
errorOfAbsorption=0.0005;
partialSpectrumData=
  CIP`DataTransformation`AddErrorToXYData[partialSpectrum,
  errorOfAbsorption];
```

The spectral data are smoothed with cubic splines (compare chapter 2)

```
reducedChisquare=1.0;
curveFitInfo=CIP`CurveFit`FitCubicSplines[partialSpectrumData,
  reducedChisquare];
```

with a convincing result:

```
labels={"Wavenumber [1/cm]","Absorption",
  StringJoin["IR spectrum with smoothing splines (",ToString[time],
  " min)"]};
CIP`CurveFit`ShowFitResult[{"FunctionPlot"},partialSpectrumData,
  curveFitInfo,CurveFitOptionLabels -> labels];
```

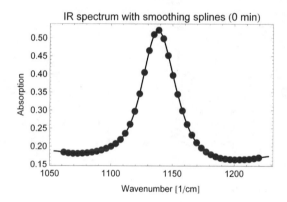

Thus the spectral data may be successfully represented by the smoothing cubic splines function:

```
argumentRange={partialSpectrumData[[1,1]],
  partialSpectrumData[[Length[partialSpectrumData],1]]};
functionValueRange={0.1,0.55};
pureFunction=Function[internalArgument,
  CIP`CurveFit`CalculateFunctionValue[internalArgument,
  curveFitInfo]];
labels={"Wavenumber [1/cm]","Absorption","Smoothing cubic splines"};
graphics=CIP`Graphics`Plot2dFunction[pureFunction,argumentRange,
  functionValueRange,labels]
```

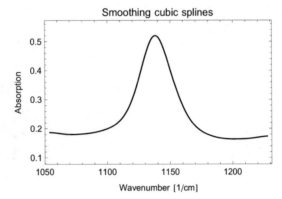

Next the locations of the minimum left of the peak, of the peak maximum and the minimum right of the peak are determined with the aid of the smoothing cubic splines model:

```
roots={};
oldPoint={partialSpectrumData[[1,1]],
 CIP 'CurveFit 'CalculateDerivativeValue[1,partialSpectrumData[[1,1]],
 curveFitInfo]};
Do[
 newPoint=
  {i,CIP 'CurveFit 'CalculateDerivativeValue[1,i,curveFitInfo]};
 If[oldPoint[[2]]*newPoint[[2]]<= 0,
  AppendTo[roots,x/.FindRoot[
  CIP 'CurveFit 'CalculateDerivativeValue[1,x,curveFitInfo],
  {x,oldPoint[[1]],newPoint[[1]]}]]]
 ];
 oldPoint=newPoint,
 {i,partialSpectrumData[[1,1]],
  partialSpectrumData[[Length[partialSpectrumData],1]],1}
 ];
```

The detected locations

```
roots
```

{1071.68, 1138.07, 1200.94}

can be used to construct a connection (linearly interpolated line) between the left and right minima to approximate the unknown baseline of the peak. The absorption value may then be corrected with the baseline value at maximum absorption

```
minimum1={roots[[1]],
 CIP 'CurveFit 'CalculateFunctionValue[roots[[1]],curveFitInfo]};
maximum={roots[[2]],
 CIP 'CurveFit 'CalculateFunctionValue[roots[[2]],curveFitInfo]};
minimum2={roots[[3]],
 CIP 'CurveFit 'CalculateFunctionValue[roots[[3]],curveFitInfo]};
```

```
baselineCorrection={
  maximum[[1]],
  minimum1[[2]]+
    (minimum2[[2]]-minimum1[[2]])/(minimum2[[1]]-minimum1[[1]])*
    (maximum[[1]]-minimum1[[1]])
};
```

which may be graphically depicted for illustration:

```
Show[
  graphics,
  Epilog -> {Thickness[0.005],PointSize[0.03],RGBColor[1,0,0],
   Line[{minimum1,minimum2}],Line[{baselineCorrection,maximum}],
   Point[minimum1],Point[maximum],Point[minimum2],
   Point[baselineCorrection]}
]
```

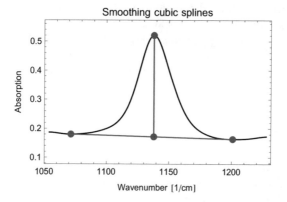

The corrected absorption value is the distance between the linearly interpolated baseline value at the maximum position and the maximum measured absorption value. This procedure is applied to all spectra with the GetAcetanhydrideKinetics-Data2 method of the CIP ExperimentalData package:

```
kineticsDataMethod2=GetAcetanhydrideKineticsData2[];
```

A graphical display of the kinetics data demonstrates the improvement in comparison to method 1:

```
labels={"Time [min]","Absorption",
  "Corrected absorption maxima around 1140 1/cm"};
CIP`Graphics`Plot2dPoints[kineticsDataMethod2,labels]
```

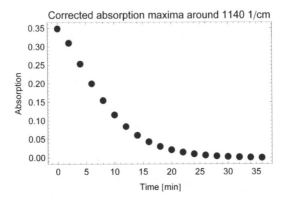

The absorption values now seem to drop to zero as it is expected for a vanished acetanhydride. Note that there is still an obvious deviation of the absorption peak value of the first spectrum at 0 minutes which could be traced to an initial measurement delay. As a final remark it should be clear that the kinetics data obtained with both methods are flawed by systematic errors due to their individual data extraction process: Whereas method 1 completely neglects all baseline issues the more elaborate method 2 performs an arbitrary linear approximation procedure only.

A.2.4 Iris flowers

The iris flower classification data set (see [Fisher 1936]) consists of measurements of the length and width of sepal and petal of 50 samples from each of the three species of iris flowers: Iris setosa (denoted species 1), iris versicolor (species 2) and iris virginica (species 3). Each input is a vector with 4 components that denote the sepal length (component 1), the sepal width (component 2), the petal length (component 3) and the petal width (component 4) in millimeter. Each output codes the corresponding species (see chapter 1 for encoding classification data set outputs). The inputs alone may be accessed for each species separately

```
inputsOfSpecies1=CIP`ExperimentalData`GetIrisFlowerInputsSpecies1[];
inputsOfSpecies2=CIP`ExperimentalData`GetIrisFlowerInputsSpecies2[];
inputsOfSpecies3=CIP`ExperimentalData`GetIrisFlowerInputsSpecies3[];
```

or joined:

```
inputs=CIP`ExperimentalData`GetIrisFlowerInputs[];
```

The complete classification data set with all I/O pairs for all three species can be accessed with

```
classificationDataSet=
 CIP'ExperimentalData'GetIrisFlowerClassificationDataSet[];
```

A.2.5 Adhesive kinetics

The adhesive kinetics data set describes the dependence of a kinetics parameter on the composition of an adhesive polymer mixture. The adhesive polymer mixture contains four components: Methyl methacrylate (MMA), poly(methyl methacrylate) (PMMA), dibenzoyl peroxide and N,N-diethylol-p-toluidine. Each input vector contains 3 components with the mass ratio of MMA to PMMA in percent (component 1), the mass of dibenzoyl peroxide in gram (component 2) and the mass of N,N-diethylol-p-toluidine in gram (component 3) - the complete mass of each mixture was about 20g. The output vector contains 1 component which is the time in seconds to the maximum temperature as a characteristic property of the exothermic adhesive hardening reaction (see [Koch 2003]). The adhesive data set comprises 73 different mixtures

```
dataSet=CIP'ExperimentalData'GetAdhesiveKineticsDataSet[];
Length[dataSet]
```

73

with a clear design of experiment which may be illustrated by a 3D display of the inputs

```
inputs=CIP'Utility'GetInputsOfDataSet[dataSet];
labels={"Ratio","C2","C3"};
viewPoint3D={0.6,-3.4,2.0};
CIP'Graphics'Plot3dPoints[inputs,labels,
 GraphicsOptionViewPoint3D -> viewPoint3D]
```

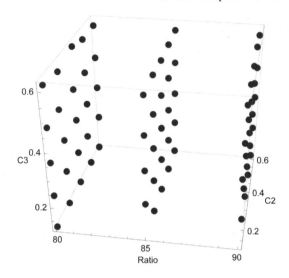

where "Ratio" denotes the mass ratio of MMA to PMMA, C2 the mass of dibenzoyl peroxide (in gram) and C3 the mass of N,N-diethylol-p-toluidine (in gram). There are 3 MMA:PMMA ratios measured: 80%, 85% and 90%. Since the full adhesive kinetics data set is four-dimensional (3 input components plus 1 output component) a three-dimensional data subset with two input components and 1 output component may be obtained for each fixed MMA:PMMA ratio, e.g. for MMA:PMMA = 80%

```
polymerMassRatio="80";
dataSet3D=
 CIP`ExperimentalData`GetAdhesiveKinetics3dDataSet[
 polymerMassRatio];
labels={"C2","C3","t"};
CIP`Graphics`Plot3dDataSet[dataSet3D,labels]
```

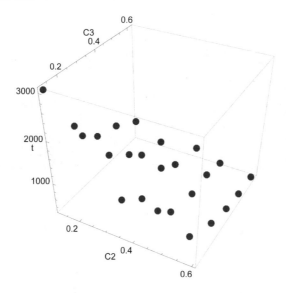

where "t" denotes the time in seconds to the maximum temperature. The experimental time-to-maximum-temperature errors were reported to be in the order of
10% to 20% with some outliers.

A.2.6 Intertwined spirals

A intertwined spiral classification data set consists of inputs of dimension 2 and
corresponding outputs that code one of the two spiral classes. The inputs of the first
spiral are calculated with

$$(x,y) = \left(2\cos(u)e^{u/10}, 1.5\sin(u)e^{u/10}\right) \; ; \; \pi \leq u \leq 3.5\pi$$

and those of the second spiral with

$$(x,y) = \left(2.7\cos(u)e^{u/10}, 2.025\sin(u)e^{u/10}\right) \; ; \; -\pi/2 \leq u \leq 2.5\pi$$

(see [Juillé 1996] and [Paláncz 2004]). The number of points along the spirals
within the defined intervals may be specified:

```
numberOfSingleSpiralIoPairs=30;
classificationDataSet60=
 CIP`ExperimentalData`GetSpiralsClassificationDataSet[
 numberOfSingleSpiralIoPairs];

classIndex=1;
inputsOfSpiral1=
```

```
CIP `DataTransformation `GetInputsForSpecifiedClass[
 classificationDataSet60,classIndex];
classIndex=2;
inputsOfSpiral2=
 CIP `DataTransformation `GetInputsForSpecifiedClass[
 classificationDataSet60,classIndex];

points2DWithPlotStyle1={inputsOfSpiral1,{PointSize[0.02],Red}};
points2DWithPlotStyle2={inputsOfSpiral2,{PointSize[0.02],Green}};
points2DWithPlotStyleList={points2DWithPlotStyle1,
 points2DWithPlotStyle2};
labels={"x","y","Intertwined spirals"};
CIP `Graphics `PlotMultiple2dPoints[points2DWithPlotStyleList,labels]
```

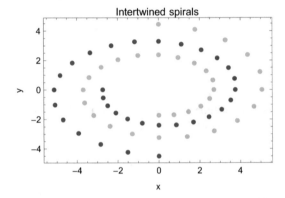

A.2.7 Faces

There are 3 faces image classification data sets. They contain 18 I/O pairs each (faces of 6 cats (class 1), 6 dogs (class 2) and 6 humans (class 3)) with the same faces but different backgrounds (see [Faces 2010]): An image classification data set with white background

```
imageClassificationDataSetWhite=
 CIP `ExperimentalData `GetFacesWhiteImageDataSet[];
imageInputsWhite=
 CIP `Utility `GetInputsOfDataSet[imageClassificationDataSetWhite];
GraphicsGrid[
 Table[
  Image[imageInputsWhite[[(i-1)*6+j]],"Byte"],
  {i,3},{j,6}
 ],
 ImageSize->300
]
```

one with gray background

```
imageClassificationDataSetGray=
 CIP`ExperimentalData`GetFacesGrayImageDataSet[];
imageInputsGray=
 CIP`Utility`GetInputsOfDataSet[imageClassificationDataSetGray];
GraphicsGrid[
 Table[
  Image[imageInputsGray[[(i-1)*6+j]],"Byte"],
  {i,3},{j,6}
 ],
 ImageSize->300
]
```

and one with black background:

```
imageClassificationDataSetBlack=
 CIP`ExperimentalData`GetFacesBlackImageDataSet[];
imageInputsBlack=
 CIP`Utility`GetInputsOfDataSet[imageClassificationDataSetBlack];
GraphicsGrid[
 Table[
  Image[imageInputsBlack[[(i-1)*6+j]],"Byte"],
  {i,3},{j,6}
 ],
 ImageSize->300
]
```

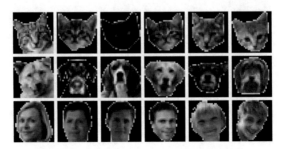

Each input of the image classification data sets is a matrix with 30 rows and columns that code 30×30 grayscale images (with a $30 \times 30 = 900$ pixels in total per image):

```
Dimensions[imageInputsWhite[[1]]]
```

{30,30}

Each pixel contains a specific shade of gray (out of 256 possible values), e.g. the first pixel

```
imageInputsWhite[[1,1,1]]
```

255.

codes "white" (255). To use an image classification data set for clustering or machine learning the (pixel) matrix must be converted to a (pixel) vector by successive concatenation of the pixel matrix rows. This is performed with a specific method of the CIP DataTransformation package, e.g. for the faces image classification data set with white background:

```
classificationDataSetWhite=
 CIP`DataTransformation`ConvertImageDataSet[
 imageClassificationDataSetWhite];
```

Each input of the resulting classification data set contains a vector with 900 components where each component contains the grayscale value of its corresponding pixel:

```
inputsWhite=
 CIP`Utility`GetInputsOfDataSet[classificationDataSetWhite];
 Length[inputsWhite[[1]]]
```

900

```
inputsWhite[[1,1]]
```

255.

A.2.8 Wisconsin Diagnostic Breast Cancer (WDBC) data

The Wisconsin Diagnostic Breast Cancer (WDBC) classification data set (see [WDBC data] in the references) consists of 569 I/O pairs

```
classificationDataSet=
 CIP`ExperimentalData`GetWDBCClassificationDataSet[];
CIP`Graphics`ShowDataSetInfo[{"IoPairs","InputComponents",
 "OutputComponents","ClassCount"},classificationDataSet]
```

Number of IO pairs = 569

Number of input components = 30

Number of output components = 2

Class 1 with 357 members

Class 2 with 212 members

where each input is mapped onto one of two classes. Every I/O pair refers to a single patient. The 30 components of each input are real-valued quantities that are computed from a digitally scanned image of a fine needle aspirate of a breast mass. Fine needle aspiration biopsy is a very safe and minor surgical procedure which is widely used in the diagnosis of cancer: A thin and hollow needle is inserted into the tumor tissue to extract cells which are then (after being stained) examined and digitally scanned under a microscope. The 30 input components describe characteristics of the extracted cell nuclei present in a digital image. Components 1-10 describe the mean, components 11-20 the standard deviation and components 21-30 the maximum ("worst" or "largest") of a feature. The 10 single features that are computed for each cell nucleus are (1) radius, (2) texture, (3) perimeter, (4) area, (5) smoothness, (6) compactness, (7) concavity, (8) concave points, (9) symmetry and (10) fractal dimension. Thus input components 1, 11 and 21 refer to the same feature (radius) and are the mean, the standard deviation and the maximum of this feature for all cell nuclei of a digitized image. The two output components code two classes where class 1 (coded $\{1.0, 0.0\}$) denotes the diagnosis of a benign tumor and class 2 (coded $\{0.0, 1.0\}$) the diagnosis of a malignant tumor.

A.2.9 *Wisconsin Prognostic Breast Cancer (WPBC) data*

The Wisconsin Prognostic Breast Cancer (WPBC) classification data set (see [WPBC data] in the references) consists of 198 I/O pairs

```
classificationDataSet=
 CIP`ExperimentalData`GetWPBCClassificationDataSet[];
CIP`Graphics`ShowDataSetInfo[{"IoPairs","InputComponents",
 "OutputComponents","ClassCount"},classificationDataSet]
```

Number of IO pairs = 198

Number of input components = 32

Number of output components = 2

Class 1 with 151 members

Class 2 with 47 members

where each input is mapped onto one of two classes and every I/O pair refers to a single patient. The first 30 components of each input correspond to those of the WDBC classification data set. Component 31 is the tumor size (i.e. the diameter of the excised tumor in centimeters) and component 32 the lymph node status (i.e. the number of positive axillary lymph nodes). The two output components code two classes where class 1 (coded $\{1.0, 0.0\}$) denotes a non-recurrent and class 2 (coded $\{0.0, 1.0\}$) a recurrent tumor.

The WPBC non-recurrent and recurrent regression data sets (see [WPBC non-recurrent data] and [WPBC recurrent data] in the references) consist of 151 I/O pairs

```
regressionDataSet=
 CIP`ExperimentalData`GetWPBCNonrecurrentDataSet[];
CIP`Graphics`ShowDataSetInfo[{"IoPairs","InputComponents",
 "OutputComponents"},regressionDataSet]
```

Number of IO pairs = 151

Number of input components = 32

Number of output components = 1

and 47 I/O pairs respectively

```
regressionDataSet=
 CIP`ExperimentalData`GetWPBCRecurrentDataSet[];
CIP`Graphics`ShowDataSetInfo[{"IoPairs","InputComponents",
 "OutputComponents"},regressionDataSet]
```

Number of IO pairs = 47

Number of input components = 32

Number of output components = 1

where each input corresponds to those of the WPBC classification data set. The single output component for the non-recurrent data denotes the disease-free time in months and for the recurrent data the tumor recurrence time in months.

Note that all WPBC data sets have missing numerical values (which are coded as NaN: Not a Number) thus a cleaning operation is mandatory prior to use.

A.2.10 QSPR data

A Quantitative Structure-Property Relationship (QSPR) is a model that maps features of chemical compounds or reactions (the input) to a quantity of interest (the output). The features are usually structural and molecular descriptors, i.e. characteristic numbers which can be calculated for a specific chemical compound like its molecular weight, dipole moment or ring count. The (e.g. physico-chemical) output quantity of interest is usually measured experimentally.

There are two alienated QSPR data sets available taken from academic and industrial research: The first data set comprises 183 input/output (I/O) pairs that each represent a single chemical compound. Each input vector consists of 155 components where each component is a calculated structural descriptor. Each corresponding output vector contains a single experimentally measured value for a compound-related physico-chemical quantity:

```
regressionDataSet=
 CIP`ExperimentalData`GetQSPRDataSet01[];
CIP`Graphics`ShowDataSetInfo[{"IoPairs","InputComponents",
 "OutputComponents"},regressionDataSet]
```

Number of IO pairs = 183

Number of input components = 155

Number of output components = 1

(Note that at first glance the usefulness of the whole data set is in doubt since the number of calculated input components is nearly equal to the number of I/O pairs, i.e. already a linear method like MLR is likely to be able to establish a perfect relationship between the molecular descriptors and the corresponding physico-chemical quantity). The second data set comprises 2169 input/output (I/O) pairs that each represent a single chemical reaction. Each input vector consists of 130 components where each component is a calculated structural descriptor. Each corresponding output vector contains a single experimentally measured value for a reaction-related physico-chemical quantity.

```
regressionDataSet=
 CIP`ExperimentalData`GetQSPRDataSet02[];
CIP`Graphics`ShowDataSetInfo[{"IoPairs","InputComponents",
 "OutputComponents"},regressionDataSet]
```

Number of IO pairs = 2169

Number of input components = 130

Number of output components = 1

A.3 Parallelized calculations

```
Clear["Global`*"];
<<CIP`Utility`
<<CIP`ExperimentalData`
<<CIP`MLR`
```

Several CIP functions (see below) support parallelized calculations which may lead to considerable performance improvements on computing devices with multiple (or multicore) processors. Note that parallelized calculations always require an unavoidable computational overhead so that performance gains may be smaller than expected up to a performance decrease for specific cases.

As an example for a performance improvement with parallelized calculations the relevance analysis of inputs components is chosen. For the WDBC classification data set

```
classificationDataSet=
 CIP`ExperimentalData`GetWDBCClassificationDataSet[];
CIP`Graphics`ShowDataSetInfo[{"IoPairs","InputComponents",
 "OutputComponents"},classificationDataSet]
```

Number of IO pairs = 569

Number of input components = 30

Number of output components = 2

the relevance of its 30 input components for a predictive MLR model is evaluated in a heuristic step-by-step fashion: In a first step all single input components are scanned for the (winner) one that leads to the best linear mapping from input to output with the best predictivity. Then this single best input component is fixed and the remaining input components are scanned for the next most relevant feature so that a pair of two relevant input components is achieved. This procedure is continued to get a relevant triple etc. It is obvious that such a procedure is well suited for parallelization since the relevance of each component is evaluated by an independent single MLR fit. For comparison the non-parallelized sequential calculation (which is the default for all CIP calculations) is initially performed

```
trainingAndTestSet={classificationDataSet,{}};
Print["Time in s: ",
 AbsoluteTiming[
```

```
    sequentialMlrInputComponentRelevanceListForClassification=
     CIP `MLR `GetMlrInputInclusionClass[trainingAndTestSet]
   ][[1]]
   ]
```

Time in s: 206.598

with a relevance result list

```
numberOfComponents=30;
sequentialInputComponentInclusionList=
 CIP `MLR `GetMlrClassRelevantComponents[
  sequentialMlrInputComponentRelevanceListForClassification,
  numberOfComponents
  ]
```

$\{28, 21, 22, 24, 15, 29, 16, 11, 30, 6, 8, 27, 17, 14, 18, 7, 1, 2, 25, 4, 13, 3, 20, 19, 23, 26, 12, 9, 5, 10\}$

that shows component 28 to be the most relevant component, component 21 to be second most relevant component etc. For the parallelized calculation the desired number of parallel Mathematica kernels has to be specified first (where argument 0 is the default value and leads to the use of all available physical processor cores)

```
numberOfLaunchedKernels=
 CIP `Utility `SetNumberOfParallelKernels[0]
```

8

(where in this case an octacore processor is used) and then the function call is repeated with the ParallelCalculation option

```
Print["Time in s: ",
 AbsoluteTiming[
  parallelizedMlrInputComponentRelevanceListForClassification=
   CIP `MLR `GetMlrInputInclusionClass[
    trainingAndTestSet,
    UtilityOptionCalculationMode -> "ParallelCalculation"]
  ][[1]]
  ]
```

Time in s: 71.6658

to come to the same component relevance result as before

```
parallelizedInputComponentInclusionList=
 CIP `MLR `GetMlrClassRelevantComponents[
  parallelizedMlrInputComponentRelevanceListForClassification,
  numberOfComponents
  ]
```

$$\{28,21,22,24,15,29,16,11,30,6,8,27,17,14,18,7,1,2,25,4,13,3,20,19,23,26,12,9,5,10\}$$

but with improved performance. The following CIP functions may be called with the ParallelCalculation option:

- GetSilhouettePlotPoints

  For METHOD = Mlr, Mpr, Perceptron, Svm:

- GetBestMETHODClassOptimization
- GetMETHODInputInclusionClass
- GetMETHODInputInclusionRegress
- GetMETHODInputRelevanceClass
- GetMETHODInputRelevanceRegress
- ScanRegressTrainingWithMETHOD

  For METHOD = Cluster, Mlr, Mpr, Perceptron, Svm:

- ScanClassTrainingWithMETHOD

  For METHOD = Mpr, Perceptron, Svm:

- FitMETHODSeries

  For METHOD = Perceptron, Svm:

- FitMETHOD
- GetMETHODTrainOptimization

Note that a parallelization of function FitMETHOD is only available for Perceptrons and SVMs. Function GetMETHODTrainOptimization is also available for MLR and MPR but GetMlrTrainOptimization and GetMprTrainOptimization do not contain a ParallelCalculation option since Perceptrons and SVMs may use internal parallelization for a single fit but MLR and MPR do not.

References

[Andrade 1934] E. N. da C. Andrade, *A Theory of the Viscosity of Liquids. - Part I*, Philosophical Magazine 17 (112), 497-511, 1934.

[Baggott 2005] J. Baggott, *A Beginner's Guide to Reality*, New York 2005, Penguin Books.

[Barlow 1989] R. J. Barlow, *Statistics: A Guide to the Use of Statistical Methods in the Physical Sciences*, Chichester 1989, Wiley VCH.

[Barrat 2013] J. Barrat, *Our Final Invention: Artificial Intelligence and the End of the Human Era*, New York 2013, Thomas Dunne Books.

[Bevington 2002] P. Bevington, D. K. Robinson, *Data Reduction and Error Analysis for the Physical Sciences*, New York 2002, McGraw-Hill, 3rd Edition.

[Bishop 2006] C. M. Bishop, *Pattern Recognition and Machine Learning*, New York 2006, Springer.

[Blue Brain Project] Blue Brain Project, see http://bluebrain.epfl.ch. Found at 2016/01/27.

[Box] Quotation (Chapter 1.1) from Wikiquote: http://en.wikiquote.org/wiki/George_E._P._Box. Found at 2010/01/27.

[Brandt 2002] S. Brandt, *Data Analysis: Statistical and Computational Methods for Scientists and Engineers*, New York 1998, Springer, 3rd Edition.

[Carpenter 1991] G. A. Carpenter, S. Grossberg, D. B. Rosen, *ART 2-A: An Adaptive Resonance Algorithm for Rapid Category Learning and Recognition*, Neural Networks 4, 493-504, 1991.

[Chatterjee 2000] S. Chatterjee, A. Hadi, B. Price, *Regression Analysis by Example*, New York 2000, John Wiley & Sons, 3rd Edition. Chapter 3: *Multiple Linear Regression*, pages 51-84.

[Cherkassy 1996] V. Cherkassy, D. Gehring, F. Mulier, *Comparison of adaptive methods for function estimation from samples*, IEEE Trans. Neural Networks 7 (4), 969-984, 1996.

[Churchland 1996] P. M. Churchland, *The Engine of Reason, The Seat of the Soul: A Philosophical Journey into the Brain*, Massachusetts 1996, MIT Press.

[CIP] Computational Intelligence Packages (CIP), Version 2.0. Open source library for Mathematica 10 or higher designed by Achim Zielesny. Internet: http://www.gnwi.de. Installation instructions for the CIP Mathematica packages are provided within the ZIP container available for download at this internet site.

[Clark 2010/2015] Tim Clark, private communication at the 2010 Beilstein Symposium on Nanotechnology in Bolzano, Italy, and at the 2015 German Conference on Chemoinformatics in Fulda, Germany.

[Cristianini 2000] N. Cristianini, J. Shawe-Taylor, *An Introduction to Support Vector Machines and other kernel-based learning methods*, Cambridge 2000, Cambridge University Press.

[Cyc 2010] The Cyc project, Internet: http://www.cyc.com

[Dirac 1929] P. A. M. Dirac, *Quantum mechanics of many-electron systems*, Proceedings of the Royal Society (London) A 123, 714-733, 1929.

[Dreyfus 1992] H. L. Dreyfus, *What Computers Still Can't Do: A Critique of Artificial Reason*, Cambridge 1992, MIT Press.

© Springer International Publishing Switzerland 2016

A. Zielesny, *From Curve Fitting to Machine Learning*, Intelligent Systems Reference Library 109, DOI 10.1007/978-3-319-32545-3

[Dyson 2004] Attributed to John von Neumann by Enrico Fermi, quoted by: F. Dyson, *A meeting with Enrico Fermi,* Nature 427, 297, 2004.

[Edwards 1976] A. L. Edwards, *An Introduction to Linear Regression and Correlation,* San Francisco, CA 1976, W. H. Freeman.

[Edwards 1979] A. L. Edwards, *Multiple Regression and the Analysis of Variance and Covariance,* San Francisco, CA 1979, W. H. Freeman.

[Faces 2010] The face images are a courtesy of Rebecca Schulz, University of Applied Sciences Gelsenkirchen, Germany.

[Fan 2005] R. E. Fan, P. H. Chen, C. J. Lin, *Working set selection using the second order information for training support vector machines,* Journal of Machine Learning Research 6, 1889-1918, 2005.

[FindMinimum/FindMaximum] The FindMinimum and FindMaximum commands of the Mathematica system provide an unified access to different unconstrained and constrained local optimization algorithms (FindMinimum for minimization, FindMaximum for maximization: Both commands essentially use the same algorithms since minimization of a function f means maximization of $"-f$ or $f^{-1}"$). Constraint optimization is chosen if a constraint is defined in the command signature otherwise an unconstrained optimization is performed. The default unconstrained local optimization algorithm used is the BFGS variant of the Quasi-Newton methods. If the function to be optimized is detected to be a sum of squares the Levenberg-Marquardt algorithm is used as a default. Other unconstrained local minimization methods like Conjugate-Gradient or Newton and their variants may be specified with the method option (see [Press 2007] for algorithmic details). For constrained local optimization there is the Interior Point method used as a default (see [Mehrotra 1992] and [Forsgren 2002] for algorithmic details).

[FitMlr] The CIP FitMlr method is build on top of Mathematica's Fit command which performs least squares fits with linear combinations of functions. See [Edwards 1976], [Edwards 1979], [Chatterjee 2000] and [Press 2007] for details.

[FitModelFunction] The FitModelFunction method is build on top of the NonlinearModelFit command of Mathematica which uses the Levenberg-Marquardt method for iterative unconstrained local minimization of $\chi^2 (a_1,...,a_L)$ since this quantity is a sum of squares (compare chapter 2). See [Bevington 2002], [Brandt 2002] and [Press 2007] for algorithmic details.

[FitMpr] The CIP FitMpr method is build on top of Mathematica's Fit command which performs least squares fits with linear combinations of functions. See [Edwards 1976], [Edwards 1979], [Chatterjee 2000] and [Press 2007] for details.

[FitPerceptron] As a default the CIP method FitPerceptron is build on top of Mathematica's FindMinimum command for an unconstrained local minimization with the (Polak-Ribiere variant of the) Conjugate-Gradient method (see [FindMinimum/FindMaximum]). Other optimization methods available through FitPerceptron are an evolutionary algorithm based global minimization with NMinimize (which uses Differential Evolution, see [NMinimize/NMaximize]), a Backpropagation plus Momentum local minimization or a genetic algorithm based global minimization (see [Freeman 1993]). The different minimization techniques may be selected with option OptimizationMethodOption. The (default) option MultiplePerceptronsOption allows to fit a perceptron for every single output component of a data set's output which can be especially accelerated by parallelized calculation with the ParallelCalculation option (see Appendix A).

[FitSvm] The CIP methods FitSvm uses Mathematica's NMaximize command with an evolutionary algorithm based constrained global maximization method (Differential Evolution, see [NMinimize/NMaximize]). The SVM code itself is based on the implementation in [Paláncz 2005]. The fit of data sets with multiple output components may be especially accelerated by parallelized calculation with the ParallelCalculation option (see Appendix A).

[Fisher 1936] R. A. Fisher, *The Use of Multiple Measurements in Taxonomic Problems,* Annals of Eugenics 7, 179-188, 1936.

[Ford 2015] M. Ford, *Rise of the Robots: Technology and the Threat of a Jobless Future,* New York 2015, Basic Books.

[Forsgren 2002] A. Forsgren, P. E. Gill, M. H. Wright, *Interior Methods for Nonlinear Optimization*, SIAM Rev. 44 (4), 525-597, 2002.

[Freeman 1993] J. A. Freeman, *Simulating Neural Networks with Mathematica*, Boston 1993, Addison-Wesley Longman Publishing Co.

[Frenkel 2002] D. Frenkel, B. Smit, *Understanding Molecular Simulation: From Algorithms to Applications*, San Diego 2002, Academic Press.

[Gaussian 2003] Gaussian 03, Revision B.05, M. J. Frisch, G. W. Trucks, H. B. Schlegel, G. E. Scuseria, M. A. Robb, J. R. Cheeseman, J. A. Montgomery, Jr., T. Vreven, K. N. Kudin, J. C. Burant, J. M. Millam, S. S. Iyengar, J. Tomasi, V. Barone, B. Mennucci, M. Cossi, G. Scalmani, N. Rega, G. A. Petersson, H. Nakatsuji, M. Hada, M. Ehara, K. Toyota, R. Fukuda, J. Hasegawa, M. Ishida, T. Nakajima, Y. Honda, O. Kitao, H. Nakai, M. Klene, X. Li, J. E. Knox, H. P. Hratchian, J. B. Cross, C. Adamo, J. Jaramillo, R. Gomperts, R. E. Stratmann, O. Yazyev, A. J. Austin, R. Cammi, C. Pomelli, J. W. Ochterski, P. Y. Ayala, K. Morokuma, G. A. Voth, P. Salvador, J. J. Dannenberg, V. G. Zakrzewski, S. Dapprich, A. D. Daniels, M. C. Strain, O. Farkas, D. K. Malick, A. D. Rabuck, K. Raghavachari, J. B. Foresman, J. V. Ortiz, Q. Cui, A. G. Baboul, S. Clifford, J. Cioslowski, B. B. Stefanov, G. Liu, A. Liashenko, P. Piskorz, I. Komaromi, R. L. Martin, D. J. Fox, T. Keith, M. A. Al-Laham, C. Y. Peng, A. Nanayakkara, M. Challacombe, P. M. W. Gill, B. Johnson, W. Chen, M. W. Wong, C. Gonzalez, and J. A. Pople, Gaussian, Inc., Pittsburgh PA, 2003.

[Gasteiger 2003] J. Gasteiger, T. Engels, *Chemoinformatics*, Weinheim 2003, Wiley-VCH.

[Glasmachers 2006] T. Glasmachers, C. Igel, *Maximum-Gain Working Set Selection for SVMs*, Journal of Machine Learning Research 7, 1437-1466, 2006.

[GetClusters] CIP GetClusters is build on top of Mathematica's FindClusters command with a method specification for partitioning around medoids (see [Kaufman 1990] for details). If the number of resulting clusters k is not defined in advance the silhouette test is chosen to obtain the best k value (see [Rousseeuw 1987] and chapter 3).

[Grant 1998] This general inability of computational chemistry to quantitatively predict rate constants for chemical reactions may be regarded as the *single biggest unsolved problem in chemistry* (from: G. H. Grant, W. G. Richards, *Computational Chemistry*, Oxford 1998) - with severe impacts on modern systems biology in form of an overall lack of kinetics data which are necessary for a realistic dynamical study and thus an understanding of biological systems.

[Gunn 1998] S. R. Gunn, *Support Vector Machines for Classification and Regression*, Technical Report, University of Southampton, Faculty of Engineering, Science and Mathematics, School of Electronics and Computer Science, 10 May 1998.

[Hamilton 1964] W. C. Hamilton, *Statistics in the Physical Sciences*, New York 1964, Ronald Press.

[Hampel 1986] F. R. Hampel, E. M. Ronchetti, P. J. Rousseeuw and W. A. Stahel, *Robust Statistics: The Approach Based on Influence Functions*, New York 1986, Wiley.

[Hawkins 2005] J. Hawkins, S. Blakeslee, *On Intelligence*, New York 2005, Times Books/Henry Holt and Company.

[Hertz 1991] J. A. Hertz, A. S. Krogh, R. G. Palmer, *Introduction To The Theory Of Neural Computation*, Redwood City, CA 1991, Addison-Wesley.

[Hofstadter 1981] D. R. Hofstadter, D. C. Dennett, *The Mind's I: Fantasies and Reflections on Self and Soul*, New York 1981, Basic Books.

[Human Brain Project] Human Brain Project, see https://www.humanbrainproject.eu. Found at 2016/01/27.

[IBM Watson] For information about IBM Watson see websites http://www.ibm.com/smarterplanet/us/en/ibmwatson/ and http://www.research.ibm.com/. Found at 2016/01/12.

[Intelligence 2010] Intelligence, from Wikipedia: http://en.wikipedia.org/wiki/Intelligence. Found at 2010/06/02.

[Jensen 2007] F. Jensen, *Introduction to Computational Chemistry* (2nd edition), Chichester 2007, John Wiley & Sons Ltd.

[Joachims 1999] T. Joachims, *Making large-Scale SVM Learning Practical*, chapter 11 in B. Schölkopf, C. J. C. Burges, A. J. Smola (editors), *Advances in Kernel Methods - Support Vector Learning*, Cambridge, MA 1999, MIT Press.

[Juillé 1996] H. Juillé, J. B. Pollack, *Co-evolving Intertwined Spirals*, Proceedings of the Fifth Annual Conference on Evolutionary Programming, San Diego, CA, February 29 - March 2, 1996, MIT Press, 461-468.

[Kaku 1998] M. Kaku, *Visions: How Science Will Revolutionize the 21st Century*, New York 1998, Anchor Books.

[Kaufman 1990] L. Kaufman and P. J. Rousseeuw, *Finding Groups in Data: An Introduction to Cluster Analysis*, New York 1990, John Wiley & Sons.

[Keerthi 2002] S. S. Keerthi, E. G. Gilbert, *Convergence of a Generalized SMO Algorithm for SVM Classifier Design*, Machine Learning 46, 351-360, 2002.

[Koch 2003] The adhesive kinetics data were provided in 2003. They are a courtesy of Prof. Dr. Klaus-Uwe Koch, University of Applied Sciences Gelsenkirchen, Germany. The data were measured in his polymer laboratory. See also: K.-U. Koch, A. Zielesny, *Neuronale Netze verkuerzen die Klebstoffentwicklung*, Adhaesion 1-2, 32-37, 2004.

[Koch 2004] C. Koch, *The Quest for Consciousness - A Neurobiological Approach*, Englewood, Colorado 2004, Roberts & Company Publishers.

[Kurzweil 2005] R. Kurzweil, *The Singularity Is Near: When Humans Transcend Biology*, New York 2005, Viking Penguin.

[Kurzweil 2012] R. Kurzweil, *How to Create a Mind: The Secret of Human Thought Revealed*, New York 2012, Viking Penguin.

[Leach 2001] A. R. Leach, *Molecular Modelling: Principles and Applications*, Harlow 2001, Prentice Hall.

[Leach 2007] A. R. Leach, V. J. Gillet, *An Introduction to Chemoinformatics*, Dordrecht 2007, Springer.

[MacQueen 1967] J. B. MacQueen, *Some Methods for Classification and Analysis of Multivariate Observations*, Proceedings of 5th Berkeley Symposium on Mathematical Statistics and Probability, 281-297, 1967, University of California Press.

[Mathematica] Wolfram Mathematica, Version 10. Mathematica is a registered trademark of Wolfram Research, Inc., Internet: www.wolfram.com.

[Mehrotra 1992] S. Mehrotra, *On the Implementation of a Primal-Dual Interior Point Method*, SIAM J. Optimization 2, 575-601, 1992.

[Meyer 2010] The acetanhydride hydrolysis IR spectra were provided in 2010. They are a courtesy of Prof. Dr. Gerhard Meyer, University of Applied Sciences Gelsenkirchen, Germany. The data were measured in his analytical laboratory.

[Mitchell 2009] M. Mitchell, *Complexity: A Guided Tour*, New York 2009, Oxford University Press.

[Murphy 2012] K. P. Murphy, *Machine Learning. A Probabilistic Perspective*, Cambridge (Massachusetts) 2012, MIT Press.

[NMinimize/NMaximize] The NMinimize and NMaximize commands of the Mathematica system provide access to the Differential Evolution method for constrained global optimization via their method option (for algorithmic details about Differential Evolution see [Price 1997], [Storn 1997], [Price 1999] and [Price 2005]). A local refinement of the constrained global optimization result with the Interior Point method (see [Mehrotra 1992] and [Forsgren 2002]) is performed by default as a post process.

[Nobel Prize 1998/2013] Royal Swedish Academy of Sciences, http://www.nobelprize.org. *Additional background material on the Nobel Prize in Chemistry 1998*, from http://www.nobelprize.org/nobel_prizes/chemistry/laureates/1998/advanced-chemistryprize1998.pdf. Found at 2016/01/26.

[Paláncz 2004] B. Paláncz, L. Völgyesi, *Support Vector Classifier via Mathematica*, Periodica Polytechnica Civ. Eng 48 (1-2), 15-37, 2004.

[Paláncz 2005] B. Paláncz, L. Völgyesi, Gy. Popper, *Support Vector Regression via Mathematica*, Periodica Polytechnica Civ. Eng 49 (1), 59-84, 2005.

[Pascal] Quotation (Preface) from *Pascal's Pensees*, Introduction by T. S. Eliot, New York 1958, E. P. Dutton & Co., Inc..

[Penrose 1991] R. Penrose, *The Emperor's new Mind*, New York 1991, Penguin Books.

[Penrose 1994] R. Penrose, *Shadows of the Mind: A Search for the Missing Science of Consciousness*, Oxford 1994, Oxford University Press.

[Platt 1999] J. Platt, *Fast training of support vector machines using sequential minimal optimization*, in B. Schölkopf, C. J. C. Burges, A. J. Smola (editors), Advances in Kernel Methods - Support Vector Learning, Cambridge, MA 1999, MIT Press.

[Press 2007] W. H. Press, S. A. Teukolsky, W. T. Vetterling, B. P. Flannery, *Numerical Recipes: The Art of Scientific Computing,* Cambridge 2007, Cambridge University Press, 3rd Edition.

[Price 1997] K. Price, R. Storn, *Differential Evolution: Numerical Optimization Made Easy,* Dr. Dobb's J. 264, 18-24, 1997.

[Price 1999] K. Price, *An Introduction to Differential Evolution,* 77-106, in: D. Corne, M. Dorigo, F.Glover (editors), New Ideas in Optimization, London 1999, McGraw-Hill.

[Price 2005] K. Price K., R. Storn, J. Lampinen, *Differential Evolution - A Practical Approach to Global Optimization,* Berlin 2005, Springer.

[Reinsch 1967] C. H. Reinsch, *Smoothing by Spline Functions*, Numer. Math. 10, 177-183, 1967.

[Reinsch 1971] C. H. Reinsch, *Smoothing by Spline Functions II*, Numer. Math. 16, 451-454, 1971.

[Rojas 1996] R. Rojas, Neural Networks: *A Systematic Introduction*, Berlin 1996, Springer.

[Rousseeuw 1987] P. J. Rousseeuw, *Silhouettes: A Graphical Aid to the Interpretation and Validation of Cluster Analysis*, J. Comput. Appl. Math. 20, 53-65, 1987.

[Rousseeuw 2003] P. J. Rousseeuw, A. M. Leroy, *Robust Regression and Outlier Detection*, Hoboken 2003, Wiley.

[Schank 1977] R. C. Schank, R. P. Abelson, *Scripts, Plans, Goals and Understanding. An Inquiry into Human Knowledge Structures*, New York 1977.

[Schneider 2008] G. Schneider, K.-H. Baringhaus, *Molecular Design: Concepts and Applications*, Weinheim 2008, Wiley-VCH.

[Schölkopf 1998] B. Schölkopf, A. J. Smola, *A Tutorial on Support Vector Regression*, Neuro-COLT2 Technical Report Series, NC2-TR-1998-030, 1998.

[Schölkopf 1999] B. Schölkopf, C. J. C. Burges, A. J. Smola (editors), *Advances in Kernel Methods - Support Vector Learning*, Cambridge, MA 1999, MIT Press.

[Schölkopf 2002] B. Schölkopf, A. Smola, *Learning with Kernels: Support Vector Machines, Regularization, Optimization, and Beyond*, Cambridge, MA 2002, MIT Press.

[Storn 1997] R. Storn, K. Price, *Differential Evolution: A Simple and Efficient Adaptive Scheme for Global Optimization over Continuous Spaces*, J. Global Optimization 11, 341-359, 1997.

[Turing 1950] A. Turing, *Computing Machinery and Intelligence*, Mind 59 (236), 433-460, 1950.

[Vapnik 1995] V. Vapnik, *The Nature of Statistical Learning Theory*, New York 1995, Springer.

[Vapnik 1998] V. Vapnik, *Statistical Learning Theory*, New York 1998, Wiley.

[Vogel 1921] H. Vogel, *Das Temperaturabhängigkeitsgesetz der Viskosität von Flüssigkeiten*, Physikalische Zeitschrift 22, 645, 1921.

[WDBC data] Wisconsin Diagnostic Breast Cancer (WDBC) data set. Taken at 2011/01/30 from the UCI (University of California at Irvine) machine learning repository at http://archive.ics.uci.edu/ml. Repository citation: A. Frank, A. Asuncion (2010), UCI Machine Learning Repository (http://archive.ics.uci.edu/ml), Irvine, CA: University of California, School of Information and Computer Science. First citation in medical literature: W.H. Wolberg, W.N. Street, and O.L. Mangasarian, *Machine learning techniques to diagnose breast cancer from fine-needle aspirates*, Cancer Letters 77 (1994), 163-171.

[WPBC data] Wisconsin Prognostic Breast Cancer (WPBC) data set. Taken at 2011/01/30 from the UCI (University of California at Irvine) machine learning repository at http://archive.ics.uci.edu/ml. Repository citation: A. Frank, A. Asuncion (2010), UCI Machine Learning Repository (http://archive.ics.uci.edu/ml), Irvine, CA: University of California, School of Information and Computer Science.

[WPBC non-recurrent data] Wisconsin Prognostic Breast Cancer (WPBC) non-recurrent data set. Taken at 2011/01/30 from the UCI (University of California at Irvine) machine learning repository at http://archive.ics.uci.edu/ml. Repository citation: A. Frank, A. Asuncion (2010), UCI Machine Learning Repository (http://archive.ics.uci.edu/ml), Irvine, CA: University of California, School of Information and Computer Science.

[WPBC recurrent data] Wisconsin Prognostic Breast Cancer (WPBC) recurrent data set. Taken at 2011/01/30 from the UCI (University of California at Irvine) machine learning repository at http://archive.ics.uci.edu/ml. Repository citation: A. Frank, A. Asuncion (2010), UCI Machine Learning Repository (http://archive.ics.uci.edu/ml), Irvine, CA: University of California, School of Information and Computer Science.

[Weast 1975] R. C. Weast (editor), *Handbook of Physics and Chemistry*, 56th edition, CRC Press, 1975.

[Weizenbaum 1966] J. Weizenbaum, *ELIZA - A Computer Program for the Study of Natural Language Communication between Man and Machine*, Communications of the ACM 9 (1), 36-45, 1966.

[Wienke 1994] D. Wienke, Y. Xie, P. K. Hopke, *An adaptive resonance theory based artificial neural network (ART -2a) for rapid identification of airborne particle shapes from their scanning electron microscopy images*, Chemometrics and Intelligent Laboratory Systems 25, 367-387, 1994.

[Zupan 1999] J. Zupan, J. Gasteiger, *Neural Networks in Chemistry and Drug Design*, Weinheim 1999, Wiley-VCH.

Index

© Springer International Publishing Switzerland 2016   467
A. Zielesny, *From Curve Fitting to Machine Learning*, Intelligent Systems
Reference Library 109, DOI 10.1007/978-3-319-32545-3

Printed in the United States
By Bookmasters